Papers presented at the

International Conference on

flow induced vibrations

in Fluid Engineering

Reading, England
September 1982

Organised and sponsored by
BHRA Fluid Engineering
England

Editors: H.S. Stephens
Mrs. G.B. Warren

NOTES

The Organisers are not responsible for statements or opinions made in the papers.

The papers have been reproduced by offset printing from the authors' original typescripts to minimise delay.

When citing papers from this volume, the following references should be used:-

Title, Author(s), Paper No., Pages, International Conference on Flow Induced Vibrations in Fluid Engineering, Reading, U.K. BHRA Fluid Engineering, Cranfield, Bedford, U.K. 14-16 September, 1982.

ISBN 0-906085-73-X
ISSN 0263-4198

The entire volume can be purchased from BHRA Fluid Engineering for $82.00

Printed and published by
BHRA Fluid Engineering
Cranfield, Bedford MK43 0AJ, UK

ACKNOWLEDGEMENTS

The valuable assistance of the Organising Committee and Panel of referees is gratefully acknowledged.

ORGANISING COMMITTEE

Prof. D. J. Johns (Chairman)	Loughborough University of Technology
Dr. N. Cook	Dept. of Environment Building Research Station
D. A. Crow	BHRA Fluid Engineering
Dr. J. D. Hardwick	Imperial College of Science and Technology
Dr. C. L. Kirk	Cranfield Institute of Technology
R. C. T. Rainey	Atkins R & D
C. Scruton	Consultant
Dr. T. L. Shaw	Sir Robert McAlpine & Sons Ltd
H. S. Stephens	BHRA Fluid Engineering
D. E. Walshe	National Maritime Institute
Mrs. G. B. Warren	BHRA Fluid Engineering

LIST OF PAPERS PRESENTED

FLUID ELASTIC RESPONSE STUDY OF
THE NAKDONG BARRAGE GATES

R.J. de Jong and Th.H.G. Jongeling

Delft Hydraulics Laboratory, The Netherlands

Summary

This paper deals with the vibration investigation that was carried out by the Delft Hydraulics Laboratory for the radial regulating gates of the Nakdong Barrage (Republic of Korea). The barrage was designed by NEDECO (Netherlands Engineering Consultants) with D.H.L. participating. The project was commissioned by the Industrial Sites and Water Resources Corporation, Republic of Korea.

The Barrage is to be built in the mouth of the Nakdong River and will create a fresh-water reservoir. The water-level of this reservoir will be controlled by regulating gates, which will mainly be used as overflow gates.

Single radial gates are often applied but experience mostly deals with underflow. In case of overflow, strong suction forces under the nappe and strong excitation forces due to the diving jet are possible. Dependent on the shape of the gate, feedback with the movement of the gate can occur. For these reasons it was decided to investigate the dynamic behaviour of the gates in a 1:20 scale model. Because of the preliminary design stage of the gates, a schematized mass spring system model was chosen. Such a model has the advantage that shape and stiffness can be changed independently.

The test showed that strong regular vibrations could occur in the tangential direction. When a spoiler at the crest of the gate was applied, thus aerating low pressure zones under the nappe, some of the vibrations could be prevented, but in two situations strong vibrations still manifested themselves. Firstly at submerged overflow conditions (the spoiler did not aerate) and secondly at free overflow, when the falling jet hits the lower horizontal main girder.

During the investigations plausible arguments were established for both vibrations. Although fundamental research was not the aim of the investigation, this paper presents the theory developed as well as the measures for preventing vibrations, which were resolved on the basis of this theory. The model tests showed these measures to be successful.

Organised and sponsored by
BHRA Fluid Engineering, Cranfield, Bedford MK43 0AJ, England

1

NOMENCLATURE

F	= hydrodynamic force	$kg\ m\ s^{-2}$
g	= gravitational acceleration	$m\ s^{-2}$
h	= gate opening	m
h^*	= (upstream water-level) − (gate-crest level)	m
ΔH	= local head	m
N_t	= transitional frequency	s^{-1}
P_{A-B}	= pressure difference between A and B	$kg\ m^{-1}\ s^{-2}$
Q	= discharge	$m^3\ s^{-1}$
R_v	= vertical response of gate model converted to a 47.5 m wide prototype gate	$kg\ m\ s^{-2}$
$R_{v\ stat}$	= static part of R_v	$kg\ m\ s^{-2}$
$R_{v\ dyn}$	= dynamic part of R_v	$kg\ m\ s^{-2}$
S	= Strouhal number	−
t	= time	s
$\bar{v}$	= mean velocity of the flow in the discharge opening	$m\ s^{-1}$
y	= tangential gate displacement	m
$\dot{y}$	= tangential gate velocity	$m\ s^{-1}$
$\ddot{y}$	= tangential gate acceleration	$m\ s^{-2}$
μ	= contraction coefficient	−
μh	= effective flow cross section	m
A'	denotes fluctuating part of variable A	
$A::B$	means: a decrease (increase) of variable A is accompanied by a decrease (increase) of variable B	
$A::-B$	means: a decrease (increase) of variable A is accompanied by an increase (decrease) of variable B	

1   INTRODUCTION

The Government of the Republic of Korea proposes to build a barrage in the mouth of the Nakdong River. The erection of this barrage forms part of a broad, long-term plan to improve the water management of the Nakdong River basin.
The preliminary design of the Barrage consists of ten, 47.5 m wide radial gates between concrete piers (Fig. 1). The gates (radius 15 m) rotate by means of four arms, hinged from trunnions fixed to the box-girder type concrete bridge across the Barrage (Fig. 2). The gates are positioned by means of hoisting cables on either side.

During high river discharges all gates will be raised, while at low discharges the gates are mainly closed, thus providing a freshwater reservoir for agricultural, industrial and municipal purposes. Four gates will then control the water-level of the reservoir. These regulating gates will be mainly operated as overflow gates. Behind these gates a stilling basin is provided. A second target of the Barrage is to prevent salt water penetration from the sea.
The design discharge of the Nakdong River has been fixed at 18,300 m^3/s with a return period of 500 years. Normally however the discharge is much smaller (up to about 300 m^3/s in the dry season).
It is aimed to keep the reservoir level between SMSL + 1.5 m and SMSL - 0.5 m (SMSL = datum). The river discharges via the regulating gates, mostly over the crests of the gates, but flow underneath is not precluded. The sea level ranges between SMSL - 0.75 m and SMSL + 0.95 m (extreme tides, excluding wind effects of about 1 m). Wave heights of about 1.3 m are to be expected at the sea side. In order to prevent leakage flow, rubber seals will be applied at the sides and lower edge of the gates. This is an important factor when preventing salt penetration.

In Fig. 2 the basic design of the regulating gates has been plotted. The gates consist of a framework of horizontal and vertical girders covered by a curved skin plate at the riverside. The hoisting cables at both sides of the gates are provided with buffer cylinders in order to prevent unequal stresses in the separate cables. The four arms of the gates are hinged to the concrete bridge. This bridge is supported at the piers both in a horizontal and vertical direction by means of rubber bearing consoles. The bearing consoles influence the radial stiffness of the gates.
The natural frequencies of the gates are dependent on gate position and amount of submergence. Basic modes are a tangential mode and a radial mode. The calculated natural frequency in case of tangential mode ranges between 3.3 and 1.5 Hz, in case of the radial mode between 9.0 and 4.0 Hz (this includes the added mass effect). The mass of one gate amounts to about 15.10^4 kg.

2   EXPECTATIONS ON VIBRATIONS

Radial gates are widely applied, however experience with single radial gates mostly deals with underflow. The radial gates of the Nakdong Barrage were designed as overflow gates as well and therefore a desk study was undertaken concerning the possibility of the occurrence of vibrations.
In case of overflow, strongly curved streamlines around the crest of the gate will occur. This will be accompanied by low pressure zones under the nappe. Dependent on the shape of the gate, suction forces will arise, which will move the gate in a tangential direction, thus influencing the discharge opening above the crest of the gate. Because of this, vibrations related to feedback phenomena are possible. When a rounded crest is applied the flow separation point is not well-defined, especially not at submerged flow conditions. This can lead to unstable flow separation. However, when a sharp crest is provided and the flow is stably separated from the crest, unstable flow reattachment phenomena can occur at the downstream part of the gate. This phenomenon was found at the Thames Barrier gates (Ref. 1).
The stability of the flow can also be influenced by added mass flow around the moving structure, thus probably causing feedback phenomena, as described for an emergency flap gate in (Ref. 2).
When, in case of free overflow, the nappe is falling down on the downstream part of the gate, an important excitation force exists, especially in the tangential direction. Under certain conditions feedback between the enclosed air cushion and the water curtain is possible, causing vibrations in the gate structure (Ref. 4).

It will be clear that the dynamic behaviour of the radial gates of the Nakdong Barrage is an uncertain factor in case of overflow, the more so as a coupling between radial and tangential phenomena is likely. Also the gap between gate and sill requires special attention, because gap flow may lead to local vibration problems, as was found for a submersible tainter gate (Ref. 3). Because of these uncertainties it was decided to study the dynamic behaviour of the regulating gates in a scale model.

3   SCALE MODEL OF REGULATING GATES

A fair reproduction of the dynamic behaviour of gates is obtained when using an elastically similar model. Such a model is usually built in the final design stage, when the shape and stiffnesses of the structure are well-known and drastic modifications are not probable. This is of importance, because an elastically similar model can hardly be changed and the building of a new model of a complex structure is an expensive and time-consuming process.
Because of the preliminary design stage of the Nakdong Barrage gates, it was decided to build an oscillator type model. Such a model is a stiff, geometrical reproduction of the prototype and is suspended by springs. Shape and stiffness can be changed independently.

The basic modes must be well-reproduced in a model. In this case the basic modes are a tangential mode (bending of gate, strain of hoisting cables) and a radial mode (bending of gate and concrete bridge, torsion of bridge, rotation at bearing consoles).
Basic modes are characterized by their natural frequencies. When reproducing these frequencies it is not always possible to reproduce both mass and stiffness correctly, and a compromise has to be found. For the gates concerned this meant that the mass of the bridge was not reproduced and that the stiffness in the radial direction was adapted.

Because of the uniform stream pattern along the crest of the gates, it was not necessary to build a complete model of the gate. Thus only a 10 m wide portion of the gates was built on scale 1:20. See Photograph 1. A small gap between gate model and flume walls was kept open. The gate model was in radial direction supported by two adjustable springs at the end of the gate arms and was vertically suspended by one adjustable spring at the end of a suspension rod, thus providing both a radial and tangential mode at a limited range of natural frequencies. See Fig. 3. The response of the gate model could be measured with the aid of force meters between springs and measurement frame.

In the first instance attention was paid to overflow and tangential vibrations (the radial springs were replaced by stiff dummy blocks); combined mode tests then were carried out, also for underflow conditions.

4   TEST RESULTS

4.1   Original design

The original gate design was tested in a wide range of natural frequencies and boundary conditions. All tests were carried out with a controlled downstream water-level, while the upstream water-level was swept from gate-crest level up to at most 2.3 m above this level (overflow tests), by increasing the flume discharge.
It appeared that strong vibrations occurred at every gate position (between SMSL + 2.0 m and SMSL - 0.5 m) and at all the tested natural frequencies of 4.0 Hz, 2.8 Hz and 2.0 Hz (tangential mode, dry gate, prototype values).

These vibrations could roughly be described as follows:
remark : $h^* = $ (upstream water-level)-(gate-crest level)
Situation 1a h^* between 0.2 - 0.4 m. Free overflow. Strong vibrations at adjusted "dry" natural frequency $f_n = 4.0$ Hz. The triangular stiffener which was designed at the skin plate edge strengthened these vibrations.
Situation 1b h^* between 0.5 - 0.6 m. Free overflow. High gate positions. Strong vibrations at $f_n = 4.0$ Hz, weak vibrations at $f_n = 2.8$ Hz. An air pocket along the crest of the gate was present. The air was sucked in.
Situation 1c h^* between 0.8 - 1.1 m. Free overflow. Strong vibrations at $f_n = 2.0$ Hz (stronger vibrations at higher gate positions).

4

Situation 1d h* between 1.4 - 2.0 m. Free overflow. Relatively low downstream water-
          levels. See Fig. 5a. Vibrations with a mostly irregular nature (irregular
          amplitude but regular resonance frequency) occurred at all adjusted "dry"
          natural frequencies of 4.0, 2.8 and 2.0 Hz.
Situation 2 h* between 1.2 - 2.2 m. Submerged overflow conditions (relatively low gate
          positions and high downstream water-levels). See fig. 5b. Strong vibra-
          tions at all adjusted "dry" natural frequencies of 4.0, 2.8 and 2.0 Hz.
In Table 1 for each situation the maximum expierenced vibrations are given. The mea-
sured vertical response R_V is divided in a static part and a dynamic part. See Fig 4.

4.2 Aeration

It was thought that a number of strong vibrations was caused by fluctuating suction
forces, due to fluctuating pressures under the nappe. Aerating these low pressure zones
might prevent the vibrations.
Therefore the gate model was provided with a spoiler (See Fig. 6) at the crest of the
gate and the effect on vibrations was tested. Indeed this measure appeared to be successful under free overflow conditions; but a new vibration phenomenon manifested itself at a relatively high discharge and a relatively low downstream water-level (referred to as situation 3 vibrations). Though these vibrations occurred under the same
conditions as situation 1d vibrations, the excitation source was thought not to be the
same, because fluctuating suction forces under the nappe were not possible. Compare
Fig. 5a and 5c. It appeared too, that the spoiler gave some reduction of situation 2
vibrations (submerged overflow), though aeration was not possible. Apparently the
spoiler was able to disturb the vibration-initiating stream pattern.
Some examples of the measured response R_V in case of situation 2 and situation 3 vibrations have been given in Table 1.

4.3 Additional tests with aerator

So far some of the vibrations could be prevented by aerating the nappe. Two situations
with vibrations however still remained (situation 2: submerged overflow conditions,
situation 3: free overflow conditions). In order to locate the excitation source in
both cases, some special tests were carried out. First the lower main girder was provi-
sionally screened at submerged overflow conditions. This was done by clasping a shelf
between the flume walls. It appeared that situation 2 vibrations still occurred (and
were even stronger). Therefore it was deduced that the excitation source had to be
found near the upper main girder. A low pressure zone under the nappe was present; this
was shown by an upward flow through the holes in the upper main girder.
A second series of tests were carried out with a gate from which the upper main girder
was removed. At submerged overflow conditions no vibrations occurred, but at free overflow conditions very strong vibrations manifested themselves.
It appeared that the diving jet hit the lower main girder. When screening the lower
main girder the vibration was eliminated. Therefore it could be deduced that situation
3 vibrations were caused by excitation forces on the lower main girder due to the
diving jet.

5 SOME THEORETICAL CONSIDERATIONS

5.1 Vibrations with drowned gate (situation 2 vibrations)

In the same way as Kolkman did with underedges of gates, various arguments were pro-
pounded to explain and avoid the problems. Kolkman found first an explanation for low
frequency perpendicular to flow vibrations of gate underedges, based on inertia effects
in the flow (Ref. 4). High frequency in flow vibrations of gate underedges were also
explained by assuming added mass flow effects (Ref. 2). In (Ref. 2) low and high frequencies have already been defined, therefore we confine ourselves here to indicating
the transitional frequency N_t

$$N_t = \frac{S}{h} \sqrt{2g\Delta H}$$

S = in the order of 0.2 (0.1 till 0.4)

5

At lower frequencies than N_t the gap discharge follows the gap fluctuation and at higher frequencies a constant discharge is assumed. These gap phenomena are used to theorize on the vibrations with drowned crest conditions. With measured resonance frequencies of 1.3 till 2.8 Hz, head differences of 0.5 till 0.7 m (and hence velocities of about 3.5 m/s, by assuming $\bar{v} = \sqrt{2g\Delta H}$) and nappe thicknesses of about 0.5 till 1 m, N_t results in:

$$0.6 \text{ Hz} < N_t < 2.6 \text{ Hz}$$

N_t has to be compared with the measured resonance frequencies to see whether one should assume fluctuating discharges or not. What is the consequence of the assumption?

Assumption 1. The discharge Q fluctuates with the effective flow cross-section, which in its turn follows the gate position y.

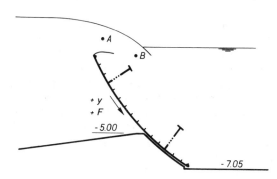

Thus $Q' :: y'$ (' denotes fluctuating part) The water mass between A and B will be accelerated by a pressure difference.

$$p'_{A-B} :: \frac{dQ'}{dt}$$

Because $\frac{dQ'}{dt} :: \frac{dy'}{dt}$ and $F' :: p'_{A-B}$ (F = hydrodynamic force on the gate) it follows $F' :: \frac{dy'}{dt}$, or : a hydrodynamic force acts in phase with the velocity of the gate movement, which means that negative damping can occur. In this situation gate vibration is self-amplifying until other (non-linear) effects result in an equilibrium amplitude. This explanation also includes the presence of discharge fluctuations, which show up in the model by surface waves, but they did not occur. An explanation with constant discharge assumption therefore will be more probable.

Assumption 2. The discharge Q is constant. As a consequence of the vibrating gate,

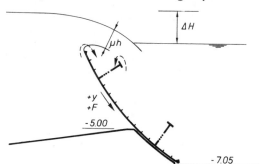

added mass flow will occur. In the figure this flow is indicated with dashed arrows for the downward (positive) gate movement y. The free boundary layer at the lower side of the nappe will be influenced by this added mass flow, but how is not easy to predict. Three assumptions are made on the effective flow cross section μh:
A $(\mu h)' :: y'$
B $(\mu h)' :: \dot{y}'$
C $(\mu h)' :: \ddot{y}'$
(The water surface streamline will follow the gate movement to a much lesser extent).
When one assumes h to be about constant, and because $Q = \bar{v} \mu h$ (with $\bar{v}$ is mean velocity) it follows:
$$\bar{v}' :: -(\mu h)'$$
It is further assumed $\bar{v} = \sqrt{2g\Delta H}$. This needs explanation because it looks as though free overflow conditions exist. Because the nappe is drowned the pressures underneath the nappe will influence the discharge. These pressures are related to the downstream water-level and therefore one can imagine that at least a local dependency of the velocities on the difference in water-levels exists. If this is accepted than $\bar{v}' :: \Delta H'$ and therefore $-(\mu h)' :: \Delta H'$.
A positive value of H' or $\bar{v}'$ will cause a lower pressure above the girder, thus $F' :: -\Delta H'$.

Combining this with the A, B and C possibilities of the $(\mu h)'$ dependency on the gate movement leads to $F' :: \dot{y}$ in case B, which means that a hydrodynamic force acts in phase with the velocity of the gate movement (negative damping). Strong vibrations will then occur.

Assumption 3. A third cause for vibrations in drowned situations can be sought in instable reattachment on the downstream part of the upper main girder. In a low frequency situation this will cause pressure and (thus) force fluctuations in phase with the position of the gate. This means that a negative hydrodynamic stiffness is present.

6

The gate will have a lower resonance frequency and will stronger respond on the fluctuating load. In a high frequency situation a phase shift will occur between the displacement and the force and a negative damping may result. This effect will go together with the vibration phenomenon of assumption 2.

The intention of this reasoning process is to look for a plausible explanation rather than to prove things. More and more detailed investigations are necessary to understand the model results. This investigation was to reach as soon as possible a reasonable design and therefore these explanations were helpful. It was believed that assumptions 2 and 3 were the most likely.

Avoiding vibrations, which follow from assumption 2, requires that pressure variations on top of the girder have to be reduced (or the girder surface has to be reduced) or that a counter working force has to be introduced.
This can be done by making the crest of the gate roof-shaped. Underneath this roof

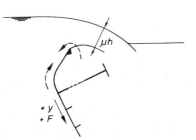

pressure fluctuations will act on the girder as well on the roof itself, this resulting in a zero total force (or a damping force in case the roof has a bigger surface as the girder). Above the roof, pressure fluctuations may (contrary with assumption 2 effect) cause a damping force. This can be seen as follows; considering during the vibration the moment of lowering the gate, this results in an upward added mass flow around the crest, causing a smaller μh and higher velocities, thus $\dot{y}' :: -(\mu h)' :: \bar{v}'$.

Higher velocities cause lower pressures above and near the crest. This results in a negative force, that fluctuates in phase with the positive gate movement. Thus
$-F' :: \dot{y}'$, which means a damping force.
Finally the roof will throw the nappe farther from the girder edge, reducing possible effects of assumption 3.

5.2   Vibrations with free-falling jet (Situation 3 vibrations)

Tests without the upper main girder indicated that these vibrations were due to a coupling between the impact forces on the lower main girder and the gate movement. This phenomenon can be compared to a father pushing the swing, his son is sitting on, at the right times. The phenomenon is called negative damping. Here on can imagine that the (aerated) free-falling nappe gets an alternating wavy shape due to the movement of the gate. The shape is influenced by the nappe thickness and the height of fall of the nappe.
Feedback phenomena possibly can be disturbed by preventing regular excitation by the falling nappe. When designing an alternative edge at the top of the gate two effects may be achieved:
- varying the thickness of the nappe along the gate crest will result in a variation of the wave amplitude of the water curtain. This is due to a variation of inertia forces in the nappe along the gate crest
- varying the nappe falling height will result in a time shift in the impact forces on the lower main girder, which are due to the falling water masses.
It is expected that mainly the second effect will disturb the feedback phenomenon.

6   PREVENTATIVE VIBRATION MEASURES TESTED

In the previous section various measures were discussed, at a theoretical level, to prevent or reduce both major vibrations. These measures were tested in the model. The results are discussed here.

As a measure to reduce situation 2 vibrations it was thought that the application of a roof above the upper main girder, in order to generate counter forces, would be suitable. Fig. 7a shows the modified crest. In the beginning this measure appeared to be not very successful, because at submerged flow conditions heavy vibrations still occurred. But when the roof was somewhat enlarged, thus equalizing the roof surface and the upper main girder surface, vibrations were no longer generated. However at free overflow this concept could not prevent situation 3 vibrations: the nappe was not thrown far enough away from the lower main girder.

Strong vibrations occurred, even stronger and more regular than those found for the original aerated gate design. As mentioned in Par. 5.2 a phase shift in the impact forces due to the falling jet had to be applied. This effect was created with crest blocks (Fig. 7b.). The measure gave good results, but it was thought that the blocks were problematic in case of floating debris. A notched upper edge was then tested with regular teeth (Fig. 7c). This appeared a satisfactory solution, but the results could be improved by applying one bigger tooth in five smaller ones (see Fig. 8). Some other tooth configurations were tested, but they gave no better results. It should be noticed that the lower main girder was not safe-guarded against flow excitation by the falling jet with these measures. The response of the gate model on this excitation however was irregular and the amplitude of the response kept within acceptable bounds. In Table 2 the vertical response of the gate model (final design) in case of submerged overflow and free overflow has been indicated (some maximum values).

7   FINAL REMARKS

After completing the overflow tests (tangential mode), the gate model was tested in underflow situations. These tests demonstrated that no danger exists for tangential vibrations under these conditions.

The model was then suspended by adjustable leaf springs in the radial direction only. After that is was suspended both in the tangential and radial direction by springs. In both cases the model appeared to be free of regular vibrations. Changing the width of the gap between gate and sill (0.20 m instead of 0.10 m) did not influence these results. It should be noticed however, that local plate vibrations in case of flow in the gap were not examined in the model (the model was not appropriate for this kind of test).

Finally quasi-static and dynamic wave loads were deduced from former investigations for the Eastern Scheldt storm surge barrier. With the aid of response calculations design loads were determined for the horizontal main girders and the skin plate of the gates.

REFERENCES

1. Hardwick, J.D. "Hydraulic Model Studies of the Rising Sector Gate conducted at Imperial College". Conf. on Thames Barrier Design, Oct. 1977, London.
   Publ. Inst. of Civ. Eng. Paper 12 pp. 133-134.

2. Kolkman, P.A. "Development of vibration-free gate design: learning from experience and theory". Symp. on Practical Experiences with Flow Induced Vibrations, Sept. 1979, Karlsruhe.
   Publ. nr. 219 of Delft Hydraulics Laboratory (1979).

3. Pickett, E.B. "Experience with flow-induced vibrations". Symp. on Practical Experiences with Flow Induced Vibrations, Sept. 1979, Karlsruhe.

4. Kolkman, P.A. "Flow induced gate vibrations". Thesis Delft University of Technology; also publ. nr. 164 of the Delft Hydraulics Laboratory (1976).

TABLES

Table 1.   Vertical response of original gate in case of overflow.

	situation	adjusted tangential frequency* (dry gate) (Hz)	gate-crest level (m to SMSL)	upstream water-level (m to SMSL)	downstream water-level (m to SMSL)	vertical response R_V		
						static part (kN)	dynamic part (maximum amplitude) (kN)	resonance frequency (Hz)
without spoiler	1a	4.0	+ 0.50	+ 0.80	− 0.75	− 50	620	3.1
	1b	4.0	+ 2.00	+ 2.65	− 0.75	− 380	540	3.5
	1c	2.0	+ 2.00	+ 2.95	− 0.30	− 1260	925	1.6
	1d	4.0	+ 1.00	+ 2.80	− 0.75	− 2180	525	2.9
		2.8	+ 1.00	+ 2.80	− 0.75	− 1990	410	1.8
		2.0	+ 1.00	+ 2.80	− 0.75	− 1990	510	1.3
	2	4.0	− 0.50	+ 1.80	+ 0.10	+ 190	1330	2.8
		2.8	− 0.50	+ 0.90	+ 0.10	+ 290	750	1.9
		2.0	− 0.50	+ 1.70	+ 0.10	+ 290	760	1.3
with spoiler	2	4.0	− 0.50	+ 1.80	+ 0.10	+ 320	655	2.8
	3	4.0	+ 1.00	+ 2.90	− 0.75	− 240	715	3.1

Table 2.   Vertical response of modified gate in case of overflow.
         (final design, aerator applied)

	situation	adjusted tangential frequency* (dry gate) (Hz)	gate-crest level (m to SMSL)	upstream water-level (m to SMSL)	downstream water-level (m to SMSL)	vertical response R_V		
						static part (kN)	dynamic part (maximum amplitude) (kN)	remarks
with spoiler	2	4.0	− 0.50	+ 1.70	− 0.30	− 1070	95	irregular response
	3	4.0	+ 1.00	+ 2.20	− 0.75	− 1050	145	irregular response

* Tangential frequencies were adjusted with the crest of the gate at SMSL + 2.00 m.
When lowering the gate, the "dry" natural frequencies decreased with a factor up
to at most 1.12 at a gate position of SMSL − 0.50 m.

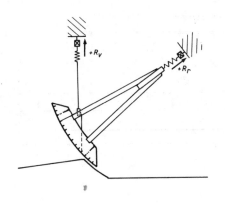

R_V = vertical response of the gate model converted to a 47.5 m wide prototype gate.

R_r = radial response of the gate model converted to a 47.5 m wide prototype gate.

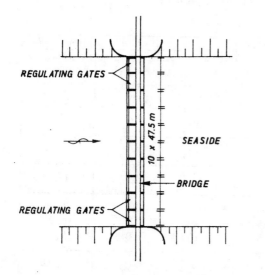

FIG. 1 BASIC DESIGN OF
NAKDONG BARRAGE

FIG. 2a PLAN OF ONE OPENING WITH REGULATING GATE

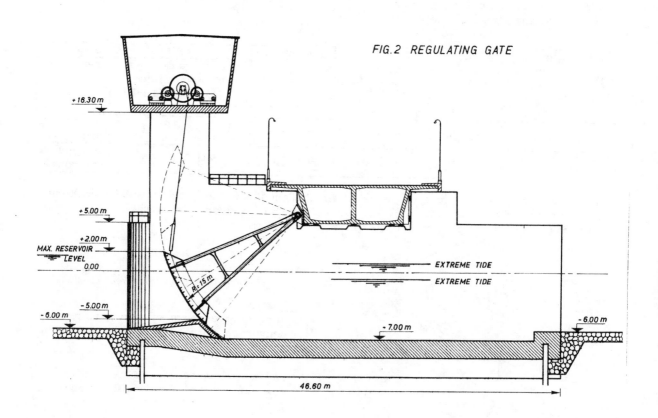

FIG. 2 REGULATING GATE

FIG. 2b CROSS SECTION AA

10

FIG. 3 SCALE MODEL OF REGULATING GATE

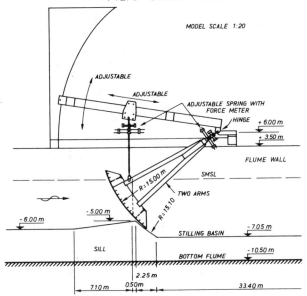

MODEL SCALE 1:20

FIG. 3a CROSS SECTION OF SCALE MODEL

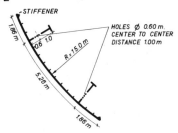

FIG. 3b DIMENSIONS OF SCALE
MODEL (ORIGINAL DESIGN)

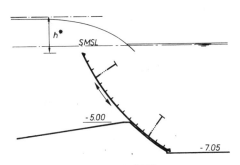

FIG. 5b SITUATION 2 VIBRATIONS

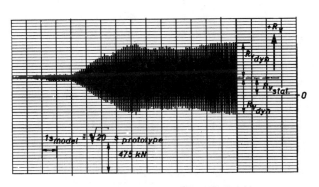

FIG. 4 EXAMPLE OF MEASURED VERTICAL
RESPONSE R_v (SITUATION 2 VIBRATIONS)

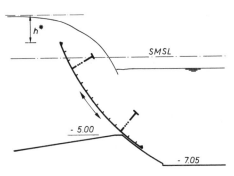

FIG. 5a SITUATION 1d VIBRATIONS

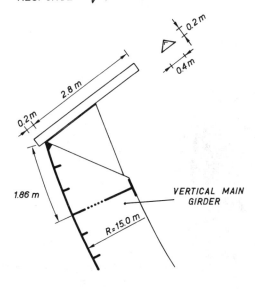

FIG. 6 GATE MODEL (ORIGINAL DESIGN)
WITH SPOILER

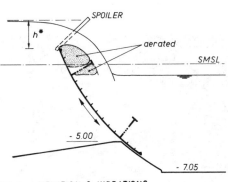

FIG. 5c SITUATION 3 VIBRATIONS

FIG. 5 VIBRATION CIRCUMSTANCES

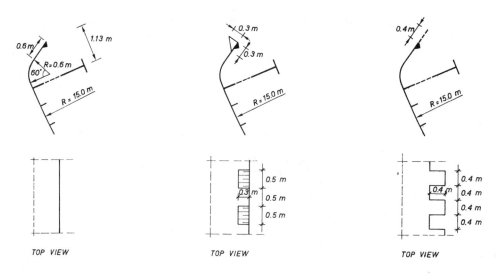

FIG.7a GATE WITH ROOF

FIG.7b GATE WITH ROOF
AND CREST BLOCKS

FIG.7c GATE WITH ROOF
AND REGULAR TEETH

FIG. 7 MODIFIED DESIGN OF GATE MODEL

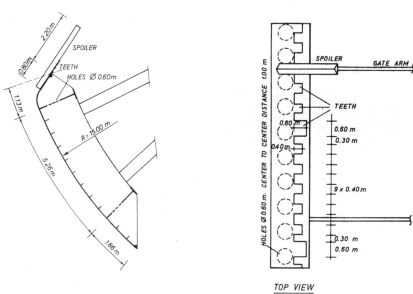

FIG. 8 FINAL DESIGN OF GATE MODEL

PHOTOGRAPH 1
MODEL OF REGULATING GATES
(ORIGINAL DESIGN)

FLOW INDUCED VIBRATIONS IN FLUID ENGINEERING

Reading, England: September 14-16, 1982

ADDED MASS BEHAVIOUR AND ITS CHARACTERISTICS AT SLUICE GATES

Nguyen D. Thang

University of Karlsruhe, Germany

Summary

This experimental study is to provide a systematic account of added mass behaviour and of important influences upon its characteristics at hydraulic gates under more realistic boundary and submergence conditions.

In the first part, a brief survey is presented on the main features affecting the added mass of submerged structures. In particular, the applicability of still-fluid added mass data in real flow situations will be discussed.

In the second part, added mass results for vibrating gate models are presented. By progressively changing boundary conditions and measuring the decaying response to plucked excitation in still water, a relatively broad range of vertical-lift gate types and service conditions can be covered.

The results are shown to suit general tendencies set out at the beginning. Comparisons with added mass values of some vibrating gates in field studies and with dynamic measurements at a hydroelastic gate model in a parallel study show a good agreement, especially in the condition of maximal resonance.

Organised and sponsored by
BHRA Fluid Engineering, Cranfield, Bedford MK43 0AJ, England

NOMENCLATURE

C_m = added mass coefficient

C_{mb} = added mass coefficient of bottom plate

C_{ml} = in-phase fluid force coefficient

C_{mr} = reduced added mass coefficient (without 75% contained water mass)

C_{mt} = added mass coefficient of top-plate

D = gate depth or characteristic body dimension

H = height of body

L = spanwise width of body

Re = Stokes number

U = reduced velocity

U_r = reduced velocity at resonance

V = characteristic flow velocity

Y_o = reduced vibration amplitude

a = lateral gap distance between gate and gate slot

f = vibration frequency in flow

f_n = natural frequency in air

f_w = vibration frequency in still water

h = water depth

m' = added mass

m_s = structural mass in air

n = number of waves

p = pressure ($\bar{}$ = mean..., $'$ = of perturbation)

s = gate opening

s_o = conduit height

y_o = vibration amplitude

u, v, w = velocities in x, y and z direction ($\bar{}$ = mean ..., $'$ = of perturbation)

α = added mass coefficient

ε = submergence degree

ζ_s = structural damping ratio in air

ν = kinematic viscosity

ρ = fluid density

ω = circular vibration frequency

1. INTRODUCTION

In safety considerations of hydraulic structures with respect to flow-induced vibrations, the knowledge of added mass is essential in determining the system resonant frequency. In many high-head installations, where gates and valves are deeply submerged and highly confined by conduit boundaries, the added mass of these could attain values many times greater than the structural mass and thus considerably affect the dynamics of the system. However, very scarce quantitative information exists hitherto hereupon. Most available data on added mass, whether theoretically or experimentally gained, are based upon strongly idealized conditions of unconfined bodies in an infinite fluid medium (Ref. 1,2) and may severely underestimate real added mass values under the diverse prototype conditions.

The aim of this experimental investigation is to provide a systematic account for the various important influences upon the added mass coefficient of sluice gates under more realistic boundary and submergence conditions, in order to facilitate quantitative estimates in specific cases.

Another objective of the study is to compare still-fluid added mass data with data measured in a dynamic model under flow conditions (Reference 3). This is a question of basic interest in hydroelasticity. A common practice is to incorporate some constant added mass and fluid-damping value measured at zero-flow condition into the structural counterparts for vibration analysis in case of flow. In fact, under dynamic conditions, added mass is a manifestation of the fluid-structure interaction and may play an important role in the delicate energy balance between damping and excitation, inherent to hydroelastic vibrations.

In a comprehensive study on fluid forces at a circular cylinder performing driven harmonic oscillations (Ref. 4), Sarpkaya noted that "added mass is one of the least understood and most confused characteristics of time-dependent flow". He further showed in Ref. 5 that the in-phase (added-mass like) lift force coefficient strongly varies with the reduced velocity, $U = V/fD$, dropping only to the classical potential-flow value of 1.0 near $U = 0$ and at resonance. Herein, V denotes some characteristic flow velocity, f the vibration frequency and D a characteristic body dimension. Comparisons between still-fluid and flowing-fluid added mass data will be presented in this paper for a flat-bottomed gate.

2. GENERAL ASPECTS ON ADDED MASS BEHAVIOUR

2.1 Validity of still-fluid added mass data in real flow

When a body vibrates in a fluid, an instantaneous pressure field is created, inducing an unsteady flow about the body, even if the fluid is stagnant. In the latter case (Fig. 1), the total force exerted on the fluid is only proportional to the body's acceleration and the resulting fictitious added mass per unit length, m', represented by the added mass coefficient:

$$C_m = \frac{m'}{\rho D^2} \qquad [1]$$

can be computed for simple configurations by potential flow theory [+] (Ref. 1). In case of flow, the perturbation due to the vibrating body induces pressure and velocity fluctuations p', u', v', w' superimposed on mean-flow conditions (Ref. 6). After subtraction of the mean flow terms from the Navier-Stokes equation, one obtains for instance for the x-direction:

[+] This definition differs somewhat from the classical added mass coefficient as used e.g. by Lamb (Ref. 1): $\alpha = 4 m'/\rho \pi D^2$, which relates the added mass to the displaced fluid mass of a cylinder of equal diameter D. The latter idea appears to historically stem from the early works of Dubuat (1786) and Stokes (1843), among others, on spherical pendulums. It is, however, somewhat misleading, since the displaced fluid mass is quasi zero in case of a flat plate but not the added mass, depending on the direction of vibration of the plate.

$$\frac{\delta p'}{\delta x} = \rho \underbrace{\frac{\delta u'}{\delta t}}_{A} + \rho \underbrace{\left[u' \frac{\delta \bar{u}}{\delta x} + \bar{u} \frac{\delta u'}{\delta x} + \bar{v} \frac{\delta u'}{\delta y} + v' \frac{\delta \bar{u}}{\delta y} + \ldots \right]}_{B} \qquad [2]$$

The A-term represents the classical inertia-type added mass, which is independent of the mean flow, i.e., measurable, for instance, in still-fluid experiments. The B-terms represent the drag-type part of the motion-induced fluid force. These are coupled with mean flow conditions. Several coupling effects related to added mass in real flow should be mentioned: the coupling of different vibration degrees of freedom due to added-mass-induced potential differences at asymmetric geometries (Ref. 7) and the coupling of inertia- and drag-type parts in the fluid force coefficient, as evident from Eqn. 2 and described in Ref. 8. In order to elucidate the limit condition for the latter problem, Kolkman (Ref. 6) suggested the following order-of-magnitude considerations: assuming $u' \sim \dot{y}$, $\delta u'/\delta t \sim \dot{y}\omega$ and $u'\delta\bar{u}/\delta x \sim \dot{y}V/D$ (where $\sim$ means "proportional to", $\dot{y}$ denotes the body velocity and $\omega = 2\pi f$ denotes the circular vibration frequency), the motion-induced fluid force will be predominantly an added-mass force if $\dot{y}\omega \gg \dot{y}V/D$, or

$$U = \frac{V}{fD} \ll 2\pi \qquad [3]$$

Thus, the use of still-fluid (or potential-flow) added mass data for the evaluation of the fluid force upon the body and of the resonant frequency appears only justified as an approximation in vibrations at low reduced velocities. Even then, the added mass (or in-phase lift force) coefficient may still vary with U, with the zero-flow case as a limit condition (Ref. 5). Another serious limitation of potential flow methods is the assumption of infinitely high velocities at sharp edges (Fig. 1), whereas the real flow will involve a quite different flow pattern, including separation, vortex shedding and synchronisation. In spite of all these limitations, still-fluid added-mass experiments are relatively easier to perform and, therefore, preferable but allowance must be made for a "calibration" of the results through comparison with similar data measured under flow conditions.

2.2   Effects influencing still-fluid added mass

Due to the fact that added mass results from the motion-induced unsteady pressure field in the ambient fluid (Fig. 1), any change in the boundary conditions of the potential-flow net may strongly affect the value of the time-averaged added mass coefficient C_m. In particular, the more the body is confined by solid boundaries, the closer the equipotential lines are and the larger the added mass. Therefore, solid-boundary confinement exerts the strongest influence upon C_m. In Ref. 9, Weaver stated that "if the confining surfaces are within two characteristic dimensions of the vibrating body, the added mass coefficient increases considerably and (α)-values from five to ten or more are not unusual". As a consequence, direct application of potential-theory added mass values (under inconfined conditions) may severely underestimate the real situation of a highly confined structure.

The proximity of a liquid free surface, where the pressure - unlike at a solid surface - must drop to zero, results in an inverse effect as confinement. Added mass increases with increasing submergence. This holds in particular for vertical immerged structures, as confirmed by many experiments, including those of Chandrasekaran (Ref. 10) at bridge piers and Hardwick (Ref.11) at hydraulic gates, depicted in Fig. 2. For totally submerged three-dimensional (or low-section ratio) bodies, however, the free surface affects only C_m if it is within about one body-diameter above the body (Ref. 12).

Furthermore, added mass may also vary with body geometry (shape, degree of two-dimensionality and exposure of structural components) and type of motion (vibration amplitude and frequency). The former effect has been investigated both analytically and experimentally by Patton (Ref.2) for simple shapes (Fig. 3). It is interesting to note that, much like the effect of confinement

16

discussed above, an increased boundary restraint is present at body shapes with higher two-dimensionality (in any direction) and results in higher C_m values as compared to similar three-dimensional shapes. This can also be seen for the different pier forms in Fig. 2a.

The effect of both vibration amplitude and frequency is generally negligible, as long as very small relative amplitudes y_o/D are involved. These are associated with the size of impulsively generated vortices. For circular cylinders, Sarpkaya (Ref. 4) reported that no vortex shedding occurs up to $y_o/D = 1$. In general, however, it appears that C_m becomes both amplitude and frequency dependent if y_o/D exceeds about 10% (Ref. 9). On the other hand, viscosity effects represented by the Stokes number

$$Re_\omega = \frac{\omega D^2}{\nu}$$

[4]

may also be involved as the frequency changes (ν being the kinematic viscosity of the fluid), with C_m generally decreasing with increasing Re_ω (Ref. 13).

3. EXPERIMENT

3.1 Gate model

Still-fluid added mass measurements at vertical-lift gate models were performed in a 0.16 m-wide and 0.66m-high water channel (Fig. 4). In order to cover a wide range of boundary conditions typical of common gate types, ranging from smooth-plate weir gates to girder-type conduit gates, two gate models were successively used, with equal depth, D = 30 mm but different heights and widths. The first, 400 mm high, 156 mm wide and made of plexiglass, was used for partial submergences as a weir-type gate. The second, 185 mm high (or 215 mm high with a 45°-inclined gate bottom), 142 mm wide, was made of bronze and provided with ten uncovered stiffening girders. It was used as the model of a fully submerged high-head gate, in conjunction with two separate concrete conduit parts. Lateral gate slots, with a maximal depth of 22 mm, were obtained by fitting side plates. The gate model was elastically supported by two horizontal cantilevered metal beams (8 x 50 mm² cross-section) and by two robust vertical profiles rigidly connected to the beams. The system was designed as to perform only vertical vibrations without additional guidance if centrically displaced. The vibration frequency was primarily about 12 Hz but could be varied between 8 Hz and 35 Hz by varying the clamping length of the metal beams. The total vibrating mass in air, m_s, including part of the cantilever beams, was determined by adding known masses and measuring the frequencies. It varied between 4286g and 5380 g.

Free vibration with controlled initial amplitudes were excited by a magnetic device and a transmitting rod. The decaying response was measured by a inductive displacement transducer (Hottinger W50). The signal was connected to a frequency counter and a ultra-violet light pen. A photograph of the set-up is shown in Fig. 5. Each measurement was repeated ten times, the result was thus averaged over about 120 oscillations. The damping ratio of the decaying vibrations $\zeta_s = (1/2\pi n) \ln(x_i/x_{i+n})$ averaged over n = 20 waves was typically 0.25% in air. The added mass m' and its coefficient C_m can be calculated from the measured frequencies f_n (in air), f_w (in water) and the structural mass m_s, as follows:

$$m' = m_s \left[\left(\frac{f_n}{f_w} \right)^2 - 1 \right]$$

[5]

$$C_m = \frac{m'}{\rho D^2 L}$$

[6]

where L is the width of the model.

The many (primarily geometrical) parameters affecting the added mass of gates, especially at high-head installations, can be readily observed in such field studies as reported by the US Waterways Experiment Station in Ref. 14 at Fort Randall Dam (the conduit and gate geometry of which was reproduced in the present model) and by Hardwick et.al, in Ref. 15. At present,

these were progressively investigated and included the following effects (see Fig. 4 for symbols):
- vibration behaviour, including vibration amplitude (y_o/D) and frequency $(Re = \omega D^2/\nu)$
- submergence degree ($\varepsilon = (h-s)/D$) and lateral confinement by gate slots and sidewalls
- gate opening ratio (s/D)
- uncovered girders, watermass within gate and top-plate added mass
- confinement due to conduit parts
- gate bottom forms.

The hydroelastic gate model used to "calibrate" the still-fluid added mass data with respect to flow effects has been described in detail in Ref. 3.

4. RESULTS AND DISCUSSION

4.1 Effect of vibration behaviour

The effect of vibration amplitude and frequency upon the added mass coefficient C_m is shown in Fig. 6 for the same boundary conditions as in Ref. 11 for the purpose of direct comparison, except for the side gate slots depicted in Fig. 4. The amplitudes were varied at a constant frequency of 13 Hz and inversely the frequencies were varied at a constant minimal amplitude of $y_o/D = 0.35\%$. The effect of amplitudes is seen to be minimal for y_o/D-values above 1%. This is in accordance with Hardwick's results, which give slightly smaller C_m values. This can be traced to the presence of the side gate slots in our study as described in the next section. At very small frequencies $(Re_\omega < 7\times 10^4)$, increasing viscosity effects result in strongly increased C_m values, in accordance with Chen's findings (Ref. 13). In the subsequent sections, the amplitude and frequency were kept invariant, at $y_o/D = 2\%$ and f = 13 Hz (or $Re_\omega = 7.3 \times 10^4$).

4.2 Effect of submergence and lateral confinement by gate slots and sidewalls

The effect of submergence was investigated for s/D = 0.67 and different side-confinement conditions, i.e. with and without gate slots; in the former case, also, the gap distance between gate and sidewalls has been varied. As clearly shown in Fig. 7, added mass increases linearly with the submergence degree, $\varepsilon = (h-s)/D$ in accordance with the tendency at Fig. 2. Furthermore, the presence of typical gate slots on both sides of the model increases the absolute value of C_m by about 0.25 (or 20% in the average). On the other hand, the gap distance between gate and sidewalls, designated by a, is seen to exert only a minor effect on C_m. In all the following, the gate slots were maintained along with a mean side gap distance of a = 5 mm.

4.3 Effect of gate opening

The effect of relative gate opening s/D or confinement due to channel bottom has been investigated for two submergence degrees, as shown in Fig. 8. Both curves reveal that C_m is only affected by the channel bottom if the gate is within one characteristic gate dimension from it, in accordance with the equally included results of Hardwick (Ref. 11) for a gate model without lateral confinement by slots and with Weaver's finding cited in Section 2.2.

4.4 Effect of uncovered girders and topplate

In view of the added mass of deeply submerged girder-type high-head gates (see Fig. 4), some important questions arise: 1. How much of the contained water mass within the vibrating gate is to be added like a rigid ballast to the structural mass? 2. Does the added mass coefficient of a fully submerged gate tend asymptotically to some constant value (see Fig. 2 and 3)? The results, obtained at a large gate opening, for a closed-type and an open girder-type gate (Fig. 9), reveal some interesting facts:

- both gates show a quite similar behaviour, except for the larger increase rate of C_m with increasing ε in the partly-submerged range of the girder-type gate, due to the increasing contained water mass and additional added masses of the girders. In the fully-submerged range, the two curves become practically parallel. The added mass excess, due to contained water and girder effects, is $\Delta C_m = 3.3$, corresponding to about 75% of the total contained water mass. Since the latter value depends upon specific gate dimensions, it is suggested as a generalized approximation for girder-type gates that 75% of the gate-dependent contained water mass be added to some universal "reduced" added mass coefficient C_{mr}, obtained at a corresponding closed-type gate.

- the topplate contributes a distinct added mass part, C_{mt}, (obtainable by linear extrapolation of the curve for partial submergence), which is uniformly larger than that of the bottom plate, C_{mb}, at a similar submergence.

- the curves appear to increase continuously with ε, even at full submergence. Patton's theoretical value for a 1:10 rectangular cylinder, included for comparison, appears to be only applicable at some smaller submergence degree.

4.5   Effect of confinement due to conduit parts

A further substantial increase of C_m is due to the confining effect of both conduit parts upstream and downstream of the gate. This is shown in Fig. 10 for a flat-bottomed gate at $s/D = 4$, placed in a conduit with a height $s_o = 7D$. The curve for the inconfined configuration of Fig. 9 is included in Fig. 10 for comparison and is seen to lie distinctly below the curves with the conduit parts present. Equally included is the curve for a similar closed-type gate in a conduit obtained through subtraction of 75% of the contained watter mass. This curve provides, as mentioned before, a basis for generalisation in view of girder-type gates with variable volumes of contained water.

4.6   Effect of gate bottom form

Fig. 11 shows the variation of C_m with gate opening at $\varepsilon = 15$ for girder-type gates with two typical gate bottom forms (both with 45-degree bottom inclination). Also, results for a face-type open-girder gate, i.e. without front conduit part, are shown, showing distinctly the extra-confinement effect due to that conduit part. The geometry of the gate bottom is seen to exert only a minor effect upon C_m. The present data can thus be extended to any gate geometry, provided that the dominant effects of confinement and submergence are properly accounted for. Again, through subtraction of a value corresponding to 75% of the contained water mass, a universal "reduced added mass" curve, C_{mr}, can be obtained for conduit gates. The accuracy of this assumption will be considered in the next section. The effect of the conduit height itself (s_o/D) may be expected at similar gate positions, s/s_o, to be irrelevant.

5.   COMPARISON WITH VIBRATION DATA IN REAL FLOW

As mentioned in the introduction, still-fluid added mass data are not necessarily valid at flow conditions, in spite of much experimental counter-evidence reported in the literature (e.g. Ref. 11 and 16). In particular, added mass varies with the reduced velocity. The condition of maximal resonance appears, however, to be an exception. For circular cylinders, Sarpkaya (Ref.5) found indeed that the inphase fluid force coefficient, C_{ml}, equals the classic (still-fluid) added mass value at resonance, i.e. $U_r = 1/Sh$, where Sh denotes the Strouhal number at a stationary cylinder. Obviously, further evidence is needed for other body shapes. At present, comparisons of the results will therefore be made both with some field data on vibrations at high-head gates and with our own measurements at a hydroelastic low-head gate model (Ref. 3).

5.1   Comparison with some field data

The Fort Randall gates (geometrically similar to the present case) have been reported (Ref. 14) to perform maximal vibrations (up to $\pm$ 3.8 mm) at opening ratios around $s/D = 5.7$ and a submergence degree (in the gate-well) of $\varepsilon = 16$. At an average vibration frequency of 3.3 Hz

19

a natural frequency of 4.2 Hz and a structural mass of 42 tons, the added mass amounts to 25.7 tons from Eqn.5 (including 18 tons contained water), yielding C_m = 7.7. Its reduced value, C_{mr} = 3.6, agrees well with the present data (Fig. 11). If the whole contained water mass were subtracted, a C_{mr}-value of 2.3 would result, which would be even smaller than the measured C_{mr}-value at a completely inconfined gate (Fig. 9).

Another vibration case has been reported in Ref. 15. The vibrations were observed in the fully-opened position (s/s_o = 1) at a submergence degree around ε = 18 and at frequencies around 1.7 Hz. The total added mass was estimated by Hardwick at 280 tons, including 170 tons contained water and 110 tons external added mass. The corresponding C_m and C_{mr} values are 10.4 and 5.6, respectively. The latter, depicted in Fig. 11, lies somewhat above the present C_{mr}-curve but still within acceptable ranges, considering the different conditions of confinement and submergence, among others.

5.2 Comparison with measurements at a hydroelastic gate model

The dynamic vertical fluid force and response at a flat-bottomed hydroelastic gate model (gate depth and width: D = 50 mm, L = 250 mm, mass-damping parameter $2\pi \zeta_s m_s / \rho D^2 L \simeq 2.0$) were measured at an opening ratio s/D = 0.5 and an averaged submergence ε = 6.0. The reduced amplitude $Y_o = y_o/D$, vibration frequency f/f_n (where f_n is the natural frequency in air) and C_m (according to Eqn. 5 and 6) are plotted versus the reduced velocity $U = V/f_n D$ over the resonance or lock-in range around U_r = 2.16. An interesting analogy to Sarpakya's finding at circular cylinders can be observed in the present case, with the added mass coefficient decreasing continuously to the value C_m = 1.28 at maximal resonance as U_r = 2.16 is approached from below. This value agrees well with the corresponding still-fluid added mass coefficient of Fig. 7 (without gate slot) as well as with Hardwick's data (Fig. 8). At reduced velocities slightly above U_r, the frequency increases sharply and C_m becomes negative. Outside the resonance range, the vibrations are very small and aperiodic, so that C_m cannot be properly defined.

6.   CONCLUSIONS AND ACKNOWLEDGEMENTS

The various effects affecting the added mass of hydraulic gates have been progressively investigated in this study. The results are shown to suit general tendencies known about the added mass behaviour of immersed bodies. It is hoped that the emerging physical picture would much facilitate quantitative estimates and corrections for diverse specific prototype conditions. Also, comparisons with real-flow data have shown that added mass values measured in still-fluid can be utilized for the prediction of vibration frequencies at the condition of maximal resonance.

The investigation was part of a research program on "Structural Flow-Induced Vibrations" at the Institute of Hydromechanics, University of Karlsruhe. The financial support of the Volkswagen Foundation is gratefully acknowledged. Special thanks are also due to Mr. M. Scherberger for his devoted help in carrying out all the measurements.

7.   REFERENCES

1. Lamb, H.: "Hydrodynamics". 6th ed. Cambridge University Press, 1932, Art. 71 and 102.

2. Patton, K.T.: "Tables of hydrodynamic mass factors for translational motion". Winter Annual Meeting of ASME (Chicago, Illinois, USA, Nov. 1965), Paper No. 65-WA/UNT-2.

3. N.D. Thang: "Force and vibration measurements on a hydroelastic gate model". (Kraft- und Schwingungsmessungen an einem hydroelastischen Schützmodell). In: Proc. International Conference on Transducer Technology and Temperature Measurement, Sensor '82, (Essen, Germany, Jan. 12-14, 1982), AMA, SIMA and IAHR Committee on Hydraulic Laboratory Instrumentation, 1982, Vol. 1, pp. 53-69 (in German).

4. Sarpkaya, T.: "Transverse oscillations of a circular cylinder in uniform flow, Part I". Technical Report No. NPS-69 SL 77071-R, 1977, Naval Postgraduate School, Monteray, California, USA.

5. Sarpkaya, T.: "Vortex-induced oscilattions, a selective review". Journal of Applied Mechanics, 46, 2, June 1979, pp. 241-258.

6. Kolkman, P.A.: "Flow-induced gate vibrations". Ph.D.-thesis, Delft Univ. of Technology, Publ. No. 164 of Delft Hydr. Lab., 1976.

7. Kolkman, P.A.: "Development of vibration-free gate design: Learning from experience and theory". IAHR/IUTAM Symposium on Practical Experiences with Flow-Induced Vibrations, ed. by E. Naudascher and D. Rockwell, (Karlsruhe, Germany, Sept. 3-6, 1979), Springer Verlag, Berlin, 1980, pp. 351-385.

8. Savkar, S.D.: "A survey of flow-induced vibrations of cylindrical arrays in cross-flow". Journal of Fluids Engineering (Trans. ASME), 99, 3, 1977, pp. 517-519.

9. Weaver, D.S.: "On flow-induced vibrations in hydraulic structures and their alleviation". Second Symposium on Applications of Solid Mechanics (McMaster University, Hamilton, Ontario, Canada: June 17-18, 1974), sponsored by CSME, CSCE and ASME.

10. Chandrasekaran, A.J., Saini, S.S. and Malhotra, M.M.: "Virtual mass of submerged structures". Journal of Hydraulics Division, Proc. ASCE, HY5, May 1972, pp. 887-896.

11. Hardwick, J.D.: "Periodic vibrations in model sluice-gates". Ph.D.-thesis, Imperial College of Science and Technology, London, U.K., May 1969.

12. Todd, F.H.: "Ship hull vibrations", Edward Arnold Publishers, London, 1961.

13. Chen, S.S., Wambsganss, M.W. and Jendrzejczyk, J.A.: "Added mass and damping of a vibrating rod in confined viscous fluids". Journal of Applied Mechanics, Transactions of the ASME, June 1976, pp. 325-329.

14. U.S. Waterways Experiment Station: "Vibration and pressure-cell tests - Flood-control intake gates, Fort Randall Dam, Missouri River, South Dakota". Technical Report No. 2-435, Vicksburg, Mississippi, June 1956.

15. Hardwick, J.D., Kenn, M.J. and Mee, W.T.: "Gate vibration at El-Chocon hydro-power scheme, Argentina". IAHR/IUTAM Symposium "Practical Experiences with Flow-Induced Vibrations" (Karlsruhe, Germany, Sept. 3-6, 1979), Springer Verlag, Berlin, 1980, pp. 439-444.

16. King, R.: "The added mass of cylinders". BHRA-Report TN-1100, 1971.

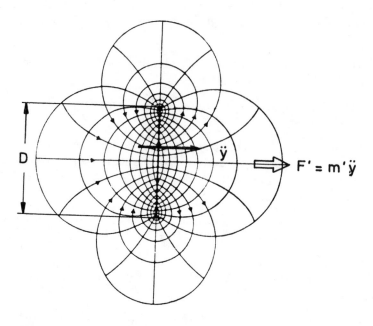

Fig. 1: Added mass flow at an accelerated flat plate

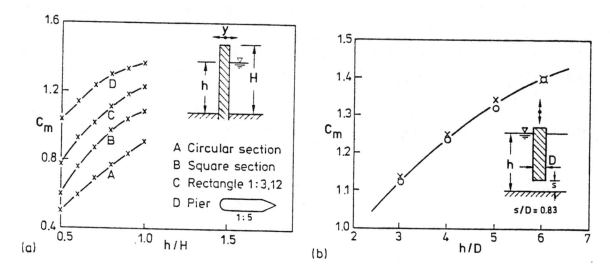

(a)

(b)

Fig. 2: Variation of added-mass coefficient C_m with surbmergence for
(a) a cantilevered pier (after Chandrasekaran et al.) and
(b) a flat-bottom gate (after Hardwick)

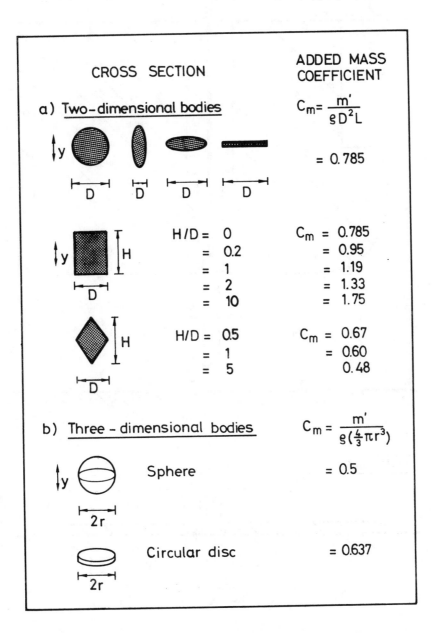

Fig. 3: Effect of body shape on C_m
(after Patton, Ref. 2)

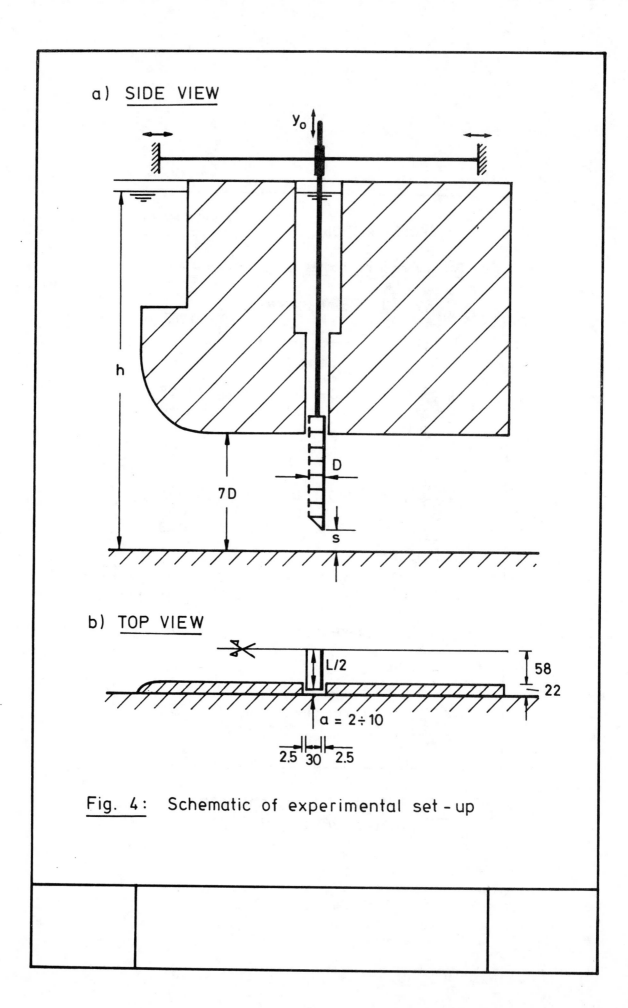

Fig. 4: Schematic of experimental set-up

Fig. 5: Photograph of the experimental set-up

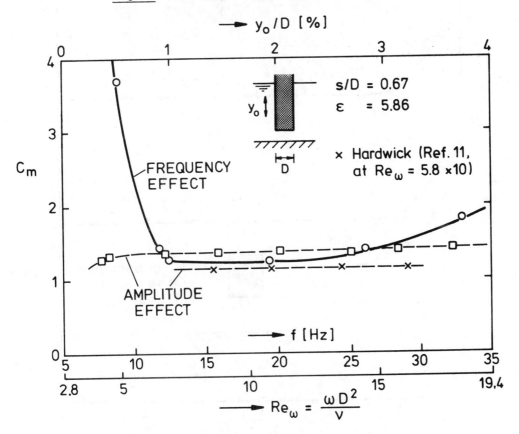

Fig. 6: Effect of vibration behaviour on C_m

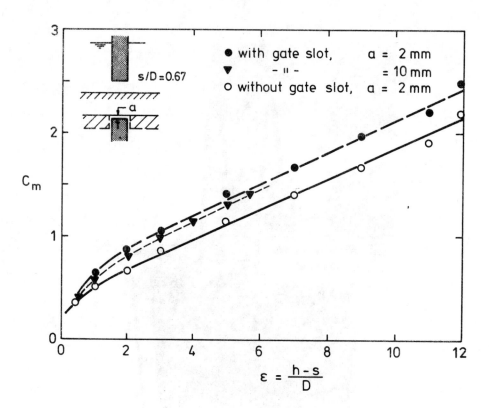

Fig. 7: Effect of submergence and lateral confinement by gate slots and side walls on C_m

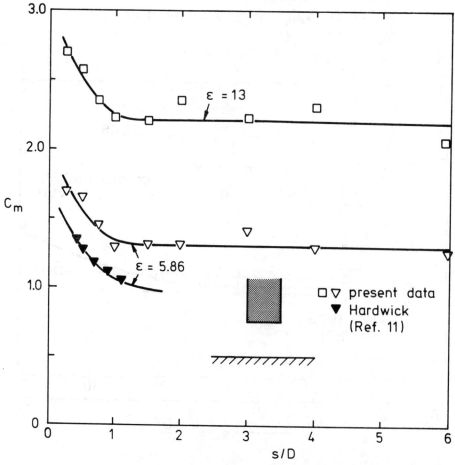

Fig. 8: Effect of gate opening on C_m

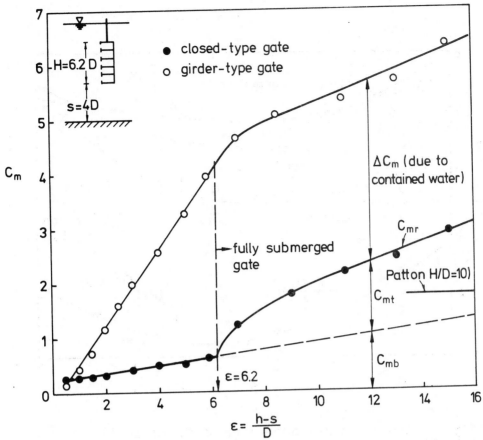

Fig. 9 : Effect of uncovered girders and topplate on C_m

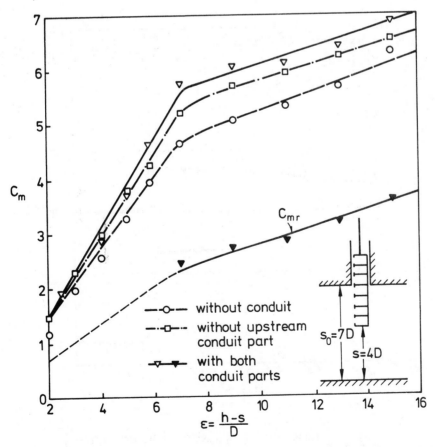

Fig 10 : Effect of confinement due to conduit parts on C_m

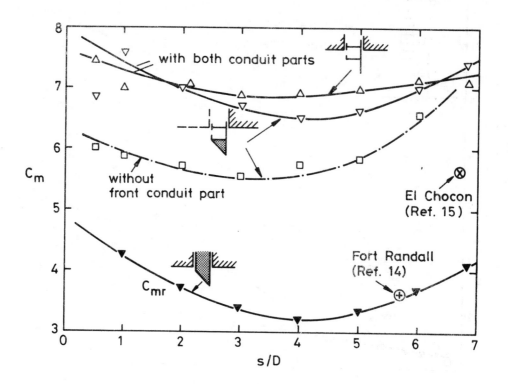

Fig. 11 : Effect of gate bottom form on C_m at $\varepsilon = 15$

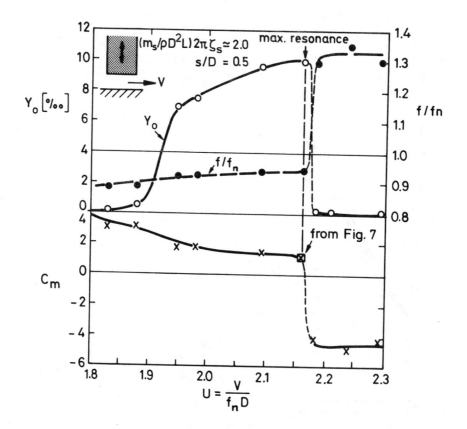

Fig. 12: Variation of C_m with reduced velocity over the resonance range of a hydro-elastic gate model

RESPONSE STUDIES OF THE STORM BARRIER
OF THE EASTERN SCHELDT

R.J. de Jong, R.M. Korthof and H.W.R. Perdijk

Delft Hydraulics Laboratory, The Netherlands

Summary

The storm surge barrier in the Eastern Scheldt, The Netherlands, is now under construction. The barrier will only be closed at extremely high tide and wind set-up but normally the barrier is opened and 75% of the former tide will still reach the Eastern Scheldt. It is built in the tidal channels using prefabricated main parts: concrete piers, concrete beams, steel gates.

Hydraulic investigations have taken several years to complete. This concerned first quasi-static loadings from waves and water level differences, and then the wave impacts and flow-induced vibrations were investigated. The special nature of the structure resulted in several complex phenomena.

The girder structures of the gates face the sea, thus inviting heavy wave loads and heavy wave impacts. Both gates and beams are flexible structures with natural frequencies of the same order, thus when dynamic responses have to be expected coupling phenomena will be present.

The experimental research is done using rigid models equipped with pressure cells, schematized mass spring models with force meters and elastic scale models with strain gauges, accelerometers and total force meters. The latter models were firstly used as research models for vibration phenomena. When the main design appeared to be reliable on this point the models were used as final models for a series of tests including various flow and wave conditions.

The paper concerns the use of the elastic scale models and gives some typical results concerning vibrations, wave impacts and quasi-static wave forces.

Organised and sponsored by
BHRA Fluid Engineering, Cranfield, Bedford MK43 0AJ, England

NOMENCLATURE

c	= damping factor	kg s^{-1}
c_w	= damping factor due to water	kg s^{-1}
E	= modulus of elasticity	N m^{-2}
f	= frequency	s^{-1}
F_w	= hydrodynamic force	kg m s^{-1}
F_H	= total horizontal impact load	kg m s^{-2}
F_v	= total vertical impact load	kg m s^{-2}
g	= gravitational acceleration	m s^{-2}
H_s	= significant wave height	m
I	= moment of inertia	m^4
k	= spring rigidity	N m^{-1}
k_w	= spring rigidity due to water	N m^{-1}
l	= length	m
m	= mass of structure	kg
m_w	= added water mass	kg
n_l	= scale factor of model related to length, etc.	−
$P_{I,II}$	= height of schematization triangle I and II	N m^{-2}
R_H	= total horizontal impact response	kg m s^{-2}
R_v	= total vertical impact response	kg m s^{-2}
t	= time	s
t_I, t_{II}	= impact time of schematization triangle I and II	s
x	= distance	m
τ	= pressure rising time	s

All quantities referred to are prototype values.

1. INTRODUCTION

In recent years there is a trend towards increasingly light civil engineering structures. As a consequence of this development the weight of the structure will, in comparison, contribute less towards the material stress than the stresses caused by the external static and dynamic load. In case of relatively high dynamic forces (compared to static forces and own weight) knowledge about these forces is of growing importance with respect to the strength of the structure. Statistical analysis of loads and knowledge about structure strength including fatigue play an important role. In this article emphasis is put on the hydrodynamical aspects viz.: flow induced vibrations and wave impact responses. Material stresses due to these aspects are to be reduced to an acceptable level by changing the geometry and mechanical properties of the design. But, besides that, a large amount of data on the hydraulic loads were necessary for the determination of the design loads, using the probalistic approach method. In this case a semi-probabilistic approach is practised in which the exceeding probabilities of hydraulic loads are estimated. The design loads then follow from the chosen failure probability. This yields a more balanced design in comparison with the deterministic method in which the external design conditions are chosen resulting in design loads with unknown exceeding chances. The semi-probabilistic method requires knowledge of load data in all relevant external conditions. This method is applicable for static and quasi-static loads, but less feasible for responses (Ref. 1).
The purpose of the storm surge barrier is a double one. Firstly it is designed to act as a water retaining structure during storm flood tides and secondly during ordinary conditions it has to allow through an average tidal range of about 75% of the existing range for environmental reasons. Attention had to be given to quasi-static and dynamic wave loads and flow-induced vibrations. Vibrations are especially treacherous and can initiate the failure of the structure. Moreover, during closure situations, wave impacts can cause high local pressures resulting in strong responses of the structure.

2. DESCRIPTION OF THE PROTOTYPE

The storm surge barrier has been built in such a way as to safeguard the low lying land adjacent to the Eastern Scheldt from flooding and to maintain the salt marine ecosystem in the estuary (see Fig. 1). The structure effectively embraces three openings in an otherwise earthen dam. The barrier with a length of about 3000 m will be closed during storm flood tides by lowering 63 steel gates suspended between 66 concrete piers (gravity structures). The prefabricated concrete piers will be placed with a centre-to-centre distance of 45 m. The piers offer a guide for the hydraulically operated steel gates. To minimize the height of the gates and thus minimize costs, upper and sill (lower) beams of concrete are placed between the piers (see Fig. 2). The sill beams form the top of a stony rockfill sill structure. For the construction of the concrete beams, closed box girders were chosen which combine the bearing and retaining function of the structure. The bearing construction of the gate consists of 2 or 3 main girders, which are provided with perforations (10%). The retaining function is performed by vertical shells which are mounted on the main girders via T-shaped girders also providing an extra through flow possibility. Close to the vertical shells a vertical truss construction is designed. This gate design answers the design requirements of minimal torsion stiffness (to allow unequal settlement of the adjacent piers). In order to achieve slight torsion stiffness for the torsion-stiff box girder type beams, they are supported on rubber bearings.
The shell construction of the gates forms, together with the upper beam and the sill beam a continuous vertical retaining plane. The upper beam has been projected at the Eastern Scheldt side of the gates in order to prevent heavy wave impacts on the horizontal bottom of the upper beam. As a consequence, the main girders of the gate have been positioned at the sea side so wave impacts on the gate will be of great importance to the gate design. Alternative designs, where this problem is absent, were either technically or economically less feasible.

3. BOUNDARY CONDITIONS

In designing the storm surge barrier, a semi-probabilistic method for the design load calculation has been used (chosen probability of occurrence of $2.5.10^{-4}$ per year). This requires knowledge of the three-dimensional probability density function of the storm surge level, the wave energy and the Eastern Scheldt level. Measured data of these parameters and their correlations were extrapolated into regions of low probability of occurrence. Fig. 3 illustrates the wide range of natural conditions. Rough indications are:

water level differences with open structure 1-3 m
closed structure 3-6 m

incoming wave heights from the sea upto H_s = 3.80 m
from the Eastern Scheldt upto H_s = 1.50 m

4. EXPECTATIONS CONCERNING THE RESPONSE OF THE STRUCTURE ON STREAM AND WAVES

Dynamic responses were expected in the gates and beams of the structure. The heavy piers (about 18.10^6 kg) act (by the soil properties) as strongly damped rigid structures. The steel gates vertically slide in guideways which are mounted in recesses of the concrete piers. When operating the gates, dynamic effects, (so-called stick-slip phenomena) can occur at the slideways. In order to prevent horizontal rattling of the gates the supports are locked into the recesses with pre-stressed rubber profiles.
The bending and rotation frequencies of the gates and beams are mainly determined by the mass and stiffness of the structure itself, the suspension stiffness of the hydraulic lifting system dependent on the position of the gate, the horizontal suspension stiffness of the rubber profile and also the rubber blocks which support the beams. But the influence of the water (mainly the hydrodynamic mass) is also important. In the pre-design stage the bending and rotation frequencies of the gate and beams were computed using a schematization of a linear-damped mass spring system which is represented by the following equation (Ref. 2).

$$(m+m_w) \; \ddot{x} + (c+c_w) \; \dot{x} + (k+k_w) \; x = F_w(t)$$

By estimating the hydrodynamic mass and ignoring the effects of hydrodynamic rigidity, the results shown in Fig. 4 were achieved. The spreading in frequency is caused by mechanical properties like variable, vertical and horizontal suspension stiffnesses, various gate heights, various dimensions of the concrete beams and boundary conditions like the submergence degree of the different elements. The low rotational frequencies caused by the necessary low torsion stiffness were of special importance. The calculations showed for different elements frequencies of the same order, which probably result in coupling effects; hence dominant modes could not be indicated. The transfer of vibrational energy to other modes due to the construction or by the water was also expected to be possible.
The occurrence of flow-induced vibrations is closely related to the shape. (See Fig. 5). During the pre-design stage a number of critical points were distinguished and 'no experience aspects' were assigned.
- Due to the lower end of the gate, flow separation is not well defined; it may separate either from the sea side flange of the main girder or from the lower end of the shell construction. Self-exciting vibrations can be expected especially in the first case (Ref. 3). Moreover during certain combinations of gate opening and water levels, conditions may occur which lead to flow instabilities resulting in dynamic excitation of the lower girder.
- High flow velocities will occur in the gaps between the gate and the beams when operating the gates. If coupling arises between the vibrating structure and the fluid forces in such a way that the vibration in the structure is amplified, a self-exciting vibration is originated.
- The influence of waves on flow-induced vibrations is unknown.
- The gate and beam constructions are closely situated to each other: as a result transfer of vibrational energy can easily occur.
It has to be noted that the long distance between the piers results in rather large girder surfaces which are exposed to vertical loads. As a result horizontal and vertical loads will be of the same order (especially at the gates in the shallow sections).

Forced vibrations due to wave excitation will not occur. The highest wave frequency of about 0.5 Hz is small in comparison with the natural frequencies of the elements. The wave loads however are expected to be structure-shape-dependent and need to be determined with models.

Wave impact loads will cause a dynamic response from the gates and the upper beams. The gates with their girders at the seaside will be loaded by wave impacts during the closure of the gates or in raised position at higher water levels, especially if one of the girders is near to the seawater level. The upper beams may also suffer wave impact loads, but they are less critical than the impacts on the gates. Therefore these will not be considered here. Data on wave impact loads are scarce and difficult to define while these loads are strongly dependent on local wave phenomena.
Thus necessary data for the final design are:
1. vibrations levels
2. wave impact loads
3. static and quasi-static loads for a broad field of circumstances.
But firstly a reasonable design has to be achieved with respect to vibrations and wave impacts.

5. SET-UP OF THE INVESTIGATIONS

To achieve a reasonable design with respect to vibration phenomena, critical points regarding self-exciting vibrations can be distinguished. In some cases theoretical approaches are possible but often in new designs new phenomena appear to occur in model or prototype (Ref. 2). Model investigations will be necessary to achieve data with some accuracy for flow-induced vibrations as well as for wave impacts. The best model for the reproduction of flow phenomena and structure response is an elastic scale model (replica). But such a model has disadvantages. Modifications are difficult to realize in such a model. Moreover wave impact phenomena present us with scaling problems. Such a model especially serves as a check of the final design. A more flexible approach was chosen with rigid section models and schematized mass spring models in which geometrical modifications do not affect the accuracy of the measurements. Rigid section models provide information on excitation (pressures and forces), while visual observations of wave impacts are also possible.
Mass spring models provide information about responses in one preferential direction. Because preferential modes could not be estimated in advance with the possibility that in this case vibrational energy could also be transferred to other modes, early investigations in a final elastic scale model still could lead to major design modifications. Hence a new element in the research programme between the schematized models and the final models had to be introduced: a changeable, schematized elastic scale model. This elastic scale model was designed in such a way that the main dimensions and overall stiffness of gate and beams were reproduced according to the pre-design incorporated results of the section models while a number of main changes were possible if inadmissable vibrations or loads should occur. Because the vertical dimensions of the gates and beams are at each opening different (related to the different depth of the tidal channels) natural frequencies are different as well. This has led to an investigation with two elastic scale models of different sections: one in a shallow section (water depth 14.60 m) and one in a deep section (water depth 30.00 m).

With the elastic scale model, responses due to wave impacts could also be determined. Impact phenomena related to the incompressible flow of water are when they belong to the type ventilated shocks (Ref. 5) correctly reproduced according to Froude's law. Some of the wave impacts, especially those related to the dynamic behaviour of air enclosures which are present during the impact and to the generation of elastic waves in the water do not follow this scale law. For the interpretation of the measured gate responses, detailed knowledge of the impact phenomena is indispensable.
This information was obtained in an investigation which comprised visual observations and impact pressure measurements in rigid transparent models.
Early conclusions from that investigation were that the impact pressures are primarily of the "ventilated shock" type with secondary effects due to oscillations of air enclosures. Therefore the following set-up for determining prototype gate responses to wave impacts has been practised (Ref. 6).

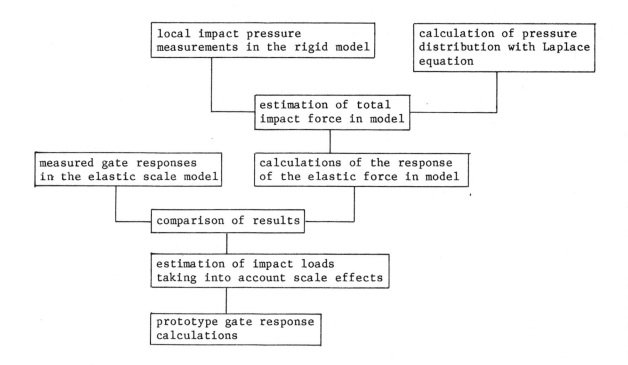

The wave loads had to be determined in the final design. The horizontal wave forces are, when the gate is closed and fully submerged in accordance to the Sainflou approach. Vertical forces can not be calcultated. For the (partly) opened gates wave forces appeared to be unpredictable due to the complex shape of the structure and the combination of waves and flow. Model investigations were necessary to show the susceptibility to different water levels and the linearity of the transfer functions (from waves to forces) for the wave height. The final elastic scale model was used to obtain these functions.

6.   DESIGN OF THE ELASTIC SCALE MODEL

At free surface flow it is essential for the reproduction of the flow pattern to scale correctly inertial and gravity forces so Froude number reproduction is attained. It can be deduced that for correct rididity the modulus of elasticity E has to be reduced on length scale $n_E = n_\ell$ (Ref. 7). This demand strongly shows the interdependence between the choice of the model scale and the kind of material. In case of bending of composite structures (like the steel gates) some freedom in the E-value is obtained when some deviation in the plate thickness is allowed such that the bending stiffness EI is scaled correctly ($n_{EI} = n_1^5$). This means a compromise concerning the geometric similarity. PVC ("Trovidur") was chosen as construction material for this scale model. In view of a correct scale relation with the E modulus a scale factor 65 should have been preferable for the gate. However, together with other aspects e.g the attainable accuracy of manufacturing a scale factor of $n_1 = 40$ was chosen. In the design of the gate with a low torsion rigidity the remaining magnitude was dependent on the local bending rigidity of the shell-shaped retaining plate. The reproduction of this element was an extra difficulty. The finite element computer programme set up for the prototype design was also used to calculate the model rigidity properties. The calibration of the elastic scale model now could also serve as calibration of the f.e.m. model. The model in the deep section is shown in Fig. 8. With the application of an elastic model hydrodynamic mass, damping and rigidity will be reproduced automatically apart for some secondary effects.

A disadvantage of the chosen material PVC is the too large structural damping.An extra damping occurred due to the changeability demand of the model. This kind of model is normally glued, but in this case major parts were fixed by small screws. The damping of the model should not exceed the prototype damping (which is however unknown) otherwise weak self-exciting vibrations are not detectable.

However the friction free support of the model gates with horizontal wires instead of

producting the pre-stressed rubber slide bearing in the prototype guideways resulted that point in a damping lower than the prototype. In this way the total damping c the dry gate model appeared to be 1.5 to 2.5% of the critical damping. In order to termine the wave loads in the final design the screwed connections were glued to wer the damping.
e model was established in a flume with sufficient length and depth and a width ual to the centre-to-centre distance of the concrete piers.

MEASUREMENT SYSTEM AND DATA HANDLING

was tried to obtain separate information on static and dynamic loads and the res-nse at each of the natural modes. To obtain this information special measurement and alysis techniques were used. The scheme is shown in Fig. 6. Normally strain gauges are plied in elastic scale models; they only affect a little the geometry and weight of e model. With regard to the changeable elastic scale models of the gates, there were sadvantages to apply only strain gauges, so small scale accelerometers were applied o (frequency range 0-100 Hz, natural frequency 500 Hz, range $\pm$ 10 g). Both devices ve their limitations. Strain gauges do not give direct information about the displa-ment and even local stresses are only obtained when one knows how strain gauges fect the local elasticity of the model. A special problem is the temperature compen-ation when the strain gauges are sometimes into or out of the water. Accelerometers – ositions can easily be changed. At the other hand accelerometers do not give direct nformation about stresses. The total vertical and horizontal forces were separately easured by replacing the gate supports by a taut wire system fixed to force meters. r the analysis of the signals a Hewlett Packard dual channel digital analyzer has een used during and after the measurements to analyze signals in the frequency domain nd time domain and to correlate two different signals. Transients like wave impacts ere also analyzed. This direct analyzing provided quick insight in the phenomena and ade an easy course for the tests possible. The wave height and wave forces were also nalyzed with a wave analysis computer programme to determine the transfer functions nd the probability distributions.

EXECUTION OF THE TESTS

n view of the wide range of combinations of natural conditions (see Fig. 3) to which he barrier is exposed the flume was specially equipped. There was an automatic steer-ng system for the water levels and an electronically-controlled wave generator (irre-ular and regular waves). Together with its measuring frame the gate could be moved lectrically. This system enabled a closing manoeuvre of the gate, while water levels ere adapted with an adaptation of the needed discharge.
he tests can be divided into three main blocks:
. dynamic behaviour of the gates and beams under flow conditions
. wave impact forces on gates and beams
. quasi-static wave load on the gates and beams.

ad A. The conditions can be described by three main variables: the sea level, the Eastern Scheldt level and the gate position. Every combination affected the mechanical system due to the vertical suspension stiffness (a function of the gate position) and the magnitude of the added mass, while the flow pattern it-self again was strongly coupled with these circumstances. To ensure that no severe circumstances for the vibrations would be missed, the whole field of circumstances had to be taken in sufficiently small steps. This was achieved by continuously varying the seawater level for several discrete gate positions and Eastern Scheldt water levels. Problematic ranges of circumstances were submitted to further investigations. In those cases the wave influence was also taken into account.
ad B. The most critical situations for wave impact loads were established by means of impact pressure measurements in the stiff models (see section 5). The pertaining findings were checked by tests with the elastic scale model, in which the clo-sing procedure of the gates was reproduced. In the most critical situations wave impact responses of the gates were measured using irregular waves.

35

ad C. The wave load tests were performed with irregular waves and a stationary current. For these tests only a fictive prototype wave spectrum on stagnant water
could be estimated without the influence of the barrier. In the model the prototype wave spectrum including reflection components from the barrier was first
established in stagnant water (for the procedure see Fig. 7). Next the same wave
generator adjustment was used for flowing water conditions. It was expected that
the current in the model transforms the wave spectrum in the same way as the
current does in the prototype. The complete procedure is shown in Fig. 7.

9.   RESULTS OF THE INVESTIGATION

9.1 Concerning the flow-induced vibrations:
The tests showed three vibration phenomena:

1. In the shallow section a self-exciting vibration of 3 Hz was found in the gate
   and beams. This was strongest in the following conditions:
   sea level at NAP + 2.75 m, Eastern Scheldt level at NAP - 1.90 m and gate
   opening 0.80 m (vibration amplitude $1.1 \cdot 10^{-2}$ m). The cause of the vibration
   had to be sought in the gap flow between the gate and the upper beam because
   it was there, with a small change of the flow pattern, that the vibration
   disappeared.
   The schematization of Fig. 10 is used to obtain an explanation. In case of
   relative low frequencies (compared to the natural frequency of the fluid which
   is probably true here (Ref. 7)) the discharge follows the gap width: with a
   positive displacement in y-direction the discharge increases ($y :: Q'$). Due to
   the inertia of the water column between gate and beam the local head difference $\Delta H'$ (downstream of the smallest cross-section) now increases with Q'.
   Because F is related to the pressure needed to accelerate the main gap flow F
   is proportional with $\Delta H'$ and with Q' and hence also with $\dot{y}$. This gives forces
   in phase with the vibration velocity which leads to negative damping (self-
   excitation) like in case of a plug valve (Ref. 8).
   The amplitude of a vibration in general will increase until the vibration is
   stabilized as a result of non-linear effects or until the construction collapses. In this particular case non-linear effects did limit the vibration amplitude. It has to be noted that there will be another stabalization point in the
   prototype because of the difference in damping. Fig. 9 shows the vibration
   mode of the gate and beams in the middle of the span.
   By fixing a little nose on the upperbeam (see Fig. 11) the vibration can be
   prevented. The critical flow section now is more downstream, by which the
   force F changes of sign ($\Delta H'$ increases upstream of the smallest cross-section). This gives forces in counter phase with the vibration velocity: this
   leads to a positive damping. Because the probability of occurrence of this
   condition was not small the nose was indeed applied in the design. The vibrations appeared to be sensitive to the presence of waves. Regular as well as
   irregular waves with related wave heights eliminated the vibration.
2. Self-exciting vibrations were also found at a condition with flow direction
   towards the sea: seawater level at NAP - 2.00 m. Eastern Scheldt water level
   at NAP + 2.00 m and a closed gate position. The vibration mode with a frequency of 2 Hz is shown in Fig. 12. A complete explanation for the vibration has
   yet to be found but it occurs probably because of a variation in pressure at
   the downstream side of the flange of the upper main girder. If the overflow
   nappe is aerated, the vibration almost disappears. Assume the discharge varies
   with the discharge opening: the discharge increases with a wider opening which
   will lead to higher negative pressures on the flange, resulting in a larger
   discharge opening, etc. This phenomenon agrees with the principal of the instability indicator (Ref. 8). Thus a self-exciting vibration is possible.
   Although this vibration can be prevented by mounting nappe spoilers at the
   flange of the upper main girder, constructive measures will not be taken because of the small probability of occurrance of this condition (less than 2.5
   10^{-4} per year). Realistic waves from the sea side reduced the vibration amplitude to about 50%, but the vibration did not disappear.
3. Finally a low frequency, high amplitude fluid oscillation was observed (f =
   0.065 Hz) in the deep section, at a fully-raised gate and at a sea level of

NAP +3.50 m and an Eastern Scheldt level of NAP - 0.20 m. The low frequency oscillation was found in the gate and in the upper beam and was accompanied by wave radiation of the same period. The wave height at the sea side was H = 0.60 m and at the Eastern Scheldt side H = 1.20 m (see Fig. 13). The phenomenon could arise because of an unstable flow separation at the underside of the gate viz. the flange of the main girder or the underside of the shell construction. In (Ref. 4) are indicated some configurations which are sensitive to unstable flow phenomena. For this case one finds conditions which are on the limit of stability conditions. Another explanation for this case is sought by Kolkman (Ref. 2).

The load on the gates due to the fluid oscillating was small in comparison with wind waves, and the danger of gate resonance did not exist. Because the circumstances at which this oscillation occurred, were not very realistic, no measures have been taken to prevent it. It was remarkable that the oscillation appeared when regular waves were present while irregular waves had a fully disturbing effect.

From the investigations concerning the dynamic behaviour of the gates and beams by flow it was concluded that the pre-design is satisfactory. Modifications which were important for the overall behaviour were not necessary. Hence the elastic scale models were maintained as the final model after glueing the vertical truss construction. In that function it is used to determine the wave loads and responses to the gates and beams.

9.2   Wave impacts:

From the tests in the rigid model a qualitative description of the wave impact phenomena evolved (see Photo 1). This was most important for the estimation of the pressure scales to be applied in the model. If air enclosures are present this will be recognizable in the signal by a typical vibration. For the interpretation of the pressure measurements and the establishment of pressure scales (see Ref. 6). The position of the gate during closing, with the upper girder at the instantaneous mean water level, provided the most severe condition for wave impacts.

From both the pressure-cell measurements and the motion pictures it was shown that two clearly defined wave impact phenomena occurred during one wave cycle:

type a   impact against the lower side of the upper girder (girder located at the mean water level). Mostly the instant of impact is when the wave front has reached a location at the back of the upper girder (see Fig. 14). From here wave pressure radiates to other parts of the gate.

type b   impact on the topside of the upper girder, occurring when the water mass overtopping the upper girder collides with the water mass rising between upper girder and vertical plate construction. It will be clear that type b occurs after type a (see Fig. 15).

The wave impacts of type b produce a less critical situation than those of type a, even if the most pessimistic scale relations are used, since:
- the resulting vertical impact force is directed downwards while the quasi-static wave force is directed upwards at the moment of the type b impact
- the horizontal impact force is smaller because of the small area involved.
Therefore only the type a impacts have been analyzed further here.

The general time history of the type a impacts is shown in Fig. 16. Two parts have been distinguished; a non-oscillating part with an impact time of 30-70 ms in the model and an oscillating part with an oscillation frequency of 200-500 Hz (in the model). The non-oscillating part has been schematized in two triangles as shown in Fig. 16. The smaller triangle can however still be related to dynamic phenomena where enclosed air can play a role. From scaling considerations this should have been more favourable than a Froude scaling. This possibility is not further considered.

37

From a test in the deep section with 1000 irregular waves (sea water level NAP + 3.30 m, upper gate girder at sea level, H_s = 3 m, water level difference 3.00 m) initially two large impacts of the type a have been selected, but later only one is chosen for the further analysis. The pressure distribution over the girders and rear shell plating is established from a combination of pressure measurements and calculations (Ref. 6).

The maximum impact forces (F) become converted to prototype:

F_{HOR} (kN)	F_{VERT} (kN)
30500	34000

There are no scale effects between both models, therefore the same phenomena are to be expected in the elastic scale models as in the rigid model.
The gate responses R_v and R_H of the elastic scale model are calculated from the total impact forces F_H and F_v by means of model-response calculations converted to prototype values (Froude's Law).
The results were checked with the results from the elastic scale model (see Photo 2) to verify the calculation method on the presumption of constant cross-sectional pressure distribution along the gate, following the set-up mentioned in section 5.
The total support responses from the elastic scale model measured once in 1000 waves amount to:

R_{Hor} (kN)	R_{Vert} (kN)
24000	14000

The horizontal response adequately agrees with the calculated response. The measured vertical response is 20 per cent smaller than the calculated one. Although the applied calculation method with a homogeneous pressure distribution along the gate width, tends to overestimate slightly the loads and responses the conclusion is drawn, that it is still a valid assumption in case of perpendicularly incoming waves.

Fig. 17, shows measured overall responses in the elastic scale model, including the quasi-static wave load response.

9.3   Quasi-static wave load on gate and beams:
In the deep section several situations with a certain head difference and several seawater levels were investigated with significant wave heights. It appeared that the transfer functions were not very susceptable to absolute seawater levels but they were for the immersion rate of the gates. Further the loads are within realistic conditions linear with the wave height.
An example of a transfer function for the horizontal and vertical wave load on the gate in the deep section is shown in Fig. 18.
The phase relation was also determined between the horizontal and vertical forces on the gates, especially for the largest forces. Depending on the circumstances, the phase shift varies between 70° and 120° the maximum vertical loads occur before the maximum horizontal loads. The phase shift between the vertical and horizontal load has resulted in a stress reduction in the gate of about 10 to 15%.

10.   FINAL REMARK

A good insight into the wave impact process was obtained by the simultaneous use of scale models and computation. This has led to a proper design of the girders within the constructional constraints. The loadings, especially the vertical ones, appeared to be high, both overall and locally.

However a drastic reduction in the horizontal quasi-static wave force was achieved due to the probabilistic approach, and the reduction in loading enabled the use of a spatial truss girder frame.

Although vibrations, wave loads and wave impacts were relevant for this alternative design they were less critical and could be tackled with a relatively short investigation programme, using the relevant results of the investigations of the plate girder gates.

References

1. Vrijling, J.K. and Bruinsma J., "Hydraulic boundary conditions".
   Symposium on hydraulic aspects of coastal structures, 1980.

2. Kolkman, P.A., "Development of vibration free gate design: learning from experience and theory".
   IAHR/IUTAM Symposium on practical experience with flow induced vibrations, Karlsruhe 1979.

3. Vrijer, A., "Stability of vertical movable gates".
   IAHR/IUTAM Symposium on practical experiences with flow induced vibrations, Karlsruhe 1979.

4. Naudasher, E., "Entwurfskriterien für schwingungssichere Talsperre Verschlüsse.
   Institut für Hydromechanik, Universität Karlsruhe, Veroffentlichung nr. 77, 1972.

5. Lundgren, H., "Wave shock forces: analysis of deformations and forces in the wave and in the foundation".
   Proceedings research of wave action Volume II, Delft 1969.

6. Ligteringen, H., Kooman, D., Korthof, R.M. and Stans, J.C., "Wave impact forces, consequences for gate design".
   Symposium on hydraulic aspects of coastal structures, 1980.

7. Kolkman, P.A., "Models with elastic similarity for the investigation of hydraulic structures".
   Delft Hydraulics Laboratory, publication no. 49, 1970.

8. Kolkman, P.A., "Flow induced gate vibrations".
   Delft Hydraulics Laboratory, publication no. 164, 1976.

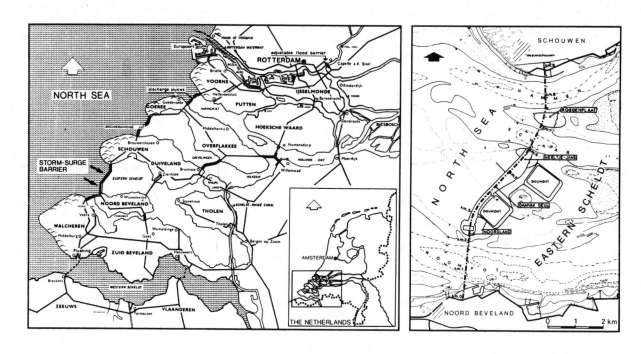

Fig. 1   Situation of the storm surge barrier

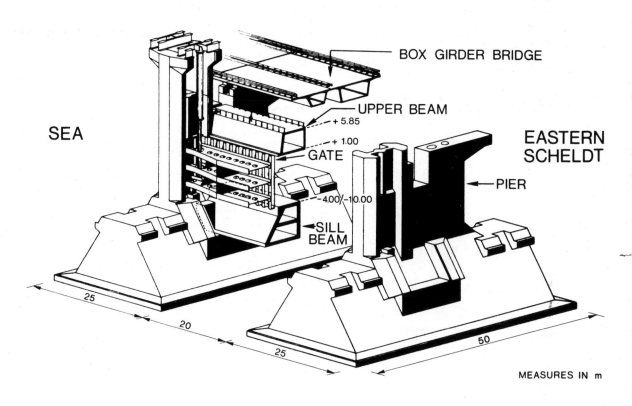

Fig. 2   Elements of the storm surge barrier

40

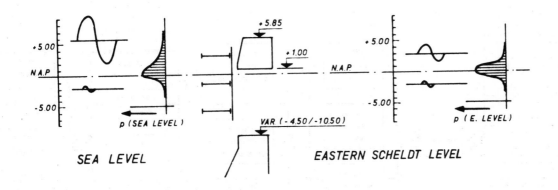

SEA LEVEL   EASTERN SCHELDT LEVEL

Fig. 3 Natural conditions

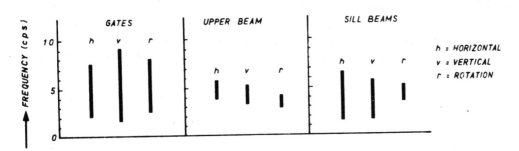

Fig. 4 Bending and rotation frequencies of gates and beams

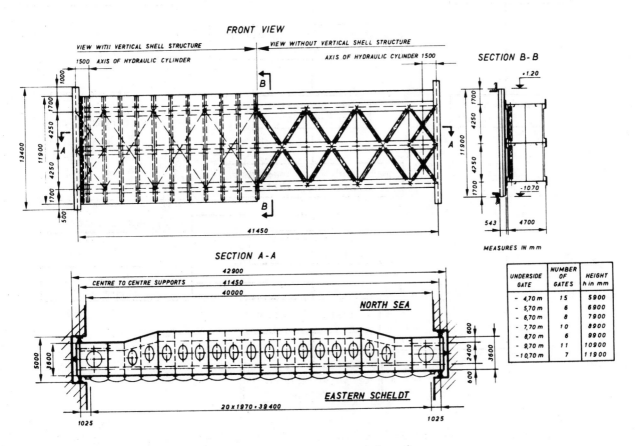

UNDERSIDE GATE	NUMBER OF GATES	HEIGHT h in mm
− 4,70 m	15	5900
− 5,70 m	6	6900
− 6,70 m	8	7900
− 7,70 m	10	8900
− 8,70 m	6	9900
− 9,70 m	11	10900
−10,70 m	7	11900

Fig. 5 Gates of the storm surge barrier

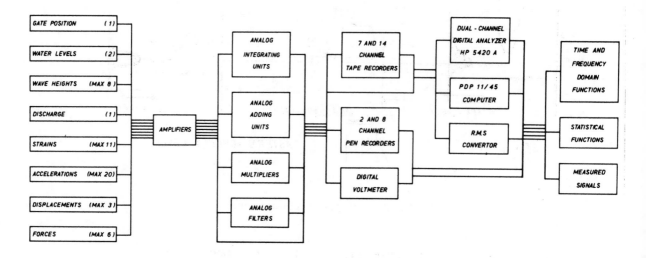

Fig. 6 Scheme of measurement and analyzing system

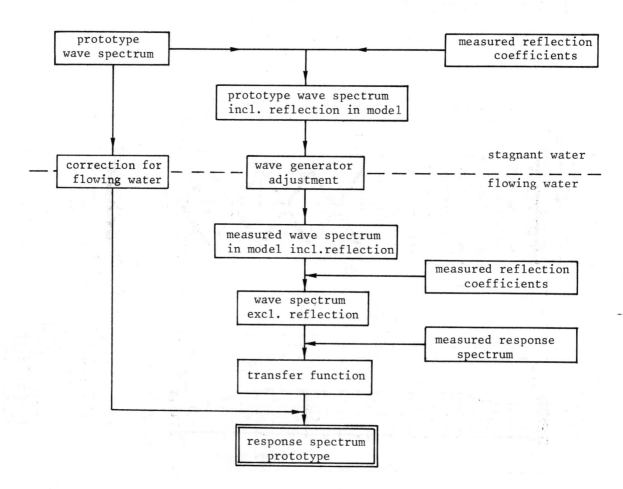

Fig. 7 Scheme for determining transfer functions from wave heights to prototype forces

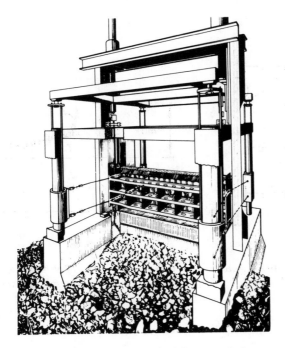

Fig. 8 Elastic scale model

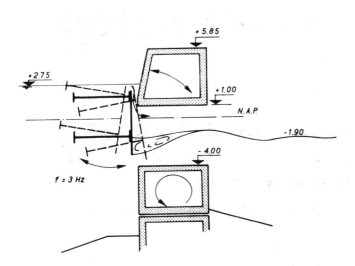

Fig. 9 Vibration due to gap flow at
        partially raised gate

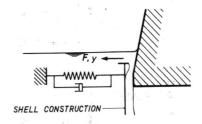

Fig. 10 Schematic mass
         spring system

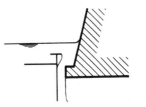

Fig. 11 Nose on upperbeam

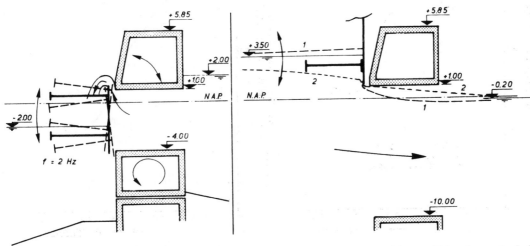

Fig. 12 Vibration due to nappe flow

Fig. 13 Fluid oscillation at fully
         raised gate

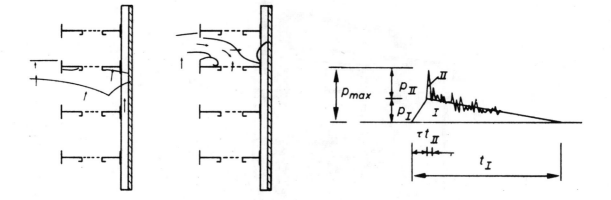

Fig. 14 Wave impact   Fig. 15 Wave impact   Fig. 16 Schematization of
        "type a"   "type b"   "type a" wave impact

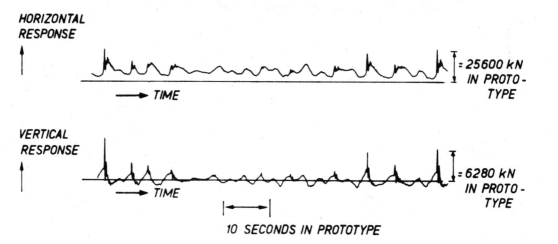

Fig. 17 Typical time history of total gate responses from e.s.m.

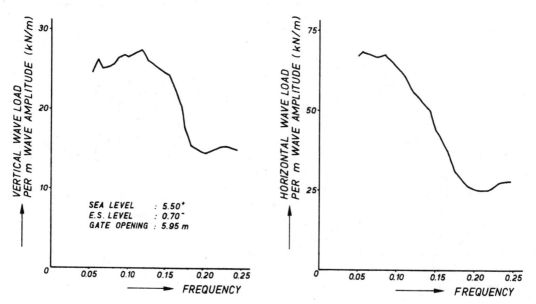

Fig. 18 Transfer function for vertical and horizontal load

44

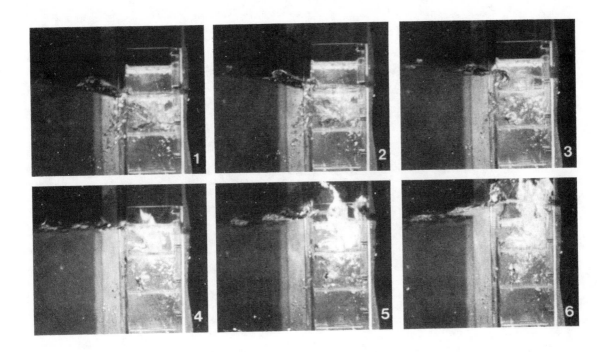

Photo 1 Photograph sequence showing
development of wave impact
type <u>a</u> (time interval 0.22 s)

Photo 2 Wave impact on the
elastic scale model

SOME LESSONS FROM HYDRO AND AERO-ELASTIC
VIBRATION PROBLEMS

S. Singh, V.S. Sakhuja and T.C. Paul

Irrigation and Power Research Institute, India

Summary

Elastic structures when immersed into fluid flows frequently, exhibit vibration phenomena. Resonance frequencies of such structures, often lead to dangerous vibrations and serious damage, if not ensured well above the dominant excitation frequencies. In this paper three prototype cases of flow induced hydro and aero-elastic vibrations, their modelling, and the lessons derived have been presented.

The first study deals with finding of causes of ripping off of steel liner from the bellmouth entrance of an irrigation tunnel at the prototype. From the data observed on the model it was inferred that pulsating differential pressure of high magnitude set the liner to vibrate and caused its ripping off. The remedy lies in appropriate anchoring of liner into concrete so as to increase stiffness. With these measures the performance of tunnel at prototype is now satisfactory.

The second study describes the dislocation of bulkhead covering the mouth of tower type intake due to the closure of emergency gate during trial tests. It was inferred from the model tests and analytical analysis that the intake shaft with top covered acted as a stand pipe or surge shaft in which under steady flow condition air was under compression. Closure of emergency gate during its trial tests induced aero-elastic vibrations in the air column which lifted the bulkhead, caused air-water spout to emerge out from the intake tower and damage to bulkhead. The mishap could have been averted if an analytical analysis had been carried out before undertaking trial tests. Mathematical modelling has shown than in such conditions even a slight disturbance (due to gate) at the interface of air and water boundary can build up strong aero-oscillations of air column sufficient enough to dislodge the bulkheads.

In the third study the case of flow induced vibrations of a masonry pier of a regulating fall has been reported. The model studies as well as field observations indicated that undular hydraulic jump fluctuating high up the glacis slopes and with asymmetrical position in adjacent bays of the fall due to location of a skew bridge downstream of it would have caused the resonance vibrations in the piers and consequently their damage. The remedy lies in raising the height of pier so as to add to their mass or extending these to join the piers of the skew bridge to have compartmentalized flow and added mass.

Organised and sponsored by
BHRA Fluid Engineering, Cranfield, Bedford MK43 0AJ, England

NOMENCLATURE

A = Cross-sectional area of liner

C = friction coefficient such that friction loss in pipe line under velocity $V = CV^2$

C_n = coefficient for end fixity and mode

d = characteristic transverse dimension of shape

E = modulus of elasticity

f_n = natural frequency of vibration in fluid in the direction of boundary

f_s = frequency of vortex shedding

g = acceleration due to gravity

I = area moment of interia about the axis of concern

L = length of liner, pipe line

ℓ = length of stand pipe/length of the pier

p = original absolute pressure

p_i = initial pressure on water under steady flow conditions

p_f = final pressure during closing of gate

Q = discharge through the pipe line

R = ratio of shaft area to the tunnel area

S = Strouhal number

T = Time period of vibration

t = time at any instant

U_m = maximum approach velocity

V = velocity of flow through pipe line at any instant

V_i = initial volume of air under compression

V_f = final volume of air under compression

w = weight per unit length of vibrating body

y = height of water in the stand pipe at any instant above the level corresponding to steady flow with Velocity V_1

y_m = maximum change in y

ρ = mass density of vibrating body

∂ = derivatives

1. INTRODUCTION

Elastic structures frequently exhibit vibration phenomena when immersed in fluid flows either wholly or partly. These are induced due to the mutual interaction of the forces due to inertia, fluid dynamics, and elasticity. Resonance frequencies of such structures, if not ensured well above the dominant excitation frequencies, often lead to dangerous vibrations and serious damages. In India, a panel(31.7 m long) of central training wall of partly completed Bhakra Dam was knocked off completely on August 7, 1958 when overflow took place(Ref. 1). This failure was attributed to the turbulence in the hydraulic jump region attended by high order($\pm$6 m)differential pressures causing forced oscillations near resonant frequency of the training wall. On August 21, 1959, after it developed serious vibrations and cracks in the floor on the upstream side of the regulation gates(measuring 7.62 m x 3.58 m) failure of the 15.24 m x 12.19 m x 21.34 m hoist chamber occurred (Ref. 2)when reservoir level against the partly completed Bhakra Dam was at El.438.91 m. In view of such a severity of the flow-induced vibrations of hydraulic structures, this paper presents a few cases from author's experience gained in the course of hydraulic model studies undertaken to ascertain the causes for mishaps in prototypes. The presentation is so oriented as to deduce some design criteria for application elsewhere as well.

2. THE CASE STUDIES

2.1 Ripping off of steel liner-Beas Dam at Pong(India).

During the construction stage of the 129.5 m high(rising to El.435.9 m)earth core-cum-gravel shell dam on the River Beas at Pong (India), five tunnels each of 9.144 m internal diameter were provided to divert the flow. In the post diversion stage, the river level intakes were closed and identical tower-type intake structures located at El.375.21 m were provided for feeding each tunnel. Three of these tunnels were later used as penstock tunnels and two designated as T1 and T2 were subsequently provided with outlet works, to be permanently used for meeting irrigation requirements. The general layout of outlet works in tunnels T1 and T2 is shown in Fig.1. Each tunnel has two rectangular con-_duits of 3.20 m x 2.14 m with bellmouth entrances. Each conduit has been provided with two high pressure slide gates(the largest ever used in India)installed in tandem.

The reservoir was filled in the monsoon of 1974. During 5-9 October, 1974, trial runs of left and right conduits in tunnel T2 were undertaken when the reservoir level was at El.402.74 m(head equal to 65.3 m)and either conduit discharging 187 m^3 s^{-1}. It was observed that although the left conduit performed satisfactorily yet in the right conduit typical noise accompanied by periodic hissing persisted. Inspection after run of about two hours of left conduit and six hours of right conduit, when the reservoir had been depleted to a level below the intake bench, revealed that the 20 mm thick steel liner in the right conduit at the bellmouth entry had been damaged and badly sheared/ripped off as shown in Fig.2. The apparent causes to which the collapse or failure of steel liner could be attributed are :i)Cavitation associated with high velocity flow, as in the case of Tarbela Dam(Pakistan)(Ref. 3), since in 187 m^3 s-1 discharge velocity of the order of 27.4 m s^{-1} was attained in this region, ii)water somehow got behind the liner and due to differential pressures during one conduit running collapse of the liner occurred, iii)the liner started vibrating either independently or together with the concrete mass on account of differential pressure pulsations and failed due to the fatigue of the material as happened in the case of divide wall of Bhakra Dam(Ref. 1).Vibration of structural elements due to high velocity flow leading to subsequent failures is not uncommon(Ref. 4), (iv)The asymmetry in the failure(ripping off of

steel liner in right conduit only)suggested that the failure phenomenon
was time dependent because left conduit was operated for only two
hours with no failure,while right conduit operated for six hours and
steel liner failed.

2.1.1 The results of model studies

With a view to ascertain the causes of the failure of the
steel liner,studies were undertaken on hydraulic model constructed to
a geometrically similar scale of 1:21.On the model of outlet works in
tunnel T2,the bellmouth entrances,outlet conduits,the emergency and
regulation gate housings and the conduit downstream of the regulation
gates upto its end was simulated in 6 mm thick plexi glass(Ref. 5).A
length of 51.2 m of 9.14 m diam.tunnel upstream of the outlet works
was fabricated out of galvanized iron sheet and its upstream end joined
to a cylindrical pressure tank for the supply of fairly stable high
pressure flow into the tunnel.Piezometers,as shown in Fig.3,were
installed at the central pier of bellmouth entrance because according
to the Inspection And Control Authorities of the project shearing of
the steel liner in the right conduit commenced at the central dividing
pier.The model was operated for reservoir levels at El.384.05 m and
El.402.74 m to observe the pressure distribution and pulsating
differential pressures at various piezometers located at stratgic
points in the bellmouth entrance so as to ascertain the possible causes
of ripping off of steel liner in this region.

The observations showed that the pressures in this reach were
no where negative contrary to the situation obtainable at Tarbela Dam
(Ref. 3)although the velocity of flow was of the same order.That this
damage was not due to cavitation is also supported by the visual
observation of pattern of damage.

The observations of differential pressures recorded at regular
interval of five seconds for a duration of five minutes are shown in
Fig.4 for the case of piezometers number 4.The trace of differential
pressure fluctuations as obtained with pressure cell(LVDT type)and
pen recorder in this case are also shown in Fig.5.The data obtained
at three piezometers numbering 2,4 and 6 are given below:

Reservoir level elevation , m	Maximum differential pressure variation,m of water	Frequency of differen-tial pressure pulsation Hz
a)One conduit running(other closed),gate opening 100 percent		
384.05 Piezometer No.2	8.39 to 10.44	Not recorded
4	23.56 to 26.57	4.4
6	10.76 to 12.74	Not recorded
402.74 Piezometer No.2	11.34 to 14.79	Not recorded
4	37.65 to 39.66	2.4
6	15.46 to 17.48	Not recorded
b)Both conduits running,gate opening 100 percent		
384.05 Piezometer No.2	2.5 to 8.08	3.3
4	-2.35 to 2.35	3.1
6	-0.67 to 11.43	3.1
402.74 Piezometer No.2	5.37 to 11.16	Not recorded
4	-1.34 to 6.71	Not recorded
6	0 to 14.79	Not recorded

It is seen that the frequency of differential pressure
pulsations and the magnitude of differential pressure are more when

one conduit is running than when both conduits are running.At the time
of failure of steel liner(reservoir level El.402.74 m)in right conduit
the magnitude of differential pressures varied from 37.65 m to 39.66 m
and the frequency of pressure pulsations was 2.4 Hz.

The natural period of vibration of the liner(unanchored to the
concrete)treating as membrane was also worked out from the following
formula(Ref. 6):

$$T = 0.281 \ L^2 \ \sqrt{\rho A/EI}$$

where T is time period, A is cross sectional area, ρ is mass density of
material, ρA is mass per unit length,EI is section stiffness,L is
length,and I is second moment of inertia of area of cross section
about the central axis.

In the case under study(Fig.2):

L = 4.04 m,A=4.04 m2, ρ = 2386/g Kgm^{-3},E = 2.1x10^{10} Kg m^{-2},

d = 0.02m, and $I = \dfrac{Ld^3}{12} = 2.69 \times 10^{-6}$ m^4

Thus T = 0.603 or Natural frequency f_n = 1/T = 1.7 Hz

2.1.2 Inferences

From the studies it is thus inferred that the only reasonable
cause leading to failure of steel liner duly supported by the experime-
-ntal data is the effect of pulsating differential pressures of the
order of 37.65 m to 39.66 m with a frequency of 2.4 Hz.The frequency as
observed on the model is,however,not of the same magnitude as the
natural frequency of vibration of steel liner(ripped off)portion,but
is of the same order.

The remedy against any such reoccurrence lay in appropriate
anchoring of liner into concrete so as to restrict the degree of
freedom of vibration to increase stiffness.At the prototype the steel
liner was subsequently properly anchored in both the tunnels T1 and T2.
Further,to counteract the unbalanced water pressure being exerted from
outside the liner,drainage holes(one per each panel)have also been
provided in both the tunnels.These are now functioning satisfactorily
and no reoccurrence of damage has been noticed.

2.2. Damage to bulkhead-Beas Dam at Pong(India).

Out of the five diversion tunnels,provided during the
construction stage of Beas Dam at Pong(India)stipulated under para 2.1,
the tunnels T1 and T2 were provided for controlled irrigation releases
after the diversion stage was over and tunnels P1 and P2 converted into
penstocks.In tunnel P3 final closure plug was to be placed upstream of
intake shaft after outlet works in tunnel T1 and T2 were completed.
This was to be achieved with the aid of closure bulkhead in 6.40 m
x 3.05 m bulkhead gate constriction.After the construction of bulkhead
constriction and emergency gate constriction,as shown in Fig.6,the
tunnel P3 functioned as P3R(constricted tunnel)to escape river flows
during monsoons of 1972.The bulkhead shaft was provided with grating
at its top and the bellmouth entrance of tower-type intake was covered
with the help cylindrical and flat bulkheads(Fig.6)weighing 58 t.

The tunnel P_3R functioned satisfactorily(with maximum
reservoir level at El.353.57 m)during the flood season of 1972.In the
year 1974,the releases were made only through tunnel P_3R under maximum
reservoir level at El.406m.After the reservoir had depleted,the
inspection revealed that the grating at the top of bulkhead shaft was
partly missing and partly damaged.Again on August 6,1975 when the
reservoir level was at El.400.8 m vertical-lift,fixed-wheel emergency
gate was lowered in 100 s for testing purposes,water surge was observed

near the bulkhead shaft location and upstream portal.Subsequent
lowering on December 15,1975(reservoir level at El.408.19 m)however,
did not give rise to this phenomenon.As tail race dried up,twisted
liner pieces and stiffners were recovered from the stilling basin.
Under-water inspection indicated that the flat bulkhead was also missing.
The plates and stiffners of the flat bulkhead were,however,recovered
from the stilling basin.

2.2.1   The results of model studies

           Model studies(Ref. 7)were undertaken to:1)determine the
possible cause of damage to grating and flat bulkhead:ii)effect of
closure of emergency gate in 100 s on pressure distribution in the
tunnel with reservoir level at El.400.81 m,and (iii)extent of disloca-
-tion of flat bulkhead and height of surge.Since the problem was
associated with pipe flow as well as formation of vortices(the possible
cause of damage to grating),two separate models referred as Model-I on
a geometrically similar scale of 1:45(in a tank without representing
boundary conditions)and Model-II on a geometrically similar scale of
1:20(representing exact boundary conditions of reservoir area around
the intake)after Anwar's criteria(Ref. 8)were constructed.The weights
of bulkheads were correctly reproduced on the models according to
modelling laws.To facilitate closure of emergency gates in stipulated
time on the models the gate shaft was connected to electric motor
equipped with gear system.

2.2.1.1.   As the reservoir level gradually rose from El.375.21 m to
El.378.56 m,air-entraining vortex started forming at the bulkhead shaft
which vanished with reservoir level at El.387.71 m.The intensity of
vortex was maximum at reservoir levels between El.384.05 m to El.386.18
as shown in Fig.7.

2.2.1.2   With lowering of emergency gate at reservoir level
El.400.81 m an upsurge(air-water spout)rising to El.402.89 m on Model I
and to El.404.16 m on Model-II emerged through the intake structure.No
rise in water level at the upstream portal was observed on   both
the models.The flat bulkhead overtop the cylindrical bulkhead covering
the intake was dislocated during the gate closure.The dislocated
position on Model-I is shown in Fig.8.The stuck in position of flat
bulkhead at the prototype in the intake shaft is also shown in the
same figure for comparison.The cylindrical bulkhead was not observed
to have dislocated.The increase in pressure at various piezometers due
to lowering of gate in 100 s is shown in Fig.9.The increase in pressure
at piezometer number 27(at the base of intake shaft)varied from 37.03m
to 37.86 m.The instantaneous pressure increase due to closure of
emergency gate was also checked with the aid of LVDT pressure cell
connected to pen recorder.

2.2.1.3   The observations indicated that the damage to the grating
previously placed at the top of bulkhead shaft was due to formation of
air-entraining vortex during the course of filling of reservoir.The
air carried down the bulkhead shaft could not obviously travel upstream
because of high pressure flow but was locked at the top of flow in the
rectangular section of bulkhead constriction reach downstream of gate
slots as well as in the stagnant zone at the abrupt expansion section.
With reservoir level at El.400.81 m and emergency gate fully open,The
flow through the tunnel P_3R is contributed by the upstream portal at
river bed level as well as the bulkhead shaft.The pressures in this
case are everywhere positive.lowering of emergency gate in 100 s caused
increase in the magnitude of pressures all along the tunnel upstream
of it.The air locked in the main intake shaft got compressed and
traversing path of least resistance escaped through the intake shaft
in the form of air-water spout thereby dislocating the flat bulkhead.

2.2.2 Analytical analysis

           For analytical analysis,the intake shaft with bulkhead

resting at its bellmouth entrance could be looked upon as a closed stand pipe or surge tank of length' ℓ ' above the original water level and the penstock tunnel drawing supplies mainly from the low-level intake and partly from the bulkhead shaft.For such a case,Gibson(Ref: 9) has shown that at any instant't'after giving change of load,the equation of motion in the pipe line may be written as:

$$\frac{p\ell}{\ell - y} - p + y - C(V_1^2 - v^2) = -\frac{L}{g}\frac{\partial v}{\partial t} \quad \text{------------(1)}$$

where p is the original absolute air pressure;L is length of pipe line, y is height of water in the stand pipe at any instant,above the water level corresponding to steady flow with velocity,V_1.V is velocity of flow in pipe line under steady flow conditions before change of load;and C is friction coefficient such that friction loss in pipe line under velocity V(during unsteady period) $= cV^2$; ℓ is the length of stand pipe above the original water level.

On the basis of Johnson's(Ref. 10)assumption that a sufficiently close approximation is attained by putting $\partial C(V_1 - v^2) = y$, the maximum value of y i.e. y_m is given by:

$$y_m^2 = 2p\ell \; \log_e \left(\frac{\ell - ym}{\ell}\right) + 2py_m + \frac{L}{gR}(V_1 - V_2)^2 + c^2(V_1^2 - V_2^2)^2 \quad \text{---------(2)}$$

Here,V_2 is the velocity of flow in pipe line under steady flow after change of load.
For the case under study,the following data are known:

Reservoir level	=	El.400.81 m
Bottom elevation of bulkhead	=	El.375.21 m
Water level in intake shaft	=	El.349.91 m
$\therefore \ell = 375.21 - 349.91$	=	25.30 m
p_1	=	14.63 m
L	=	195.68 m
Q_1,discharge through the tunnel	=	499 m³ s⁻¹
V_1	=	7.6 m s⁻¹
$\frac{V_1^2}{2g}$	=	2.94 m

$C =$ head loss/$V_1^2 = 400.81 - \frac{(370.36 + 2.94)}{(7.6)^2} = 0.476$

$R =$ Shaft area/tunnel area $= 0.563$

Weight of bulkheads per m² $= 58$ t/74.72 $= 0.78$ tm⁻²

Weight of water column above bulkhead $= 400.81 - 377.65 = 23.16$ t m⁻²

$\therefore$ Total weight acting on the bulkhead or the thrust required to lift the bulkhead $= 23.94$ t m⁻²
$= 23.94$ m of water

Let V_i and V_f be the initial and final volumes of air under compression and p_i and p_f the corresponding pressures.
Then $V_f = \left(\frac{p_i}{p_f}\right)^{1/K} \times V_i$ where K $= 1.412$

For $p_f = 23.94$ m; ℓ computes to 17.85 m

$\therefore y_m = 25.30 - 17.85 = 7.45 \, m$

For this value of y_m eqn(2)yields $V_2 = 5.96$ m s⁻¹and ratio

$V_2/V_1 = 0.78 = Q_2/Q_1$, where Q_1 and Q_2 are discharges through the tunnel before and after change of load.

2.2.3   Inferences

Thus as emergency gate is lowered to cause reduction in discharge to the extent of 22 percent or the gate is lowered by 38 percent (assuming that the coefficient of discharge of the gate does not significantly change at partial gate openings) aero-elastic oscillations so set in attain strength enough to dislodge the bulkhead.

2.3 Damage to piers of regulating fall

A Regulating Fall in a lined main canal designed for a discharge of 192.8 $m^3 s^{-1}$ has four piers each 0.70 m wide and five spans each 3.81 m width. The details of the structure are shown in Fig.10. At a distance of barely 71.93 m on its downstream a skew-type road bridge comprising four piers and five spans is also located. During 1976, two piers number 1 and 3 (from left) of the regulating fall were damaged. The damage comprised cracks both in the horizontal direction and vertical planes and 0.25 m x 0.08 m holes in the piers all in the region where hydraulic jump oscillated upstream and downstream against the piers as shown in Fig.11. On pier number 3 two vertical cracks were also noticed in the walls supporting the gates.

2.3.1   The results of model and prototype studies

2.3.1.1   To find the causes of damage the problem was examined on a model constructed to a geometrically similar scale of 1:20 as well as in the field and the following results were obtained. The observations of flow conditions obtained on the model for a discharge of 192.8 $m^3 s^{-1}$ indicated that in the region of damaged portion of piers undular hydra--ulic jump fluctuates. The position of hydraulic jump is not symmetrical in all the five bays due to the location of skew bridge on the downstream of fall. Also due to the unequal up and down fluctuations water spilled over the piers from the bay with higher level to the one with receeding jump formation. This process is reversed the next moment.

The instantaneous differential pressures on the two faces of the piers due to turbulence in the hydraulic jump region was of the order of only 0.36 m as observed with a pressure cell and so could not be considered responsible for damage to the piers. Maximum velocity of approach at the pier was observed to be 4.64 m s^{-1} its average value as 4.42 m s^{-1}. Velocity of flow just upstream of hydraulic jump was of the order of 4.90 m s^{-1}, giving value of F_1 as 1.10. This order of Froude number was responsible for the formation of undular jump.

2.3.1.2   The results thus indicated that the only cause of damage to the piers was the formation of asymmetric undular jump in the bays high up on the glacis slope which set up flow induced vibrations to the piers and consequent damage. The data were further checked up in the field where observations were made for a discharge of 173.2 $m^3 s^{-1}$. The observations of flow conditions confirmed the findings on the model, i.e. the location of hydraulic jumps was not symmetrical in all the five bays and fluctuated widely from gate line to down below causing a difference of 0.91 m to 1.22 m and thereby generating cross flow from one bay to the other. A cycle of variation in the jump position from its highest to the next highest position takes about 6 to 7 s. The positions of hydraulic jumps at two instances are shown in Fig.12. The depth and velocity of flow before the jump were observed to be 1.72 m and 5.31 m s^{-1} yielding value of $F_1 = 1.29$. The depth at the full supply discharge was expected to rise to 3.87 m at the upstream end of piers.

2.3.1.3   An attempt was also made to calculate the frequency of vortex shedding, f_s, and natural frequency of pier f_n to ascertain whether resonant frequency is achieved in this case or not. The frequency of vortex shedding $f_s = S U_m/d$ after Pennino (Ref. 11), where fs is frequency of vortex shedding;S is Strouhal number; U_m is maximum approach velocity,and d is the characteristic transverse dimension of shape.This gives value of f_s= 1.65Hz for S= 0.25 if the piers are taken to be cylindrical in shape.The piers under consideration are,however, rectangular of length 10.93 m and width 0.70 m.As such the value of f_s will actually be less than 1.65 Hz.The natural frequency of the pier when treated as a beam subjected to vibration computed from the formula

$$f_n = Cn \sqrt{gEI/wl^4}$$ after Pennino(Ref. 11)gives a value close to 1.Here f_n is natural frequency of vibration in fluid in the direction of bending; C_n is coefficient for end fixity and mode; w is the weight per unit length of the vibrating body,and l is length of the pier.Accordingly,the ratio f_s/f_n is of the order of 1 causing resonant frequency in the pier. It is thus established that flow induced vibrations causing resonance in the region of hydraulic jump are responsible for causing damage to the piers.

2.3.2 Remedial measures

The various remedial measures tried on the model suggested that at the first instance,the cross flow should be prevented.This could be achieved by raising the height of piers to El.251.46 m.This will also add to the mass of piers thereby averting sympathetic resonance and consequent damage.The other remedy lies in extension of piers downstream so as to join the piers of the skew-bridge to form compartmentalized flow.

3. CONCLUSIONS

The failures at the three prototype installations presented above due to hydro or aero-elastic vibrations as examined both in the field and on hydraulic models suggested the following lessons:

3.1   In a tunnel when outlet works comprising more than one high pressure conduits are provided their asymmetric operation needs to be avoided failing which high order differential pressures pulsating with frequency equal to that of the steel liner may consequently lead to its ripping off.In some eventualities steel liners should be appropr--iately or adequately anchored in to concrete so as to increase stiff--ness or to restrict degree of freedom to vibrate.Further,the unbalanced water pressure being exerted from outside the tunnels can be counteracted by providing drainage holes.

3.2   At the diversion stage when the flow in the tunnels is through low-level(river bed level)intakes and intakes for the ultimate stage have also been provided before any venture to test or operate gates is undertaken a hydraulic or mathematical modelling study is desirable.Mathematical modelling has shown that in such conditions even a slight disturbance(due to gate)at the interface of air and water boundary can build up aero-oscillations of air column strong enough to blow off bulkheads.

3.3   The asymmetric formation of undular hydraulic jump against piers can set piers into vibration and consequent damage.Under such conditions remedies suggest the elimination of the very cause leading to asymmetric hydraulic jump,compartmentalization of flows(to avoid cross flow)and increasing mass of piers by raising adequately so that their natural frequency of vibration does not correspond to that of differential pressures.

4. REFERENCES

1.   Uppal,H.L.,Gulati T.D. and Sharma B.D.:" A study of causes of damage to the central training wall,Bhakra Dam spillway."Journal of Hydraulics Research,IAHR, 5,No.3,1967,pp.209-224.

2.   Uppal H.L.,Gulati T.D. and Gajinder Singh.:"Plugging of right

diversion tunnel after failure of hoist chamber at Bhakra Dam."
Journal of the Institution of Engineers(India),XLV,No.5,pt.Cl 3,
Jan.1965,pp.495-514.

3. Maurice Kenn.:"Tarbela designers were warned about cavitation
   dangers.",Water World,March 1981,p.6.

4. Arya,A.S.:Garde R.J, Ranga Raju K.G. and Brijesh Chandra.:
   "Vibration studies on divide wall of Baggi Control Works."
   Irrigation and Power;journal of CBIP,New Delhi,India, 31,No.4,1974,
   pp.469-481.

5. Dhillon,G.S. and Paul T.C.:"Performance of Pong Dam T2 Tunnel Outlet
   Works." In:Proc.45th Annual Research Session of CBIP,New Delhi
   (India), II(Hyd.),1976.pp.47-59.

6. Green W.G.:"Theory of machines."Blackie and Son(India)Ltd,Chapter,
   XVI,p.916.

7. Dhillon G.S, Paul T.C., and Sakhuja V.S.:"Performance of Pong Dam
   P_2R intake at low levels-A prototype-cum-model study."In: Proc:
   46th Annual Research Session of CBIP,New Delhi(India),III,Nov.1977,
   pp.8-19.

8. Anwar H.O.:"Prevention of vortices at intakes."Journal of Water
   Power, 10, Vol.1.,Oct.1968 ,pp.393-402.

9. Gibson A.H.:"Hydraulics and its application."Asia Publishing House,
   New Delhi(India),5th Edition,pp.543-549.

10. Johnson.R.D.:"The Differential Surge Tank."In:Trans.ASCE 1324, 78,
    1915,p.760.

11. Pennino B.T.:"Prediction of flow induced forces and vibrations."
    International Water Power and Dam Construction,Special Issue,
    Hydraulics,Feb.1981,pp.19-24.

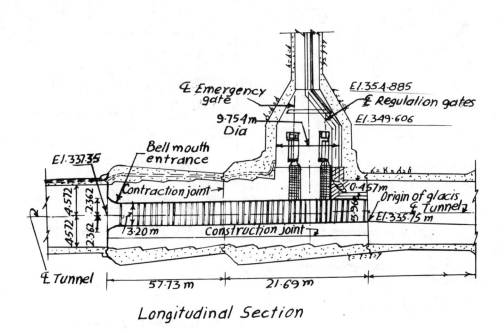

Longitudinal Section

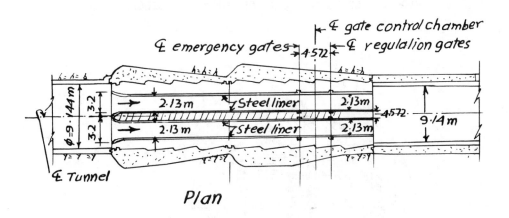

Plan

FIG. 1   GENERAL LAYOUT OF OUTLET WORKS IN TUNNELS T1 and T2

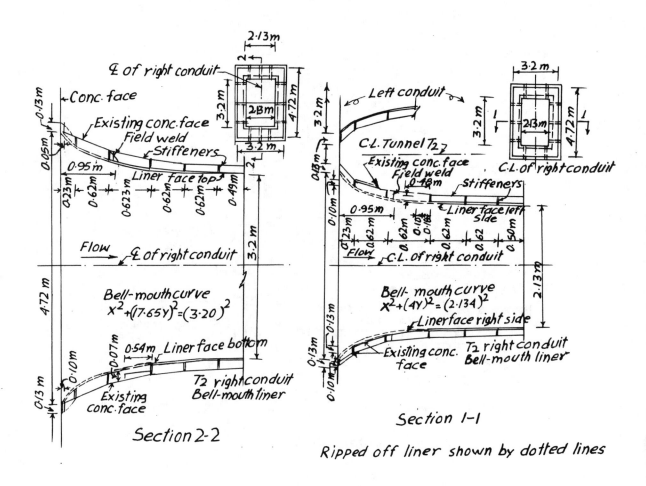

FIG. 2   SHOWING EXTENT OF DAMAGE TO STEEL LINER

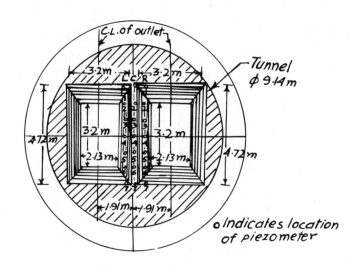

FIG. 3   LOCATION OF PIEZOMETERS AT THE CENTRAL PIER OF OUTLET CONDUITS

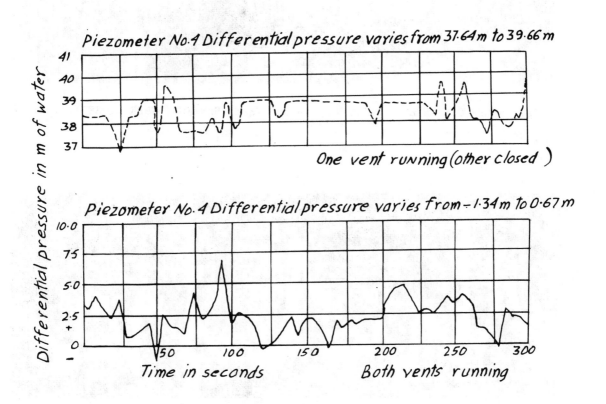

FIG. 4   DIFFERENTIAL PRESSURES OF PIEZOMETER NO 4 LOCATED ON LEFT AND RIGHT OF
CENTRAL PIER-RESERVOIR LEVEL EL. 402.74 m

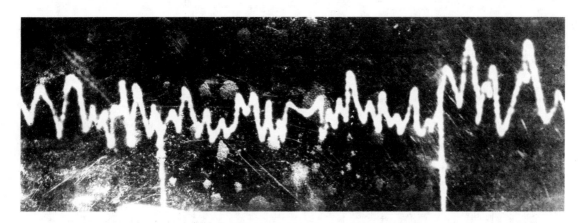

(a)   RESERVOIR LEVEL AT EL.384.05 m ONE VENT RUNNING (OTHER CLOSED), $f_n = 4.4$

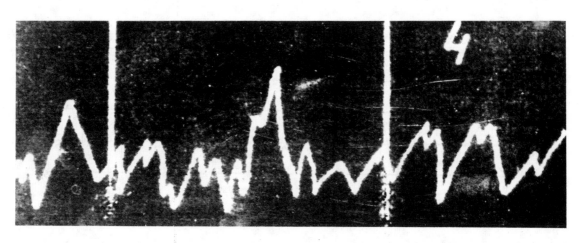

(b)   RESERVOIR LEVEL AT EL.396.24 m ONE VENT RUNNING (OTHER CLOSED), $f_n = 2.4$

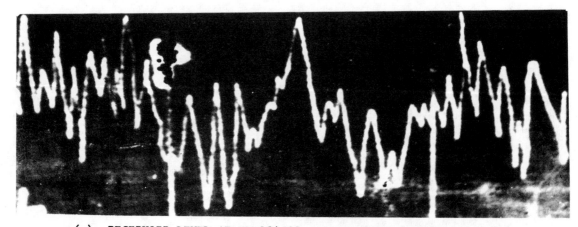

(c)   RESERVOIR LEVEL AT EL.384.05 m BOTH VENTS RUNNING, $f_n = 3.1$

FIG. 5   TRACES OF DIFFERENTIAL PRESSURE PULSATIONS IN 3 SECONDS FOR PIEZOMETER
NO 4 L-R

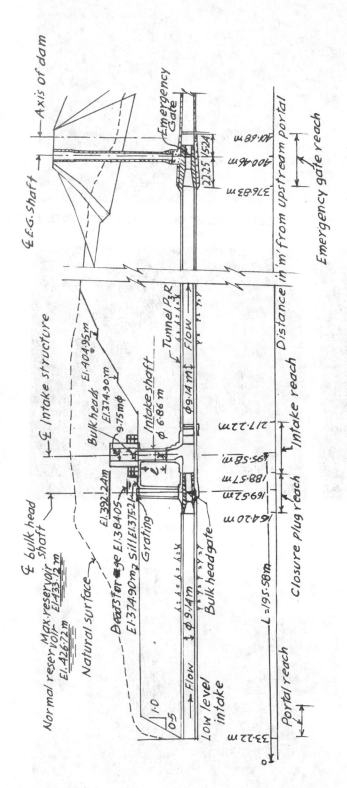

FIG. 6   LONGITUDINAL SECTION TUNNEL P₃ (CONSTRICTED)

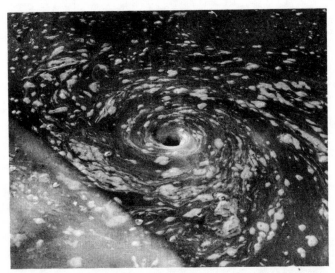

FIG. 7   VORTEX FORMING AT BULKHEAD GATE SHAFT WITH RESERVOIR LEVEL EL.385.57 m

(a)   ON MODEL I

(b)   ON PROTOYPE

FIG. 8   POSITION OF DISLOCATED FLAT BULKHEAD IN THE INTAKE SHAFT

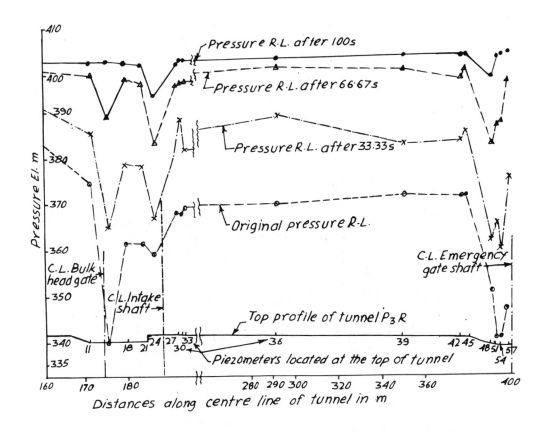

FIG. 9   INCREASE IN PRESSURE (DUE TO CLOSING OF GATE) AT TOP PROFILE OF TUNNEL

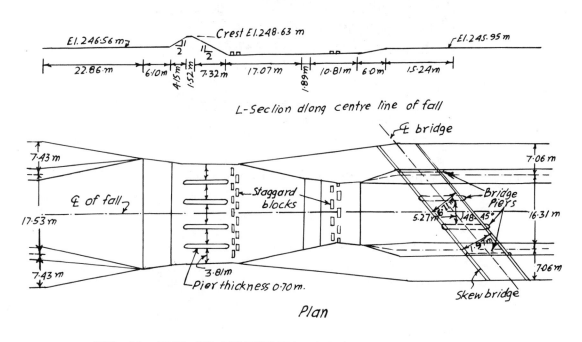

FIG. 10   PLAN AND LONGITUDINAL SECTION OF REGULATING FALL

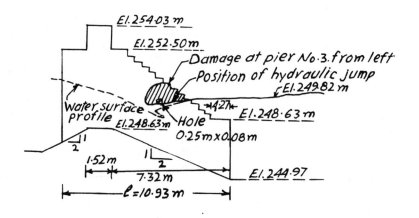

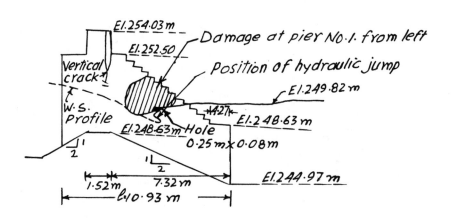

FIG. 11 LOCATION OF DAMAGE AT PIERS OF REGULATING FALL

FIG. 12 VIEWS OF FLUCTUATING HYDRAULIC JUMP IN BAYS OF REGULATING FALL AT TWO
INSTANTS FOR SAME DISCHARGE

FLOW INDUCED VIBRATIONS IN FLUID ENGINEERING

Reading, England: September 14-16, 1982

VIBRATION OF SCREEN AT LA PLATE TAILLE HYDRO STORAGE POWER STATION IN BELGIUM

R. Vanbellingen

Publics Works Ministry, Belgium

A. Lejeune, J. Marchal, M. Poels and M. Salhoul

University of Liege, Belgium

Summary

Last year in 1981, one recorded some failures in the screen equipments at La Plate Taille Hydroelectric Power Station in Belgium.

This plant is a Storage Power Station described in the paper and the characteristics are the following:
- variable head: 33 to 48 metres
- 4 groups of 35 MW each Turbine pumps

These failures have occurred about one year after the opening of the station. From diving observations, the screen steel members of the intake and the outlet were disconnected from their supports, some nuts unscrewed, some bolts were pulled out from the plugging concrete.

In the paper, general characteristics of racks are given. The good design of cross sections of vertical bars and horizontal members as far as minimum head losses are concerned is observed.

Vertical and horizontal forces due to waterflow are computed and the stability of the structure is verified.

With finite elements well known method natural frequencies of the whole structure are computed and comparisons with Von Karman vortex periods due to the flow are made. It was found that during starting phases of the turbines, flow frequency and natural frequency of the whole structure are equal, the support conditions of the members having no influence.

Details of natural and flow frequencies computations are given in the paper. Comparison with a screen of another storage power plant is made.

Organised and sponsored by
BHRA Fluid Engineering, Cranfield, Bedford MK43 0AJ, England

NOMENCLATURE

f   : natural frequency
r   : radius of gyration of bar section
h   : distance between two horizontal members
E   : Youngs Modulus of bar material
w_f : specific weight of the fluid
b   : distance between two vertical bars
s   : thickness of the bar
M_b : correction factor of the bar
f_e : frequency of shedding of a pair of vortices
v   : flow velocity
S_t : Strouhal number
M_f : correction factor function of $\dfrac{s+b}{s}$
L   : width of the longest side of the bar section
W   : specific weight of bar material

1. INTRODUCTION

In 1981,at La Plate Taille Hydroelectric Power station,some failures in the trashracks equipment were recorded.This plant,situated in Belgium,is described in figure 1 and the main characteristics are the following:

    storage hydroelectric power station;
    variable head from 33 to 48 meters;
    4 groups of 35 MW each;
    turbines pomps.

These failures were observed one year after the opening of the power plant. From diving observations,horizontal steel members of the screen were disconnected from their supports.As a rule,the screens were placed at the intake and at the outlet as showed in figure 1.It was discovered that not only were the screens insufficiently tightened up,some nuts having loosened,but also some bolts had pulled out from the plugging concrete.

2. DESCRIPTION OF SCREENS

A general description of vertical bars and horizontal members of the intake screen is given in figure 2.The screen was set in a very slight inclining position. Values of areas and inertias of each support are the following :

	Vertical bar	Horizontal member
A	3240 mm^2	2175 mm^2
I_{XX}	8748000 mm^4	-
I_{YY}	-	3810781 mm^4
I_{ZZ}	87480 mm^4	40781 mm^4
I_{twist}	327875 mm^4	152486 mm^4

Table 1

As can be observed from figure 2,the design of screen is very good,as far as minimum head loss is concerned.In fact,screen hydrodynamic forms had been studied in detail to minimize the head losses and were adopted.People know the great importance of the head losses in the power hydroelectric power station and particularly in storage one where the head losses during pumping operations have to be added to head losses during turbine operations in considering the total efficiency of the plant.
Characteristics of the outlet screen are very similar and studies have been made.
Results and conclusions are the same,but are not developed in this paper.

3. LOADS OF SCREEN MEMBERS

The maximum discharge in the intake is 107 m^3/s.Due to the forms of the intake,in that case,the velocity of the water through the screen is about 1m/s.
If loads due to the deadweight of the racks are computed, it will be found that they are unimportant in their structural design.
As far as water forces are concerned, some conclusions are made. In the case of maximum velocity,drag forces and lift forces are computed for all directions. Finally the stability of the structure is verified with a very large safety factor if those forces are taken into account.
In fact,failures are observed and the structural design of the screen should be checked for adequacy against bar vibration induced by the flow and associated shedding of vortices.

4. CLASSICAL VERIFICATION OF THE STABILITY DUE TO VIBRATIONS

For any bar thickness and velocity of flow through the racks, there is a maximum safe laterally unsupported length of bars which should not be exceeded. The design chart shown in figure 3 indicates the spacing of lateral supports or stiffeners at which the natural frequency of bar vibration in water is 2,5 times the frequency induced by the flow, to assure sufficient safety against vibration of harmful magnitude (1). The frequency of vortex shedding, due to Von Karman street, is based on Strouhal's formula, using a modified Strouhal number of 0,20 to account in part for the influence of the boundary layer thickness.
Levin's formula is used to determine the natural frequency of the bar in the water (2) i.e. (see section 8).

$$ f = M_b \frac{r}{h^2} \sqrt{\frac{g\,E}{w + \frac{b}{s} w_f}} $$

It should be notice that this formula is valid only for $\frac{b}{L} < 0,7$ where L is the width of the longest side of the bar section. If $\frac{b}{L} \geqslant 0,7$, one should use the limit value b=0,7L.
In application of those calculations, these results have been found :
        natural frequency of the bar : f = 653,7 Hz
        maximum distance of the screen members
                vertical bars (D = 18 mm)   L = 1140 mm
                horizontal members (D = 15 mm)   L = 940 mm

In conclusion, stability of the members themselves is verified because the maximum lateral length of screens members is 650 mm in La Plate Taille Power Station. So researches have to be turned to the study of the vibration of the whole screen.

5. STUDY OF NATURAL FREQUENCIES OF THE SCREEN

Determination of natural frequencies of the whole structure is not an usual problem. Some simple examples have been solved in publications and are very useful. Now to study those complex problems, finite elements method is used.
The computations have been made in the University of Liège using S.A.M.C.E.F. program of Laboratoire des techniques aérospatiales (3).
The screen has been discretized in a grid of 98 beam elements and 72 nodes (figure 4).
To examine the influence of the different support boundary conditions, the following cases have been studied
        1. bi clamped beam
        2. simply supported beam
        3. clamped and simply supported beam
        4. simply supported screen on 3 edges
        5. simply supported screen on 4 edges

The last two case have been computed to determine a future solution of the problem in adding some supplementary support members.
The results are given below taking into account the corrected values of natural frequencies $f_{water} = 0,7\ f_{air}$. (see section 8).

Support	Simply supported	Bi clamped	Simply supported clamped	Simply supported on 3 edges	Simply supported on 4 edges
1	0,70	1,01	0,90	1,48	3,16
2	1,16	2,06	1,82	1,82	6,51
3	1,33	2,64	1,85	3,92	7,18
4	1,60	2,80	1,98	4,21	7,74
5	2,69	3,94	2,92	4,64	7,81
6	3,94	4,55	4,15	4,84	8,04
7	4,63	5,23	4,67	4,94	8,18
8	4,68	5,62	5,37	5,38	8,91
9	4,84	5,63	5,57	5,67	9,08
10	5,35	5,83	5,62	5,87	9,18
12	5,62	6,36	5,88	6,89	10,53
14	6,06	7,86	6,11	7,32	11,24
16	6,91	8,37	7,10	7,33	12,58
18	8,37	9,08	8,28	9,20	13,75
20	9,41	10,11	9,21	9,70	15,24
25	15,89	20,46	13,27	12,62	22,77
30	63,71	–	25,14	17,54	69,54

Screen Natural Frequencies in Hz . Table 2

6. SHEDDING FREQUENCIES OF VON KARMAN STREET

To determine shedding frequencies of Van Karman street, Strouhal number is used in the following formula:

$$f_e = \frac{S_t\ v}{s} M_f$$

Strouhal number is given for different cases in the figure 5 and and shedding frequencies of Von Karman street are computed in figure 6.(4).
As it could be observed from the figure 6 and the table 2, in the starting phases of the turbines, flow frequency and natural frequency of the whole screen are equal, the support boundaries condition having no influence.

7. CONCLUSION

In this storage power plant, the duration of the starting phases are rather long (between 1 and 3 minutes) and the operations are very frequent on the network. So it could be conclued that this screen is in vibration and the failures observed were due to the resonance phenomena.
The very light influence of the boundary support conditions shows the weak rigidity of the screen equipment and in such conditions the potentiality of vibrations.
To complete the study, some computations have been made for another screen equipment in Belgium, at the Storage Power Station of Coo.
The main characteristics of this hydroelectric power station are the following for the total plant:
     Head from 241 m to 273 meters
     Discharge : maximum 462 m^3/s
     Power : maximum 1059 MW
Description of the screen equipment at the lower basin is given in the figure 7 and table 3 below gives the natural frequencies of the whole screen.

Number	Natural frequency Hz
1	0,88
2	1,95
3	1,96
4	2,10
5	3,04
10	6,34
15	9,24

Table 3 . Natural frequencies of screen equipment at
the lower basin in Coo.
Natural frequency in water ($f_{water} = 0,7 f_{air}$).

Computations of Von Karman shedding frequencies are made and results are very similar to
those showed in figure 6 for La Plate Taille equipments.
The conclusions should be the same for both plants but in fact there is no problem in
the Coo station power.
If we look at the distribution of the flow velocity through the screen,it is very different.
Figure 8 shows the velocity profile before the screen equipment in the lower basin
in Coo power station.This profile has been measured in scale model tested in our labo-
ratories.One of the purposes of this hydraulic model was to improve the flow in the
outlet channel (5).As it could be observed in figure 9,the velocity distribution in
La Plate Taille Power Station,the profile is uniform.
This stream flow has been obtained by integration of Navier-Stokes equations using
turbulent viscosity (6) and finite difference method.
At the end,the final conclusions show not only the importance of the relation between
natural and flow frequencies,but also the influence of the velocity profile.
The new solution adopted by the Public Works Ministry for this problem is a cable
net.This solution seems good as head looses and vibrations are concerned,but more
investigations are still necessary.

8. REMARKS

In section 4, Levin's formula has also to be applied to horizontal bars, and
in this case h and b should be interchanged in the formula itself.
In section 5, the S.A.M.C.E.F. program determines the values of natural frequencies in
the air. In fact the screen is situated in the water, and one has to take into account
the added water mass which reduces the value of natural frequencies.
The value of 0,7 is deduced from experimental studies.

9. REFERENCES

(1) Zowski, Th. "Trashracks and raking equipment - I" Water Power, vol.12, n°.9,
        pp.342-348 (September 1960)

(2) Levin, L. "Etude hydraulique des grilles de prise d'eau." Proc. 7th general mee-
        ting of I.A.H.R., Lisbon, Portugal. Vol.1, paper C11, pp.C11-1-C11-19.
        (24-31 July 1957).

(3) Sander, G. "Programme SAMCEF". Laboratoire des techniques aérospatiales. Univer-
    Geradin, M. sité de Liège. Belgium.

(4) Levin, L. "Formulaire des conduites forcées, oléoducs et conduits d'aération".
        Dunod, Paris 1968.

(5) "Etude sur modèle réduit de la phase II de la centrale de Coo". Procès-verbal
    d'essais. Laboratoires d'hydrodynamique, d'hydraulique et des constructions hydrau-
    liques. Institut du Génie Civil, Université de Liège, Belgium.

(6) Lejeune, A. "Study of the forces occuring during the movement of mitergates of
        locks". US Army Engineer Waterways Experiment Station. Corps of Engi-
        neers, n° 74.11, 1974. Vicksburg, Mississippi.

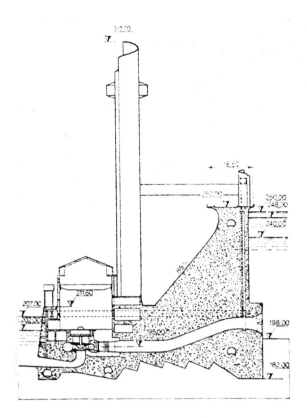

Figure 1 - La Plate Taille storage hydroelectric
Power Station

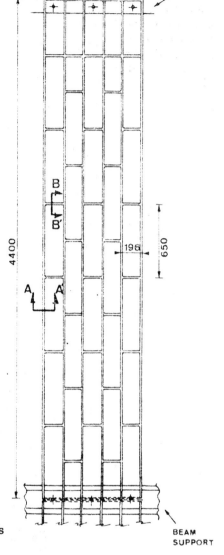

Figure 2

Screen Equipments

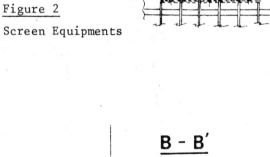

A - A'

B - B'

scale : 1/20 - 1/2

unit : mm

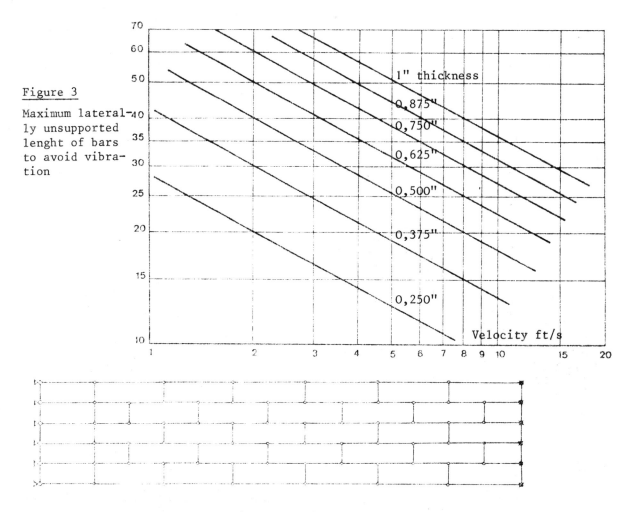

Figure 3

Maximum lateral-
ly unsupported
lenght of bars
to avoid vibra-
tion

1" thickness
0,875"
0,750"
0,625"
0,500"
0,375"
0,250"

Velocity ft/s

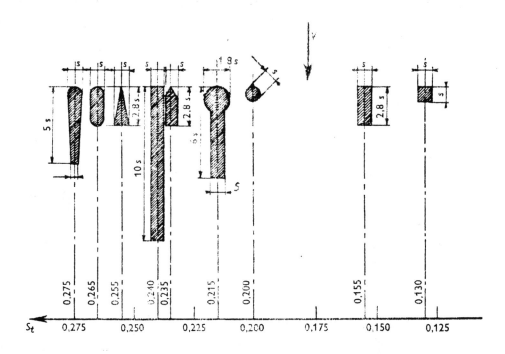

Figure 4 - Screen grid 98 beam elements - 72 nodes

Figure 5 - Strouhal number

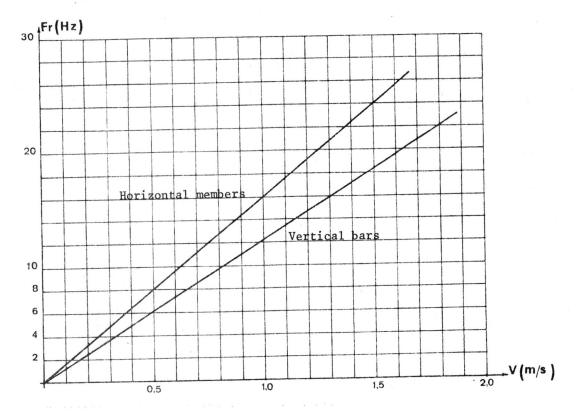

Figure 6 - Von Karman shedding frequencies

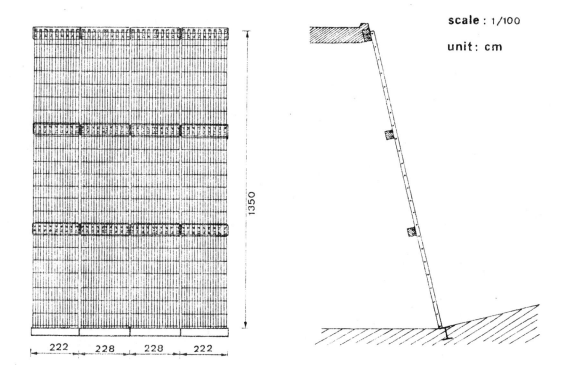

Figure 7 - Coo screen equipments

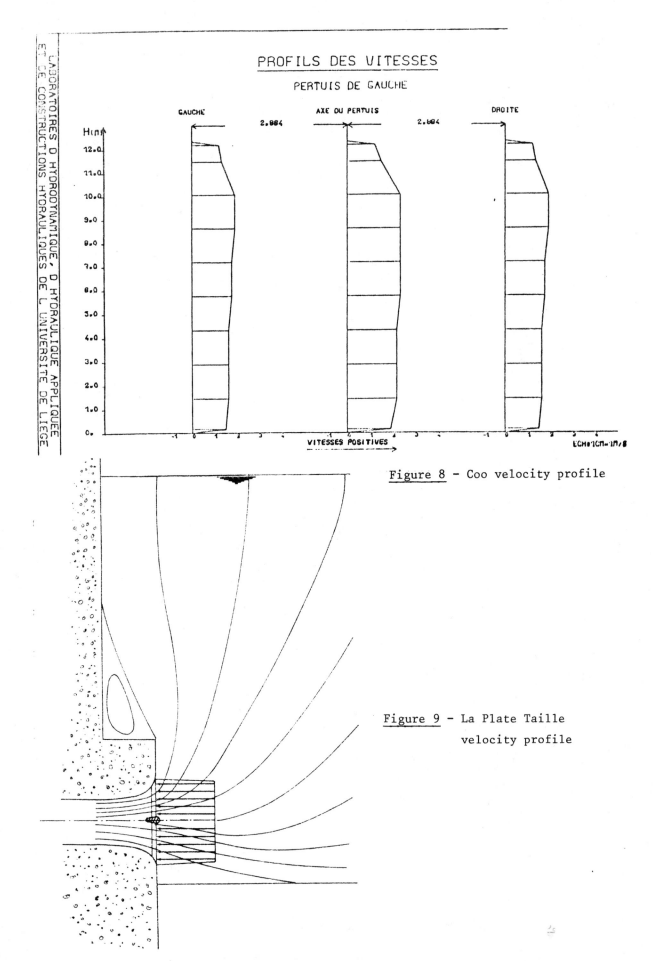

PROFILS DES VITESSES

PERTUIS DE GAUCHE

Figure 8 - Coo velocity profile

Figure 9 - La Plate Taille
velocity profile

FLOW INDUCED VIBRATIONS
IN FLUID ENGINEERING

Reading, England: September 14-16, 1982

DYNAMIC RESPONSE OF ARTICULATED PLATFORMS
IN RANDOM SEAS

C. L. Kirk and P. Bose

Cranfield Institute of Technology, U.K.

Summary

This paper presents a theoretical investigation into the dynamic response of three designs of articulated platform under the combined action of collinear wave and current forces.

The motions considered are the direct wave induced oscillations at wave frequency and the slowly varying oscillations produced by a random sea state, the two motions being coupled by nonlinear drag forces. The solution of the coupled equations of motion is achieved by means of spectral methods with the nonlinear drag force being linearised by the method of Tung and Wu (Ref. 1). The results give the r.m.s. value for angular motions of the platform, the distribution of r.m.s. shear force and bending moment distribution along the length of the platform tower and r.m.s. deck accelerations.

Comparisons of theoretical values with limited experimental results given in the literature show good agreement.

Organised and sponsored by
BHRA Fluid Engineering, Cranfield, Bedford MK43 0AJ, England

NOMENCLATURE

A = Linearising coefficient for drag force, eq.3.

A_d, B_d = coefficients in fluid drag moment, eq.15.

A_I = transfer function for fluid inertia moments, eqs.12, 13.

C_m, C_d = coefficients of inertia and drag, eq.1.

D = external diameter of tower.

d = water depth

F_I, F_d = fluid inertial and drag forces.

$F_o(\omega_k)$ = mean square slowly varying force at M.W.L.

$F_f(i\omega, y), G_f(i\omega, y)$ = complex transfer functions for fluid loading.

F_w = wave force

$G_r(i\omega, y)$ = complex transfer function for relative fluid velocity

g = acceleration due to gravity

H_s = significant wave height

H = wave height

$H_\theta(i\omega_k)$ = transfer function for slowly varying motion.

h_b = depth of ballast above ball joint.

h = effective depth of tower for slow drift forces.

I = moment of inertia of tower and added mass about ball joint

K_c = Keulegan-Carpenter number.

k = wave number

$L_g(i\omega, y)$ = transfer function for shear force due to vertical loads on tower

M_I, M_d = moment of fluid inertia and drag forces about ball joint.

M.W.L. = mean water level.

M_{wd} = moment of direct wave forces about ball joint

$\tilde{M}_d$ = moment of slowly varying forces.

$M_d(\omega_k)$ = spectral density of slowly varying moment

M_p = mass of tower deck.

m_s, m_a, m_b = mass per unit length of tower, added mass and ballast.

n = number of columns.

$R^2(\omega)$ = reflection coefficient for tower

$S(\omega)$ = spectral density function for quantity related to high frequency motion

$S_d(\omega_k)$ = spectral density of slowly varying force.

$S_H(\omega)$ = spectral density of wave height.

$S_{\bar{F}}(\omega)$ = spectral density of mean force

t = time

U_r = fluid relative velocity

$u, \dot{u}$ = fluid velocity and acceleration from linear wave theory.

V_c = current velocity

$\underline{V}_r$ = fluid relative velocity excluding current.

$\bar{V}_r$ = fluid relative velocity neglecting low frequency motion

V_t, V_w = tidal and wind induced components of current velocity

$\dot{x}$ = velocity of tower.

Y = distance from ball joint

y = distance below M.W.L.

y_p = distance from M.W.L. to deck centre of gravity.

α = $\pi \rho C_m D^2 / 4$

β = $\rho C_d DA / 2$

$\bar{\alpha}, \sigma_a \sigma_b, \gamma$ = constants in JONSWAP wave height spectrum.

β_e = equivalent linear damping ratio, eq.43.

η = distance from M.W.L.

θ = angle of dynamic rotation of tower

$\bar{\theta}$ = mean angle of rotation

λ = restoring moment on tower about ball joint

ρ = fluid density.

σ_{V_r} = root mean square relative velocity excluding current

σ = root mean square value

$\Phi(i\omega)$ = complex transfer function for tower motion, eq.19.

ψ = $4\omega^2_k / g$

ω = circular wave frequency

ω_k = frequency of slowly varying motion.

ω_θ = natural circular frequency of tower, eq.44.

ω_p = peak frequency of wave height spectrum.

1. INTRODUCTION

The future exploration for oil and gas in water depths greater than those of the present North Sea oil fields will require a new generation of structures that are technically feasible for deep water drilling and production in addition to being economic to build, instal, operate and maintain.

One type of structure for which numerous detailed design studies have been carried out by companies such as Taylor Woodrow, C.G. Doris and Elf Aquitaine is the articulated or deep water gravity tower (DWGT).

The economic advantages of the DWGT over fixed platforms is illustrated in fig.1, where it is seen that for $300 < d < 600$ m, there is a significant reduction in steel weight. Other advantages can be listed as follows.

1) Buoyant structure requires no anchor lines which can be a shipping hazard.

2) Concrete floater protects conductors and risers from wave and current action and provides good protection from ship impact.

3) Compartmented structure permits accidental flooding of cells to occur near the surface.

4) Deck motion is small and comparable to that of fixed structures.

5) From the point of view of wave forces applied to the DWGT the platform is excited at wave periods which are much shorter than the natural period. Thus the wave forces are opposed by the inertial forces of the structure and its fluid added mass, giving rise to a considerable reduction in loading.

2. DYNAMIC ANALYSIS

The present analysis considers the platform to be in an undamaged state with motions taking place about the static equilibrium position which is assumed for practical purposes to be vertical.

At the design stage of a DWGT a dynamic analysis should provide information on (1) amplitude and acceleration of deck due to direct wave forces and slowly varying forces, (2) the distribution of shear force and bending moment along the length of the structure above and below the M.W.L.

It should be recognised that significant bending moments arise from platform weight, as a result of the moment arm caused by angular motion of the tower, in addition to angular offset due to wind and current forces.

In random sea states the results of a spectral analysis yield r.m.s. values from which expected peak values can be estimated.

The three candidate towers selected for study are shown in figures 2, 3, 4 and are designated cases (a), (b), (c) for which the mean water depths are respectively 300, 250 and 250 m.

In calculating the hydrodynamic forces on a tower it is assumed that the waves and current act in the same direction. A further simplification assumes that the tower oscillates as a rigid body and that elastic mode response is negligible which is considered to be valid for water depths up to about 300 m. For greater depths however the first few flexural modes of the tower may have periods that approach those at the short period end of the wave spectrum, in which case elastic mode response could give rise to significant bending stresses similar to a T.L.P. riser (Ref.2).

Various types of articulation have been proposed for the DWGT. For example the C.G. Doris design consists of a hemispherical ball and socket joint separated by deformable laminated rubber bearing pads maintained in compression by a ballast chamber at the bottom of the tower. The ball joint resists horizontal and vertical loads and a torsion frame prevents rotation about the vertical axis. In this analysis the ball joint is assumed to offer zero rotational stiffness and damping moments.

The analysis of the high and low frequency motions are considered separately but with coupling terms due to nonlinear drag being included in both equations of motion.

2.1   Direct wave induced motion

The direct wave forces acting on the cylindrical members of a tower are calculated using Morison's formula for drag force including current and inertia force. For a unit length element of the tower at depth y below M.W.L. the element of force is given by

$$\delta F_w = \frac{\pi}{4} \rho C_m D^2 \dot{u} + \frac{\rho}{2} C_d D |U_r| U_r = \delta F_I + \delta F_d \tag{1}$$

where δF_I, δF_d = inertia and drag forces respectively

D = diameter or equivalent diameter of element.

C_m = inertial coefficient

C_d = drag coefficient,

ρ = fluid density

$\dot{u}, u$ = fluid velocity and acceleration from linear wave theory

U_r = relative fluid velocity = $u + V_c - \dot{x}$

V_c = current velocity.

$\dot{x}$ = $Y\dot{\theta}$ = velocity of tower at distance Y from ball joint

$\dot{\theta}$ = angular velocity of tower

The drag force is linearised approximately by Tung and Wu's method (Ref.1) and the time varying part is written

$$\delta F_d = \frac{\rho}{2} C_d D A V_r \tag{2}$$

where $A = \sqrt{8/\pi} \left[\sigma V_r \exp\left\{- (V_c/\sigma_{V_r})^2/2\right\} + \sqrt{2\pi} V_c \, \text{erf}(V_c/\sigma_{V_r}) \right] \tag{3}$

V_r = relative velocity excluding current = $u - \dot{x}$

σ_{V_r} = r.m.s. relative velocity for the high frequency motion to which will be added the velocity due to the slow drift motion.

$$\text{erf}(\mu) = \frac{1}{\sqrt{2\pi}} \int_0^\mu e^{-t^2/2} \, dt$$

Eq.1 is now written

$$\delta F_w = \frac{\pi}{4} \rho C_m D^2 \dot{u} + \frac{\rho}{2} C_d D A (u - \dot{x}) \tag{5}$$

$$= \alpha \dot{u} + \beta (u - \dot{x}) \tag{6}$$

In eq. 5, $\dot{x}$ only relates to the high frequency motion with the low frequency $\dot{x}$ being considered separately in the equation for slow drift motion. It can be seen that the coupling between the two motions is taken into account through σy_r augmented by the r.m.s. slow drift velocity. Hence an iterative solution of both equations of motion is required.

The equation of high frequency small angular motion of the tower neglecting potential damping is written

$$I\ddot{\theta} + \lambda\theta = M_{wd} \tag{7}$$

where I is the moment of inertia of the tower and fluid added mass about the ball joint axis and λ is the effective restoring moment due to buoyancy and weight about the ball joint.

The moment of the fluid drag and inertia forces is given by

$$M_{wd} = \int_{-d}^{0} (\delta F_d + \delta F_I)(d + y)\,dy \tag{8}$$

where it is assumed that the ball joint is at depth d below M.W.L.

For a wave height H and circular frequency ω, the horizontal fluid velocity and accelerations are given by

$$u = \omega\frac{H}{2}\frac{\cosh[k(d+y)]}{\sinh(kd)}\cos(kx - \omega t) \tag{9}$$

$$\dot{u} = \omega^2\frac{H}{2}\frac{\cosh[k(d+y)]}{\sinh(kd)}\sin(kx - \omega t) \tag{10}$$

and for deep water $k \simeq \omega^2/g$

The inertial moment is found by reference to equations 6, 8, 10 as

$$M_I = \int_{-d}^{0} \alpha(y)\dot{u}(d+y)\,dy \tag{11}$$

Substituting for $\dot{u}$ from eq.10 in eq.11

$$M_I = \frac{-H\omega^2 \pi\rho G_1(\omega)}{8\sinh(kd)}\sin\omega t \int_{-d}^{0} C_m(y)D^2(y)(d+y)\cosh[k(d+y)]\,dy \tag{12}$$

where

$G_1(\omega) = $ transfer function accounting for column spacing

Because of the variable diameter D(y) the integral in eq.12 is performed numerically by dividing the length of the tower from the sea bed to M.W.L. into a number of discrete elements, say 50. The integrand is then evaluated at the centroid of each element and summed along the length. For deep water towers it is essential to concentrate more elements near to the M.W.L. for accurate evaluation of wave forces.

Equation (12) is written as

$$M_I(\omega) = -H A_I(\omega)\sin\omega t \tag{13}$$

82

The linearised drag moment is found in a similar manner from equations (5), (8), (9) as

$$M_{d_L} = \int_{-d}^{0} \beta(y)(u - \dot{x})(d + y)dy$$

$$= \frac{\rho}{2} \int_{-d}^{0} C_d(y)D(y)A(y)\left[u - (d + y)\dot{\theta}\right](d+y)dy \qquad (14)$$

$$= H\, A_d(\omega)\cos\omega t - B_d(\omega)\dot{\theta} \qquad (15)$$

where

$$A_d(\omega) = \frac{\omega \, \rho G_2(\omega)}{4\sinh(kd)} \int_{-d}^{0} C_d(y)D(y)A(y)(d+y)\cosh\left[k(d+y)\right]dy \qquad (16)$$

and

$$B_d(\omega) = \frac{n\rho}{2} \int_{-d}^{0} C_d(y)D(y)A(y)(d+y)^2 dy \qquad (17)$$

and $G_2(\omega)$ takes account of the phase difference between the drag forces at the columns. For a single column $G_1(\omega) = G_2(\omega) = 1$.

Equation (7) can now be written

$$I\ddot{\theta} + B_d(\omega)\dot{\theta} + \lambda\theta = H\left[-A_I(\omega)\sin\omega t + A_d(\omega)\cos\omega t\right] \qquad (18)$$

Assuming a steady state solution $\theta = \bar{\theta}e^{i\omega t}$ and writing $\cos\omega t = e^{i\omega t}$, $\sin\omega t = -ie^{i\omega t}$, eq. 18 yields the complex frequency response function for direct wave excitation

$$\frac{\bar{\theta}}{H} = \frac{\left[iA_I(\omega) + A_d(\omega)\right]}{(\lambda-I\omega^2) + i\omega B_d(\omega)} = \Phi(i\omega) \qquad (19)$$

The spectral density function of the response is found from

$$S_\theta(\omega) = |\Phi(i\omega)|^2 S_H(\omega) \qquad (20)$$

where $S_H(\omega)$ is the wave height spectrum. The root mean square angular response is then given by

$$\sigma_\theta = \left[\int_0^\infty S_\theta(\omega)d\omega\right]^{\frac{1}{2}} \qquad (21)$$

To evaluate the above equations it is required to obtain the linearised drag coefficient, A, in eq.3, which contains σ_{V_r}. Hence a transfer function for V_r is derived as follows. Denoting by $\bar{V}_r(y)$, the relative velocity neglecting the low frequency motion, then

$$V_r(y) = u(y) - (d + y)\dot{\theta}$$

$$= \omega\frac{H}{2}\frac{\cosh\left[k(d+y)\right]}{\sinh(kd)}e^{i\omega t} - (d+y)i\omega\bar{\theta}e^{i\omega t} \qquad (22)$$

83

Substituting for $\bar{\theta}$ from eq.(19) gives the complex transfer function

$$\bar{V}_r(i\omega,y) = H\left[\frac{\omega\cosh.k(d+y)}{2\sinh(kd)} - i(d+y)\omega\Phi(i\omega)\right] e^{i\omega t} \tag{23}$$

or $\bar{V}_r(\omega,y) = H\ G_r(i\omega,y)$ (24)

Hence the p.s.d. of relative velocity is

$$S_{\bar{V}_r}(\omega,y) = |G_r(i\omega,y)|^2\ S_H(\omega) \tag{25}$$

from which the r.m.s. relative velocity is obtained as

$$\sigma_{\bar{V}r}(y) = \left[\int_0^\infty S_{\bar{V}r}(\omega,y)\,d\omega\right]^{\frac{1}{2}} \tag{26}$$

The influence of the slow drift velocity is now taken into account by assuming that the high and low frequency motions are statistically independent. The resultant r.m.s. relative velocity is then obtained from

$$\sigma_{Vr}(y) = \left[\sigma_{\bar{V}r}^2(y) + (d+y)^2\ \sigma_{\dot{\theta}_s}^2\right]^{\frac{1}{2}} \tag{27}$$

where $\sigma_{\dot{\theta}_s}^2$ is the mean square slow drift velocity which will be determined subsequently.

In Morison's equation the selection of C_m and C_d was based on the results of Sarpkaya (Ref.3) for a stationary cylinder and the local Keulegan-Carpenter number K_c for each element of the tower. Since the tower is oscillating it would appear to be reasonable to determine K_c in terms of the relative velocity neglecting current, thus we have assumed

$$K_c(\omega,y) = \frac{2\pi}{\omega D(y)}\ \sigma_{Vr}(y) \tag{28}$$

where ω represents the frequencies in the wave spectrum.

Other possible forms of eq.28 were considered but not pursued because of the lack of experimental data relating to K_c for oscillating cylinders under random excitation.

3. DISTRIBUTION OF R.M.S. SHEAR FORCE AND BENDING MOMENT

The resultant forces acting on the tower are found by using d'Alembert's principle and considering the forces due to wave action, buoyancy, weight and the oscillatory inertia forces due to platform structural and fluid added mass. These forces are shown in fig.5.

3.1 Shear Force

The complex transfer function for the fluid inertia and linearised drag forces are found directly from eq.14 by removing the moment arm (d+y) from equations (12), (14), (16), (17). Thus

$$F_f(i\omega,y) = H\hat{A}_d(\omega,y)\cos\omega t - \hat{B}_d(\omega,y)\dot{\theta} - H\hat{A}_I(\omega,y)\sin\omega t \tag{29}$$

where

$$\hat{A}_d(\omega,y) = \frac{\omega\cdot\rho}{4\sinh(kd)}\int_{-y}^0 C_d(\eta)D(\eta)A(\eta)\ \cosh\left[k(d+\eta)\right]\ d\eta \tag{30}$$

84

$$\hat{B}_d(\omega,y) \quad = \quad \frac{n}{2}\rho \int_{-y}^{0} C_d(\eta)D(\eta)A(\eta)(d+\eta)d\eta \tag{31}$$

and
$$\hat{A}_I(\omega,y) \quad = \quad \frac{-\omega^2 \cdot \pi\rho G_1(\omega)}{8\sinh(kd)} \int_{-y}^{0} C_m(\eta)D^2(\eta)\cosh\left[k(d+\eta)\right] d\eta \tag{32}$$

Substituting $\dot{\theta} = i\omega\theta$ from eq.19 in eq.29 gives the complex transfer function for the hydrodynamically applied shear force at section y,

$$F_f(i\omega,y) \quad = \quad H\left[\hat{A}_d(\omega,y) - i\omega\hat{B}_d(\omega,y)\Phi(i\omega) + i \hat{A}_I(\omega,y)\right] \tag{33}$$

$$= \quad H \, G_f(i\omega,y) \tag{34}$$

The shear force due to inertia, buoyancy and weight is given by
$$F_g(i\omega,y) \quad = \quad \ddot{\theta}\int_{-y}^{0}\left[m_s(\eta) + m_a(\eta)\right](d+\eta)d\eta + \ddot{\theta}\int_{-y}^{-(d-h_b)}m_b(\eta)(d+\eta)d\eta$$

$$+ g\theta\int_{-y}^{0}\left[m_a(\eta) - m_s(\eta)\right]d\eta - g\theta\int_{-y}^{-(d-h_b)}m_b(\eta)d\eta$$

$$+ \ddot{\theta} M_p(d+y_p) + \ddot{\theta}\int_{0}^{y_p}m_s(\eta)(d+\eta)d\eta - g\theta\int_{0}^{y_p}m_s(\eta)d\eta \tag{35}$$

where
$m_b(\eta)$ = ballast density per unit length
h_b = height of ballast above ball joint
$m_s(\eta)$ = structural mass density per unit length
$m_a(\eta)$ = buoyancy force per unit length (equal to added mass for $C_a = 1$)
M_p = mass of platform deck, distance y_p above M.W.L.

Substituting from eq.19 for θ and $\ddot{\theta}$ in eq.35 yields the complex transfer function

$$F_g(i\omega,y) \quad = \quad H.L_g(i\omega,y)\Phi(i\omega) \tag{36}$$

where
$$L_g(i\omega) \quad = \quad \left[-\omega^2\int_{-y}^{0} m_s + m_a (d+\eta)d\eta - \omega^2\int_{-y}^{-(d-h_b)} m_b(d+\eta)d\eta \right.$$

$$+ g\int_{-y}^{0}\left[m_a-m_s\right]d\eta - g\int_{-y}^{-(d-h_b)}m_b d\eta$$

$$\left. - \omega^2 M_p(d+y_p) - \omega^2\int_{0}^{y_p} m_s(d+\eta)d\eta - g\int_{0}^{y_p} m_s d\eta\right] \tag{37}$$

The transfer function for the total shear force at depth y below M.W.L. is then given by the difference between eq.34 and eq.36, thus

$$F_s(i\omega,y) \quad = \quad H\left[G_f(i\omega,y) - L_g(i\omega,y)\Phi(i\omega)\right] \tag{38}$$

The p.s.d. of the shear force is then

$$S_{F_S}(\omega,y) = S_H(\omega)|G_f(i\omega,y) - L_g(i\omega,y)\Phi(i\omega)|^2 \qquad (39)$$

from which the r.m.s. shear force can be found by integrating over the spectrum.

The r.m.s. bending moment is obtained in a similar manner by introducing the moment arm $(d+\eta)$ to the foregoing integrals. In the numerical integration it was essential to choose sufficiently small elements in order to satisfy the condition of zero bending moment at the ball joint.

4. SLOWLY VARYING MOTION

The slow drifting motion of a compliant offshore structure takes place primarily at the natural frequency and is due to low frequency second order wave forces from frequency differences in the wave spectrum. Since the motion is essentially resonant the amplitude will be governed by nonlinear damping, because potential damping is negligible at low frequencies.

4.1 Slowly Varying Force

The p.s.d. of the slowly varying force on a surface piercing cylinder of infinite depth suggested by Pinkster (Ref.4) is given by

$$S_d(\omega_k) = 2(\rho g D)^2 \int_0^\infty S_a(\omega)S_a(\omega+\omega_k)R^4(\omega_k/2 +\omega)d\omega \qquad (40)$$

where D = diameter of surface piercing cylinder, $S_a(\omega) = S_H(\omega)/4$, $R^2(\omega)$ = reflection coefficient due to Havelock (Ref.5), see fig.6 and ω_k = difference between any two frequencies in the wave spectrum. A typical plot of the slow drift force spectrum is given in fig.7. A derivation of eq.40 by S.O. Rice can be found in Ref.6.

4.2 Equation of Motion

The exciting moment on the tower is given by the sum of the slowly varying moment $M_d(\omega_k)$ and the nonlinear drag moment which is obtained from eq.15 by neglecting the high frequency velocity term u. The equation of motion is then given by

$$I\ddot{\theta} + \lambda\theta = \tilde{M}_d \sin\omega_k t - \dot{\theta}\,\frac{n}{2}\rho \int_{-d}^0 C_d(y)D(y)A(y)(d+y)^2\,dy \qquad (41)$$

In eq.41, $C_d(y)$ was determined in terms of $K_c(y)$ in eq.28 with ω representing frequencies in the low frequency spectrum of eq.40.

Eq.41 is now written in the form

$$\ddot{\theta} + 2\beta_e\omega_\theta\,\dot{\theta} + \omega_\theta^2\,\theta = \frac{\tilde{M}_d}{I}\sin\omega_k t \qquad (42)$$

where

$$\beta_e = B_d/2I\omega_\theta \qquad (43)$$

and

$$\omega_\theta = \sqrt{(\lambda/I)} \qquad (44)$$

The equivalent linear damping ratio β_e is evaluated from B_d, eq.17, using the r.m.s. relative velocity given by eq.27.

The steady state solution of eq.42 is

$$|H_\theta(i\omega_k)|^2 = \left|\frac{\bar{\theta}}{\tilde{M}_d}\right|^2 = \frac{1}{I\left[(\omega_\theta^2 - \omega_k^2)^2 + (2\beta_e\omega_\theta\omega_k)^2\right]} \quad (45)$$

The mean square velocity of slow drift in eq.27 is obtained from

$$\sigma_{\dot{\theta}}^2 = \int_0^\infty \omega_k^2 |H_\theta(i\omega_k)|^2 M_d(\omega_k)d\omega_k \quad (46)$$

where $M_d(\omega_k)$ is the p.s.d. of the slowly varying moment which will be determined subsequently. It can be seen that the r.m.s. slow drift velocity occurs on both sides of eq.46 hence an iterative solution will be required which involves simultaneous solution of the high frequency and low frequency equations of motion. This technique entails the selection of an assumed initial value of $\sigma_{Vr}(y)$ from which a first approximation to the r.m.s. high frequency and low frequency responses can be obtained, which then yield a new value of $\sigma_{Vr}(y)$. The procedure is repeated until satisfactory convergence is achieved. The initial distribution of relative velocity was taken as $\sigma_{Vr}(y) = \exp(-ky)$ with $k = 0.014$.

Using the converged value of r.m.s. velocity from eq.46, the r.m.s. slow drift motion is obtained from a single integration as follows

$$\sigma_\theta^2 = \int_0^\infty |H_\theta(\omega_k)|^2 M_d(\omega_k)d\omega_k \quad (47)$$

4.3   Slowly Varying Moment.

To obtain an expression for the p.s.d. of slow drift moment, $M_d(\omega_k)$, the distribution of slow drift force with depth is required. A search of the existing literature failed to uncover any treatment of the subject and hence the following approximate treatment is proposed. From the work of Havelock (Ref.5) the second order forces are assumed to vary with depth according to $\exp(2\omega_k^2 y/g)$ which is the same as the variation of the nonlinear terms in Bernoulli's equation.

Thus at depth y and frequency ω_k the mean square force corresponding to an element of frequency $d\omega_k$ is given by

$$\delta.S_d(\omega_k)d\omega_k = F_0(\omega_k) e^{4\omega_k^2 y/g} d\omega_k \quad (48)$$

where $F_0(\omega_k)$ denotes the mean square value at M.W.L. For equivalence of total mean square force on the tower on an effective length h of constant diameter D.

$$S_d(\omega_k)d\omega_k = F_0(\omega_k)d\omega_k \int_{-h}^0 e^{4\omega_k^2 y/g} dy$$

$$= \frac{gF_0(\omega_k)}{4\omega_k^2} d\omega_k(1 - e^{-4\omega_k^2 h/g})$$

from which

$$F_0(\omega_k) = \frac{4\omega_k^2 S_d(\omega_k)}{g(1-e^{-4\omega_k^2 h/g})} \quad (49)$$

The p.s.d. of slow drift moment is then given by

$$M_d(\omega_k) = F_o(\omega_k) \int_{-h}^{0} (d+y)^2 e^{4\omega_k^2 y/g} \, dy$$

$$= \frac{S_d(\omega_k)}{e^{\psi h} - 1} \left[\left\{ \frac{2d}{\psi} - (d-h)^2 + \frac{(2h\psi - 2)}{\psi^2} \right\} + e^{\psi h} \left\{ d^2 - \frac{2d}{\psi} + \frac{2}{\psi^2} \right\} \right] \quad (50)$$

where $\psi = 4\omega_k^2/g$

5.   DECK ACCELERATION

Assuming that the high and low frequency motions are statistically independent the mean square deck acceleration is determined from

$$\langle \ddot{x} \rangle^2 = (d+y_p)^2 \left[\int_0^{\infty} \omega^4 S_H(\omega) |\Phi(i\omega)|^2 d\omega + \int_0^{\infty} \omega_k^4 |H_\theta(\omega_k)|^2 M_d(\omega_k) d\omega_k \right] \quad (51)$$

For platforms with low natural periods, as shown subsequently, the acceleration due to slow drift is negligible.

6.   MEAN DRIFT MOMENT

The mean drift force on a cylinder can be found from Ref.6 and is given by

$$\bar{F} = \rho g D \int_0^{\infty} S_a(\omega) R^2(\omega) \, d\omega \quad (52)$$

where ω = wave frequency.

Since $R^2(\omega) \simeq 0.65 D\omega^2/g$ the spectrum of the mean force in eq.52 is given by

$$S_{\bar{F}}(\omega) = 0.65 \rho D^2 \omega^2 S_a(\omega) \quad (53)$$

which has the same form as the spectrum for fluid velocity . Thus $S_{\bar{F}}(\omega)$ can be assumed to vary with depth as $\exp(2\omega^2 y/g)$ and the spectrum of mean force at depth y can be expressed as

$$\bar{F}(\omega,y) = \bar{F}_o(\omega) e^{2\omega^2 y/g} \quad (54)$$

For equivalence of mean force along a surface piercing cylinder of diameter D and effective length h,

$$S_{\bar{F}}(\omega) = \bar{F}_o(\omega) \int_{-h}^{0} e^{2\omega^2 y/g} \, dy$$

which after integration gives

$$\bar{F}_o(\omega) = \frac{2\omega^2 S_{\bar{F}}(\omega)}{g(1-e^{-2\omega^2 h/g})} \quad (55)$$

The p.s.d. of the mean drift moment is then obtained from

$$\bar{M}_d(\omega) = \bar{F}_o(\omega) \int_{-h}^{0} (d+y) e^{2\omega^2 y/g} dy \tag{56}$$

$$= \bar{F}_o(\omega) \left[d - \frac{g}{2\omega^2} + \frac{h}{(e^{2\omega^2 h/g} - 1)} \right]$$

The mean angular deflection of the tower will then be given by

$$\bar{\theta} = \frac{1}{\lambda} \int_{0}^{\infty} M_d(\omega) d\omega \tag{57}$$

This static deflection will be added to the deflections obtained from wind and current forces.

7.   ESTIMATION OF MEAN PEAK VALUES

For the response variables from the direct wave induced forces it is assumed that for a 12 hour storm the mean peak values will be 4 times the r.m.s. (σ) values since the high frequency motion is caused predominantly by linear wave inertial forces, although drag forces exist but to a smaller degree. In situations where drag forces are of the same order as inertia forces it does not appear to be possible to estimate rigorously the mean peak value for an oscillating structure from the results of spectral analysis.

For the case of the slow drift motion, if it is assumed that the response is a narrow band process it would be reasonable to assume that the mean peak value is 4σ, but this cannot be substantiated analytically.

8.   STATIC BENDING MOMENTS AND SHEAR FORCES

Stresses due to current and stresses arising from bending moments caused by platform weight and tower tilt should be added to the mean peak dynamic stresses. In addition the extra bending moment caused by slow drift tilt should be taken into account. Thus the slow drift tilt will provide an increased moment arm for the platform weight which will be counteracted by an increased buoyancy restoring moment, consequently the mean peak value of slow drift can be added to the static tilt in evaluating static stresses.

9.   DATA

Wave data

The following wave height spectra were used:-

MEAN JONSWAP H_s = 15m(ω_p = 0.365 rad/s) - 100 year storm, North Sea.
  $\gamma = 3.3$
$\sigma_a = 0.07$ H_s = 7.5m(ω_p = 0.5170 rad/s) - average winter storm, North Sea.
$\sigma_b = 0.09,$
$\bar{\alpha} = .0081$

PIERSON-MOSKOWITZ H_s = 15m(ω_p = 0.3243 rad/s)

                      H_s = 7.5m(ω_p = 0.4586 rad/s)

<u>Current Data</u>

For both sea states the current profile was taken in the form

$$V_c(y) = V_t(1 + y/d)^{1/7} + V_w(1 + y/50) \tag{58}$$

where V_t = 0.75 m/s (tidal contribution)

V_w = 0.8 m/s (wind induced)

<u>Structural Details</u>

The overall dimensions and gross weights of the towers are given in figures 2, 3, 4.

10. <u>DISCUSSION OF RESULTS AND CONCLUSIONS</u>

The results given below relate to the 100 year storm (H_s = 15 m) and an average winter storm, which are considered statistical extreme values.

The distributions of r.m.s. shear force and bending moment are given in figures 9, 10, 11, 12 and it was found that the influence of drag force on the high frequency motion was negligible in comparison with inertia force. As discussed subsequently however, the drag force due to the high frequency wave motion has a considerable effect in increasing the damping for the slow drift resonant motion of the towers. The most probable extreme values are estimated as 4 x r.m.s. (4σ) by assuming all quantities have a Rayleigh distribution.

For the direct wave induced motion (or high frequency motion) the bending moments and shear forces arise from the fluid inertia forces, structural inertia forces and gravity forces due to platform weight.

The maximum bending moments in each case are observed to occur at approximately 50 to 80 m below the tower deck. Such behaviour is to be expected since the fluid loading is concentrated in about the top 100 m from M.W.L. and the inertia relief arises primarily from the large deck mass.

It is interesting to compare the positions of the maximum bending moments with the results of marine risers for tension-leg platforms (Ref.2) where the location is similarly towards the top of the riser. In the riser analysis dynamic bending moments were obtained from a normal mode analysis using riser elastic curvature. The present method could easily be modified to incorporate elastic mode bending of the tower. In such cases the results would be similar to those presented but with the superposition of additional components due to resonant modes of the tower, which may be important for water depths exceeding say 350 m.

The 4σ shear forces at the ball joint for the three cases were 280t(JONSWAP + current), 1384t(P-M) and 1680t(P-M) for H_s = 15 m.

Tables 1 and 2 present the r.m.s. results obtained with C_m = 2.2 and C_d = 0.5 over the length of the tower based on the local K-C number.

For the 'Elf' tower the S.F. and B.M. due to the high frequency motion are SF = 280t (with current), BM = 72,988t-m (no current) for the JONSWAP spectrum. The inclusion of current is seen to decrease the maximum B.M. by 3% and to increase the ball joint reaction by 7%.

In the lower sea state (H_s = 7.5 m) the B.M. is reduced by about 58% but the shear force is only slightly reduced.

It is noted that the influence of gravity forces has been included in the evaluation of S.F. and B.M. for high frequency motion. The slowly varying motion however will give additional angles of tilt which will increase the tower bending moment due to deck weight. Tables 1 and 2 show that neglect of coupling between the high and low frequency motion leads to a much larger amplitude of slow drift resonant motion than when coupling is considered. This can be explained in terms of the linearised damping coefficient β_e, eq.42 which is a function of the r.m.s. relative velocity, eq.27.

The influence of the slowly varying tilt on the bending moment at M.W.L. due to deck weight can easily be examined. Thus for H_s = 15 m and a 0.66 deg. static tilt due to wind and mean drift for case (a) the B.M. is 0.66 x 35 x 6000/57 = 2431 t-m. Addition of the slowly varying 4σ tilt of the 1.12 deg gives an increased B.M. of 9637 t-m which is a fourfold increase. If current is included the tilt due to wind, current and mean drift is now 1.52 deg and the B.M. is 5598 t-m. Thus when 4σ = 0.64 deg is added (JONSWAP + current) the B.M. is increased to 7955 t-m, i.e. 42%

It is thus seen that current increases the damping coefficient β_e and reduces the slowly varying tilt, but increases the static tilt and in case (a) the overall effect of current is to decrease the bending moment.

Table 2 shows that the larger diameter towers have a greater slowly varying motion which for case (b) is about $\pm$4 deg. Here the static tilt due to wind and mean drift is 1.69 deg. and the B.M. at M.W.L. is 40,322 t-m. With current, wind and mean drift the tilt is 2.42 deg but the 4σ slow drift has been reduced by current to $\pm$3.88 deg, giving a total maximum tilt of 6.3 deg. which is greater than the former 5.69 deg. tilt neglecting current. Therefore in this case current has given rise to a net increase in bending moment.

It is clear from the above results that slowly varying resonant motions of articulated towers can cause considerable increases in bending moment due to deck weight. The slow drift resonant motion is governed primarily by nonlinear damping whose magnitude depends on C_d. Since the slowly varying force p.s.d. can be approximated as a white noise spectrum, the r.m.s. motion with (velocity)^2 damping can be determined using the equivalent linearisation method due to Caughey (Ref.7). It can be shown that σ_θ is approximately proportional to $(1/C_d)^{1/3}$ hence a doubling of C_d will only lead to a 20% reduction in amplitude.

The most probable peak deck acceleration occurring during the 100 year storm (H_s = 15 m) is 4σ = 0.228 g and for the average storm (H_s = 7.5 m) it is reduced to 0.118 g. For the larger towers, (b) and (c) the values are somewhat smaller, being 0.175 g and 0.0978 g respectively.

The validity of the calculated results is difficult to assess in the absence of measurements of motions and bending moments from model tests. A qualitative comparison can be made with the full scale measurements and calculated results for the Statfjord 'A' Articulated Loading Platform (Ref.8). The data for the survival case is as follows: H_s = 16.5m, $\bar{T}$ = 14.5s, average float chimney diameter $\simeq$ 11m. The r.m.s. deck acceleration of 0.76m/sec^2 is similar to case (a), (0.56m/sec^2). The r.m.s. pivot reaction was 49.5t (case (a) = 60.2t). A realistic comparison of r.m.s. bending moment however was not available due to the different design details of the two A.L.Ts. The maximum r.m.s. bending moment for Statfjord 'A' was 3,091 t.m. compared to case (a) of 4500 t.m. per column.

The natural period of Statfjord 'A' is approximately T_n = 30 sec with a 4σ motion amplitude of about 7° compared to case (a) of 3.6° where T_n = 57 sec. The Elf A.L.P. in 300 m water depth has more small diameter bracing members than Statfjord 'A' and hence the quadratic damping would be greater leading to smaller slow drift motions.

11. REFERENCES

1. Wu, S.C. and Tung, C.C. "Random response of structures to waves and current forces". Sea Grant Pub.UNC-SG-75-22, Sept. 1975. Also Wu, S.C. "The effects of current on dynamic response of offshore platforms". Proc. Offshore Technology Conf. 1976, paper OTC 2540.

2. Kirk, C.L. and Etok, E.U. "Random dynamic response of a tethered buoyant platform production riser". Jnl. of Applied Ocean Research, 3, 2, April 1981, pp.73-86.

3. Sarpkaya, T. "Vortex shedding and resistance in harmonic flow about smooth and roughened circular cylinders". Proc. of BOSS '76, an international conference on the Behaviour of Offshore Structures held at the Norwegian Institute of Technology, Trondheim, Aug. 2-5, 1976, Vol.1, pp.220-235.

4. Pinkster, J.A. "Low frequency phenomena associated with vessels moored at sea". Society of Petroleum Engineers Jnl. Dec. 1975,S.P.E. paper 4837, pp.487-494.

5. Havelock, T.H. "The pressure of water waves on a fixed obstacle". Proc. Royal Soc.,175(A), July 1940, pp.409-421.

6. Rice, S.O. "Selected papers on noise and stochastic processes". Edited by Wax, N. Dover Pubs. Inc. Part III "Statistical Properties of Random Noise Currents", pp.263-266.

7. Kirk, C.L. "Random vibration with non-linear damping". The Aeronautical Jnl. of the Roy.Aero.Soc. Nov.1973, pp.563-569.

8. Korbijn, F. "Computer simulation of slow and high frequency motions and loads of the Stratfjord 'A' Articulated Loading Platform in irregular sea". Det Norske Veritas Report 79-1049-I and II, Dec. 1979, pp.1-159.

$T_n = 100$ sec.

Wave p.s.d.	H_s (m)		HIGH FREQUENCY MOTION						LOW FREQUENCY MOTION				
		θ_D (deg)	Deck Accn. (m/sec²)	β_e	Pivot S.F. (tonne)	Max.rms BM (t-m)	*θ_s(c) (deg)	+θ_s(u/c) (deg)	Deck(c) Accn. (m/sec²)	Deck(u/c) Accn. (m/sec²)	Mean Slow drift (deg)	Case
JONSWAP	15	0.55	0.43	0.019	301	84,891	1.03	3.76	0.026	0.076	0.88	(b)
		0.50	0.38	0.016	351	85,571	0.664	3.02	0.016	0.06	0.60	(c)
	7.5	0.166	0.24	0.015	207	37,227	-	-	-	-	-	
P-M	15	0.61	0.42	0.038	346	84,254	-	-	-	-	-	(b)
		0.56	0.36	0.032	420	86,623	-	-	-	-	-	(c)
	7.5	-	-	-	-	-	-	-	-	-	-	
JONSWAP + CURRENT	15	0.55	0.43	0.144	296	84,130	0.97	2.38	0.025	0.05	-	(b)
		0.50	0.37	0.113	346	85,059	0.63	1.756	0.0156	0.036	-	(c)

*c = coupled
+u/c = uncoupled

Static tilt under wind + current = 1.54 deg(b), 1.62 deg(c).

TABLE (2) r.m.s. values
Cases (b), (c)

$T_n = 57$ sec.		High Frequency						Low Frequency ($C_D = 0.5$)				
Wave p.s.d.	H_s (m)	θ_D (deg)	Deck Accn. (m/sec²)	β_e	Pivot S.F. (tonne)	Max.rms BM (tonne m)	*$\theta_s(c)$ (deg)	+$\theta_s(u/c)$ (deg)	*Deck(c) Accn. (m/sec²)	+Deck (u/c) Accn. (m/sec²)	Mean slow drift (deg)
JONSWAP	15	0.64	0.56	0.078	65	18,247	0.28	1.47	0.02	0.093	0.4
	7.5	0.18	0.29	0.084	52	7,730	-	-	-	-	-
P.M.	15	0.67	0.53	0.08	64	17,624	-	-	-	-	-
	7.5	0.22	0.29	0.085	50	7,944	-	-	-	-	-
JONSWAP + CURRENT	15	0.61	0.54	0.344	70	17,570	0.16	0.85	0.014	0.054	-

* c = coupled
+u/c= uncoupled.

Static tilt under wind + current = 1.12 deg.

TABLE (1)   r.m.s. values

Case (a)

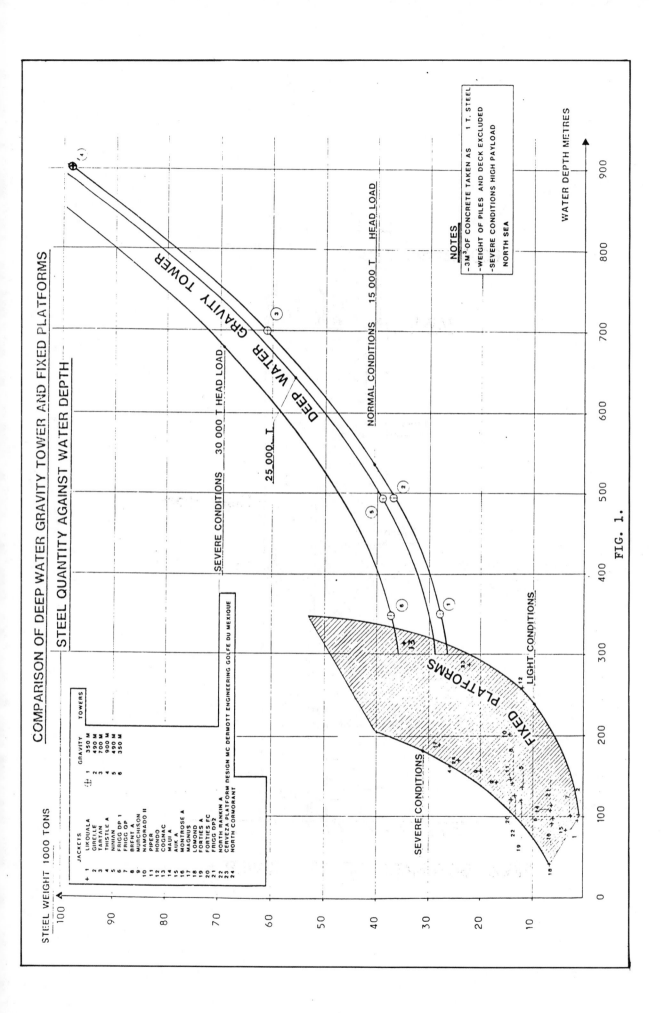

COMPARISON OF DEEP WATER GRAVITY TOWER AND FIXED PLATFORMS

STEEL QUANTITY AGAINST WATER DEPTH

FIG. 1.

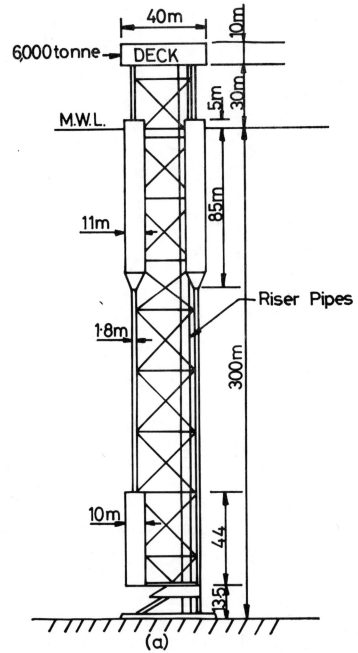

FIG.2. SCHEMATIC OF 'ELF' TOWER

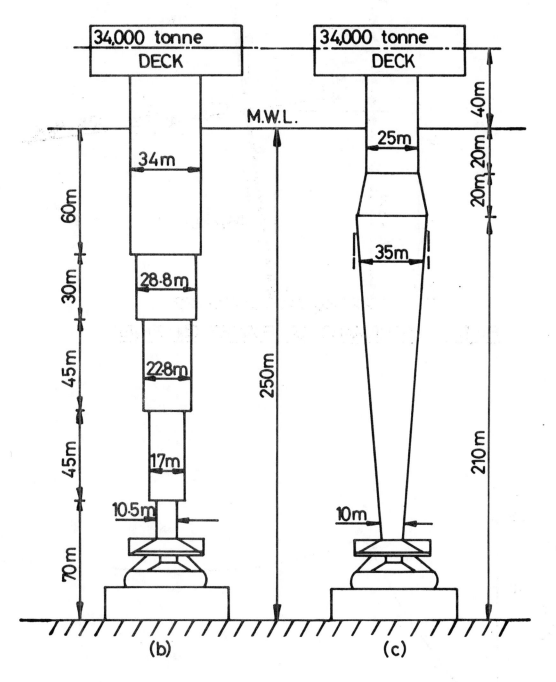

FIG. 3.   FIG. 4.

SCHEMATICS OF ARCOLPROD TOWERS.

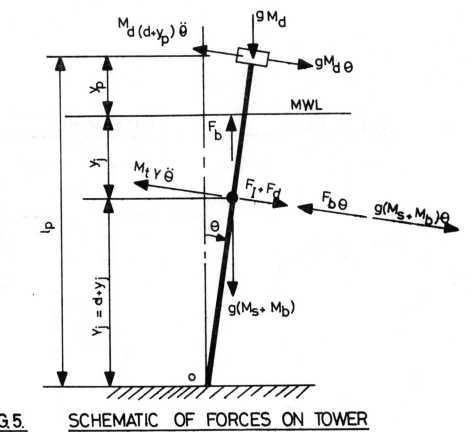

FIG. 5.   SCHEMATIC OF FORCES ON TOWER

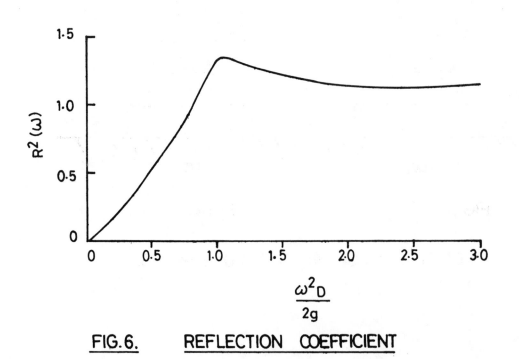

FIG. 6.   REFLECTION COEFFICIENT

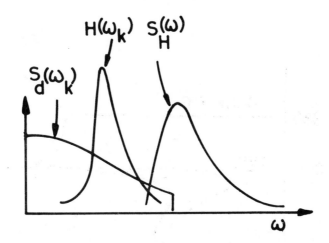

FIG.7. SPECTRA OF SLOW DRIFT
FORCE AND WAVE HEIGHT.

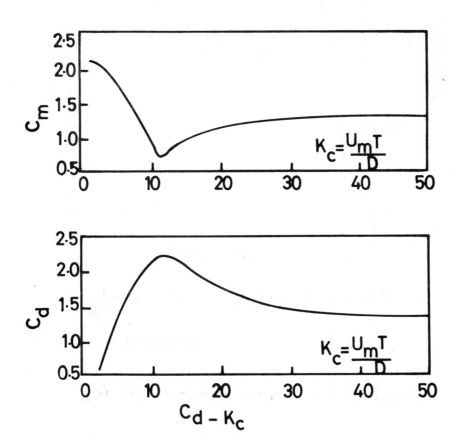

FIG.8. MEAN VALUES FOR
C_m AND C_d (REF.3).

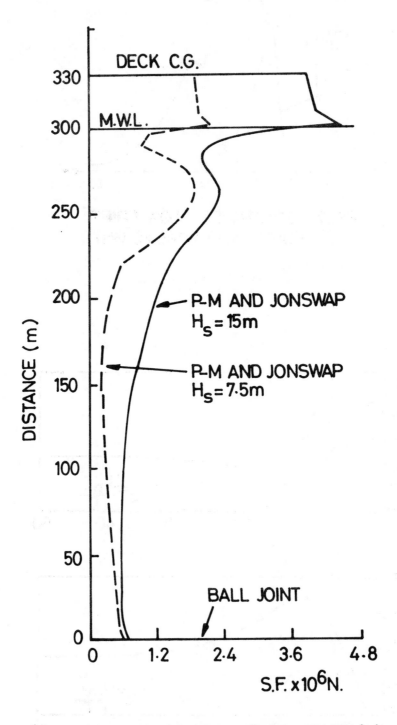

FIG.9. R.M.S. SHEAR FORCE – CASE (a).

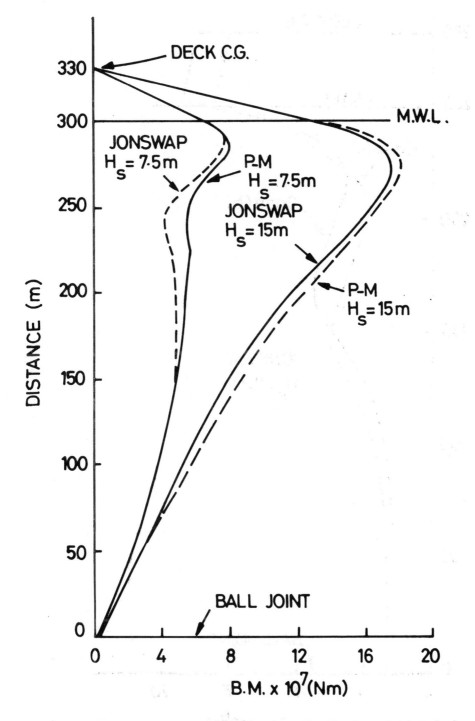

FIG.10. R.M.S. BENDING MOMENT – CASE (a).

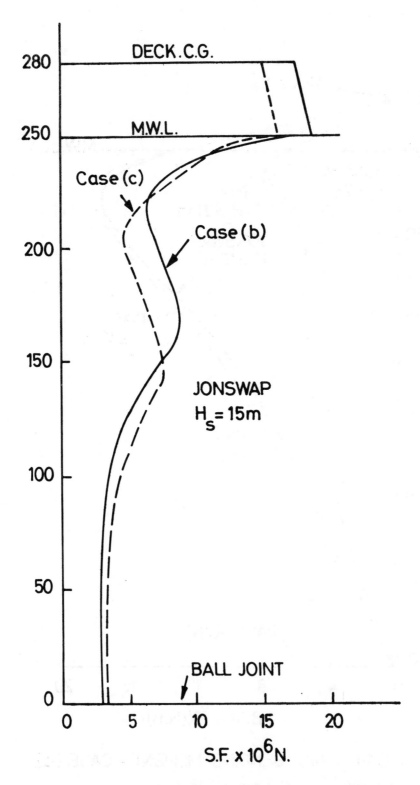

FIG.11. R.M.S. SHEAR FORCE – CASES (b) AND (c).

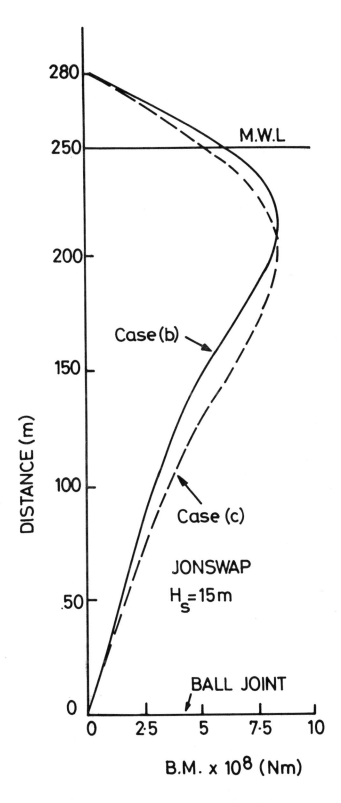

FIG.12. R.M.S BENDING
MOMENT – CASES (b) AND (c).

HYDRODYNAMIC GALLOPING OF RECTANGULAR CYLINDERS

A. R. Bokaian and F. Geoola

University College London, U.K.

Summary

The phenomena associated with transverse vibrations of rigid prisms, mounted elastically and restricting oscillations to a plane normal to the incident water flow were studied both experimentally and analytically. The prisms were all of rectangular cross sections of side ratio 1:2, of both sharp and rounded edges, with the broader side against the flow direction. A free stream turbulence of high intensity could be generated at the test section by the installation of a grid in the flume. The experiments encompassed the wake observations of stationary cylinders as well as the measurements of the lateral force coefficients at different angles of flow attack.

The variation of the lateral force coefficient in terms of the angle of flow attack was found to be highly sensitive to turbulence intensity and the corner radius ratio. The Strouhal number for a stationary cylinder was found to be independent of Reynolds number and turbulence characteristics but much dependent on corner radius ratio.

In dynamic tests, two organized forms of flow-induced vibration, namely, vortex-induced excitation and hydrodynamic galloping was observed. The response parameter was found to have a marked effect on the cylinder behaviour. As opposed to vortex resonance, turbulence was found to have strong effects on galloping vibrations. The lateral force measurements as well as the dynamic tests indicated that tendency to hydrodynamic galloping increased with an increase in turbulence intensity while it did decrease with an increase in corner radius ratio. With the highest corner radius ratio tested, the galloping was suppressed completely. For high values of response parameter, a reasonable agreement was observed between the measured galloping amplitudes and the theoretical curves based on the quasi-steady theory, whilst, for low values of response parameter, the theory led to a conservative estimation of the galloping response.

Organised and sponsored by
BHRA Fluid Engineering, Cranfield, Bedford MK43 0AJ, England

NOMENCLATURE

A = group of experiments

A_1, A_3 = coefficients of polynomial approximation to lateral force variation

B = group of experiments

B_1, B_2 = constant coefficients

C = group of experiments

C_D = coefficient of drag, (drag force)/$(\frac{1}{2}\rho hLV^2)$

C_L = coefficient of lift, (lift force)/$(\frac{1}{2}\rho hLV^2)$

C_M = coefficient of added mass, (added mass of cylinder)/$[\rho\pi L(b/2)^2]$

$C(\alpha)$ = lateral force coefficient

E = normalized power spectrum

F_y = net vertical force on the cylinder

H = a distance as defined in text

J = an integer

K_s = response parameter

L = cylinder length

L_x = longitudinal integral length scale of turbulence

Re = Reynolds number based on free stream velocity and the cylinder frontal size

S = Strouhal number

U = reduced velocity, $V/\omega_n h$

U_0 = minimum reduced flow velocity for galloping

$\overline{U}$ = twice reduced velocity, $nU/2\beta$

U_r = vortex resonance speed, $1/2\pi S$

$\overline{U}_r$ = twice reduced vortex resonance velocity, $nU_r/2\beta$

V = time mean free stream velocity

V_{rel} = relative flow velocity

Y = vibration amplitude

Y_0 = initial displacement to the system in still water

Y_i = half-cycle peak amplitude in still water

a = reduced amplitude, Y/h

$\overline{a}$ = twice reduced amplitude, $na/2\beta$

b = lateral dimension of the cylinder parallel to the flow direction

c = viscous damping coefficient of system

f_s = vortex shedding frequency

h = frontal dimension of the cylinder perpendicular to the flow direction

i = a counter

k = spring constant

m = total vibrating mass (including the added mass)

n = mass parameter

p = corner radius ratio

r = cross-sectional corner radius of cylinder

t = time

v = streamwise component of velocity fluctuation

y = vibration displacement

$\dot{y}$ = vibration velocity

$\ddot{y}$ = vibration acceleration

Ω = response frequency ratio, ω_c/ω_n

α = angle of attack of relative flow

β = reduced damping

γ = vibratory Reynolds number

η = dimensionless initial displacement to the system in still water, Y_0/h

ν = kinematic viscosity

π = 3.1415926

ρ = water density

ϕ = vibration phase angle

ω_c = system response frequency

ω_n = natural circular frequency of system (obtained in still water)

1. INTRODUCTION

A review of pertinent literature on flow-induced vibrations reveals that although work in some detail has been done on both forced and freely oscillating circular and non-circular cylinders in a steady air flow, the subject of free-transverse vibrations of non-circular cylinders in a steady water flow has received very little attention. Circular cylinders are frequently used in the form of components of marine structures. They are convenient forms of offshore piling since they have equal strength in all directions; however, their geometry is known to promote complex local flow patterns. Non-circular cylinders of various cross-sectional configurations are also used in offshore and coastal regions; examples are Rhendex piles, etc. Rectangular and square cylinders, though with rounded edges, can widely be seen as members of marine offshore structures such as four sided tethered buoyant platforms and three sided tension leg platforms. Considering that the submerged depth of these structures can be as high as 50m, the hydrodynamic stability of local bluff components as well as the entire structure under sea currents requires special consideration.

The transverse vibrations of sharp-edged bluff bodies in a steady air flow have been the subject of much study by the aerodynamicists over the past three decades. The very large amplitude oscillations associated with these structures is termed as aerodynamic galloping. These investigations have been mostly in connection with aerodynamic instabilities of towers, suspension bridges, most high-rise buildings and so forth. Den Hartog (Ref. 1) associated the galloping instability with the aerodynamic coefficients of a cross section. Parkinson and his research collaborators (Refs. 2,3,4) made extensive experimental work on galloping vibrations in a wind tunnel, and developed a quasi-steady aerodynamic theory for this type of excitation. The effects of turbulence on galloping was explored by Novak and Davenport (Ref. 5), Novak (Ref. 6), Novak and Tanaka (Ref. 7) and Kwok and Melbourne (Ref. 8) amongst others, while the combined effect of galloping and vortex resonance was investigated by Parkinson and Wawzonek (Ref. 9). However, all this research work has been restricted to sharp-edged geometries.

The primary effect of the surrounding fluid on the vibrations of a structure in a still fluid is simply to increase the effective mass of the structure. The added mass effects are ordinarily negligible in the aerodynamic analysis of buildings since the density of air is negligible compared to that of most buildings. However, since the density of water is comparable to the density of most marine structures, the added mass plays a large role in the hydrodynamic analysis of marine structures, resulting in considerably lower structural frequencies. Furthermore, fluid loading is a function of the density of the surrounding medium, which, for water, is three orders of magnitude greater than air. This means that the frequency dependent forces which may not be significant in air can be vital in water where the natural frequencies are lower.

The possibility of galloping of marine structures due to sea currents have been mentioned by many marine offshore engineers and researchers (Refs. 10, 11). Although circular sections do not gallop, sections which are not far removed from circular can. For example, it is recommended (Ref. 10) that light structural forms such as cables or bundles of cables should be wrapped in waterproof tape or enclosed in circular casing in order to prevent the possibility of galloping. This research, which is a part of a continuing programme, studying the hydroelastic instabilities of bluff cylinders, looks at the various phenomena associated with transverse vibrations of rigid prisms, mounted elastically and restricting oscillations to a plane normal to the incident water flow. The prisms were all of rectangular cross sections of side ratio 1:2, of both sharp and rounded edges, with broader side facing the flow direction. This research is concentrated on two organized forms of flow-induced oscillations, namely, the vortex-induced and galloping. In addition to sharing a transverse plane of oscillation, the two forms are nearly harmonic and occur at a frequency very close to the natural frequency of the hydroelastic system, so that from an observation at a single water speed it would be difficult to distinguish vortex-induced oscillation from galloping. However, their dependence on water speed is quite different. Vortex-induced oscillation is a form of non-linear resonance and is confined to a small range of flow speeds, at which the frequency of vortex shedding deviates from that expected for a stationary cylinder (proportional to flow speed since the Strouhal number is nearly constant) and coincides with the natural frequency of the system. The vibration of the cylinder appears to control the vortex shedding in this range, and this phenomenon has been termed "lock-on". At the upper limit of the range, the vibration can no longer control the shedding which now reverts to the frequency for the stationary cylinder, and the vibration dies out. In the second form of instability, referred to as galloping, the fluid forces which create a condition of instability are generated by the fact that the cross-section of the body is hydrodynamically unstable to small disturbances. These forces result in oscillations which grow in amplitude until the energy extracted from the fluid stream balances that dissipated through various forms of damping. The main features of galloping are that the vibration can occur in a single degree of freedom and the steady-state amplitude tends to increase with increasing water velocity, taking a linear relationship,asymptotically.

In the hydrodynamic analysis of structural members which are exposed to tidal, river or esturial flows,the effects of natural turbulence must be considered. In fact, it is the object of this research to find the transverse response of the rectangular cylinders, as described earlier, in the forms of vortex resonance and galloping under steady water flows with different turbulence characteristics. Vortex-excited oscillations result in reduced fatigue life, amplified hydrodynamic forces, and sometimes lead to structural damage and to destructive failures. If the galloping oscillations start, they can become violent as they grow to large amplitudes, severely jeopardizing the structure. To design reliably against the consequences of flow-excited oscillations, design engineers must have prior knowledge of the criteria governing these.

2. BASIC RELATIONSHIPS OF GALLOPING

Fig. 1 shows a two-dimensional linearly damped spring supported rectangular cylinder moving with the velocity $\dot{y}$ perpendicular to a two-dimensional flow having a velocity V. In this figure,y is the displacement, h, b and r are the cylinder frontal dimension perpendicular to the flow direction, the cylinder lateral dimension parallel to the flow direction and the cylinder cross-sectional corner radius respectively.

The fundamental assumption of galloping analysis is that at every instant during oscillation, the hydrodynamic force is the same as on a fixed body during a static test under the same angle of attack. Under this assumption, the lateral force F_y can be written as

$$F_y = -(C_D \sin \alpha + C_L \cos \alpha)\tfrac{1}{2}\rho h L V^2 \sec^2\alpha = C(\alpha)\tfrac{1}{2}\rho h L V^2 \tag{1}$$

in which, $\alpha = \tan^{-1}(\dot{y}/V)$ is the angle of attack of relative flow to cylinder, ρ, L, C_D, C_L and $C(\alpha)$ are the fluid density, the cylinder length, the coefficient of drag, the coefficient of lift and the lateral force coefficient respectively. When the lateral force coefficient is measured on stationary cylinder at different angles of attack α, its coefficient $C(\alpha)$ can be expressed by a polynomial in terms of $\dot{y}/V$

$$C(\alpha) = \sum_{i=1}^{j} A_{(2i-1)} (\dot{y}/V)^{(2i-1)} \tag{2}$$

in which i, J and $A_{(2i-1)}$ are a counter, an integer and constant coefficients respectively. The equation of motion of the cylinder maybe written as

$$m\ddot{y} + c\dot{y} + ky = \tfrac{1}{2}\rho V^2 h L C(\alpha) \tag{3}$$

where m, c and k are the total vibrating mass, the viscous damping coefficient and the spring constant respectively. The solution to the Eq. (3) is sought in the form of

$$y = Y \sin(\omega_n t + \phi) \tag{4}$$

where ω_n is the natural circular frequency of the system and the amplitude Y and the phase ϕ are assumed to be slowly varying functions of time t. Using the first approximation of Krylov and Bogoliulov (Ref. 12), the steady state amplitude of the Eq. (3) maybe calculated from the following relationship

$$\sum_{i=1}^{j} A_{(2i-1)} \cdot B_i \cdot \left(\frac{a}{U}\right)^{2i-2} = \frac{2\beta}{nU} \tag{5}$$

in which

$$B_i = 2 \times \frac{1 \times 3 \times 5 \times \quad \times(2i-1)}{2 \times 4 \times 6 \times \quad \times(2i)} \tag{6}$$

the reduced velocity $U=V/(\omega_n h)$, the dimensionless amplitude $a=Y/h$, the reduced damping $\beta=c/(2m\omega_n)$ and the mass parameter $n=\rho h^2 L/(2m)$. By introducing two dimensionless variables $\bar{a}=na/(2\beta)$ and $\bar{U}=nU/(2\beta)$, called twice reduced amplitude and twice reduced velocity respectively, Eq. (5) can be written as

$$\sum_{i=1}^{j} A_{(2i-1)} \cdot B_i \cdot \left(\frac{\bar{a}}{\bar{U}}\right)^{2i-2} = \frac{1}{\bar{U}} \tag{7}$$

If $A_1>0$, the onset of reduced velocity for galloping is

$$U_0=\frac{2\beta}{nA_1} \tag{8}$$

and for all flow velocities $U>U_0$, the zero position (equilibrium) is unstable and strong steady oscillations build up. However, for $A_1 \leq 0$, the zero position remains stable for all flow velocities. A detailed description of the above theory together with its stability analysis can be found in (Ref. 4).

3.   EXPERIMENTAL PROCEDURE

The experiments were conducted in a rectangular horizontal flume, 0.3m wide, 0.3m high and 18m long with a smooth aluminium bed and vertical glass side walls. A 100mm durapipe supplied water from the constant head tank to the settling chamber and distributed through a manifold. A 60mm thick layer of honeycomb baffle was positioned across the settling chamber near the pipe outlet to help dissipate disturbances in the flow, after it left the pipe outlet. In addition to this, two very fine meshes with a separation of 10cm were placed in the settling chamber just before the leading stream-lined plates of the flume. The flow rate was controlled by means of an inclined differential manometer connected to the constant head tank and the inflow pipe. The water depth in the flume could be varied by an adjustable tailgate at the downstream end of the flume. The flow from the flume was directed through a drain channel to an underground sump. The discharge from the flume could be measured volumetrically over a measured time interval with the water being collected in an underfloor volumetric tank.

Plate 1 shows the experimental apparatus designed for dynamic tests in this investigation. As can be seen in this plate, this apparatus was basically symmetrical with respect to the channel centre plane. It consisted of a horizontal square plate of dural, 600mm×600mm×10mm, being supported by the flume rails. Two 30mm×50mm×500mm blocks of dural, with two airbearings embedded in each, were bolted to the plate. The blocks were separated from each other vertically by a distance of 30mm. The airbearings were each of internal diameter 6.3mm and 30mm height. On these blocks, screws were provided to secure the parallelism of the two sets of airbearings. The rectangular cylinder under investigation which spanned approximately the whole width of the flume was held horizontally by two identical very thin arms of width 6mm and length 180mm. These arms were separated by a distance of 250mm. Each arm was attached to a shaft of external diameter 6.2mm and 200mm height. Each shaft in turn passed through two airbearings. The upper ends of the two vertical shafts were connected together by means of a horizontal cross bar and the whole system was suspended from a rigid frame by means of two identical helical springs. The horizontal component of this frame could be fixed at varying heights with respect to the square plate. Screws on the square plate allowed very fine adjustments of the plate in every direction to ensure that the cylinder could be alligned parallel to the channel bed with its broader side facing the approaching water flow. The high air pressure in the bearings ensured that the vibrating system had purely cross flow movements with a minimum viscous structural damping. However, in order to be able to change the amount of external damping artificially, a thin vertical plate immersed in oil was attached to the cross bar by means of a very slender shaft. Different types of oil or different lengths of plate could be used to generate the desired damping. The mass and the stiffness of the system could be varied by placing additive masses on the cross bar and helical springs with different stiffnesses respectively.

The whole assembly allowed a maximum peak to peak amplitude oscillation of roughly 100mm. An ultrasonic displacement measuring technique was employed to observe the cylinder movements. This device essentially consisted of a pair of transducers, one transmitting and the other one receiving. The transmitter was attached to the moving body whilst the receiving transducer was held stationary. The transmitter emitted an ultrasonic pulse which was detected by the receiver. The transit time of the pulse was proportional to the distance between them. In order to avoid the interference of any

110

ambient noise with the receiving transducer, both the receiving and the transmitting transducers were shielded, as shown in Plate 1. There was a small gap between the two shields and the shielded transmitting transducer could freely move outside the shielded receiving transducer. This device had four switches for amplitude ranges of 0-30mm, 0-100mm, 0-300mm and 0-1000mm. The instrument was coupled with a U.V. recorder which allowed a later analysis of amplitude and frequency response of the system.

In addition to the experimental rig described above, another apparatus was designed to measure the lateral force F_y and hence the lateral force coefficient $C(\alpha)$ on a stationary cylinder at different angles of flow attack α. In this case, the cylinder was held vertically in the centre plane of the channel and was mounted at the top by two identical leaf springs connected to a horizontal turntable. The cylinder broader side was perpendicular to the centre axis of the leaf springs and hence the cylinder movement was only possible in the direction normal to this axis. The base plate of the turntable was supported by the flume rails. The lower end of the cylinder was supported by a cone and socket device with the former resting on the channel bed and the latter being fixed to the cylinder lower end. There was a very small clearance between the cylinder lower end and the channel bed. By turning the turntable, the cylinder could be positioned at any desired angle of flow attack. The lateral force was inferred by measuring the displacement of the upper end of the cylinder by means of a capacitance type transducer in conjunction with a mini-computer via an analogue-digital converter and using a force-displacement calibration curve. The transducer was able to measure displacement up to ± 0.6mm. Prior to each experiment, with no water in the channel, the displacement of the upper end of the cylinder was calibrated against a horizontal force exerted to the cylinder in a direction normal to its smaller side, at a vertical distance from the channel bed equal to half water depth.

The time mean velocity of the flow was measured by means of a miniature current flowmeter. Occasionally, a Pitot-static tube was used for the same purpose. The velocity fluctuations in the free stream and in the wake of the cylinder were detected by means of a linearized constant-temperature type hot film anemometer. The output of the hot film was fed into a combined analogue-digital converter and a mini-computer. The turbulence characteristics of the water flow in which the cylinder was immersed could be varied by positioning a square mesh grid at various locations upstream of the test section. The grid which was biplanar was constructed with bars of circular cross section of 16mm diameter and a mesh to bar ratio of 5:1.

4. EXPERIMENTAL PROGRAMME

The experiments were performed at a station situated at a distance of 1.6m from the leading streamlined plates of the flume. Throughout the experiments, the water depth at this section was kept constant at approximately 290mm. With the aid of the Pitot-static tube, some checks were made of the velocity field at this section at different flow rates. It was found that the mean velocity distribution throughout the working section was essentially uniform with the exception of a region very close to the water surface (some 2mm from the water surface), where the velocity measurements showed some unsteadiness. This was probably due to the cross waves generated in the settling chamber and in the flume. The hot film probe was employed to identify the turbulence characteristics at the test section, namely, the scale and the intensity. It was set to sample the streamwise fluctuating velocity at the time interval of 0.0025s. The longitudinal integral length scale of turbulence L_x was computed from the auto-correlation function of the streamwise fluctuating velocity v and the time mean free stream velocity V based on Taylor's hypothesis. The initial or background turbulence at the test section had the intensity $(\overline{v^2})^{\frac{1}{2}}/V$, where $(\overline{v^2})^{\frac{1}{2}}$ is the r.m.s. value of v, on average, equal to 6.5%. The average value of the associated integral length scale was 16mm. This background turbulence is termed here as turbulence type 1, and it seems that it was essentially originated from the settling chamber where the inflow discharge was distributed through the manifold. A free stream turbulence of high intensity was generated by the installation of the grid at the distance of 360mm upstream of the test section. The associated turbulence intensity and integral length scale were, on average, 11.9% and 8mm respectively. Hereafter this grid generated turbulence is referred to as turbulence type 2. In both cases, the turbulence was found to be roughly homogeneous at the test section and the intensity and scale changed only slightly with the increase in the mean velocity. The frontal dimension of cylinders used in this investigation were 18.8mm, 13.3mm and 17mm. The corresponding values of turbulence length scale ratio L_x/h

were 0.85, 1.20 and 0.94 in the turbulence type 1 and 0.42, 0.60 and 0.47 in the turbulence type 2.

The assembly for the dynamic tests was adjusted so that the cylinder position in still water was half way between the channel bed and the water surface (approximately ±145mm from the channel bed). The end clearances between the cylinders and the glass side walls were kept at 1 or 2mm; these end gaps could not be made any smaller due to practical difficulties. Indicated in Table 1, are the values of the cylinder aspect ratio which were 22.41, 17.47 and 15.85. With the assembly allowing a maximum peak to peak oscillation amplitude of approximately 100mm, the nearest distance of a vibrating cylinder, in dynamic tests, from the two horizontal boundaries (the channel bed and the water surface) could be as low as H≈95mm. The values of the ratio H/h for the aforementioned cylinder frontal sizes were 5, 7.1 and 5.6 respectively, and these figures were assumed to be sufficiently high to preclude any horizontal boundary effects on the vibrating cylinders. Furthermore, the aforementioned frontal sizes occupied 6.5%, 4.6% and 5.9% of the cross-sectional area of the flow respectively. No correction was made to the results of this investigation for the effects of blockage.

As is indicated in Table 1, a total number of 13 dynamic experiments were performed in this investigation and they can be broadly classified into three groups A, B and C depending on the value of corner radius ratio r/h. In group A, the cylinders were all sharp-edged; while for groups B and C, the corner radius ratios were 0.141 and 0.187 respectively. In each dynamic test, values of the natural circular frequency ω_n, the reduced damping β and the mass parameter n were fixed and the cylinder performance was observed for about twenty different velocities ranging from 3cm/s to 45cm/s. These velocities were measured by means of the miniature flowmeter. For each velocity, the amplitude and frequency of oscillation were measured.

In addition to the aforementioned dynamic tests, two more series of experiments were carried out. In the first series, the rectangular cylinders were kept horizontal and stationary at half water depth, with the broader side against the flow direction and the frequency of vortex shedding at different flow velocities was determined. In the second series of experiments, the lateral force on stationary cylinders at different angles of flow attack was measured. The details of these two series of experiments, which were performed in both types of turbulence, will be described in Sections 5.1 and 5.2 respectively.

5.   RESULTS AND DISCUSSION

5.1   Wake observations of stationary cylinders.

The hot film probe was positioned in the centre plane of the flume, just outside the near wake of cylinder. The exact position varied and was chosen to give the steadiest signal. Figs. 2, 3 and 4 represent the normalized spectra of wake fluctuations past rectangular cylinders in the range Re≈500-7500, in both the low and high free stream turbulent flows. They were computed using Fast Fourier Transform technique (Ref. 13). As can be seen in these figures, roughly 70% of the energy in the part of the wake sampled is concentrated within a narrow band which peaks at the vortex shedding frequency. The variation of the Strouhal number defined as $S=f_s h/V$ in relation to the Reynolds number for both sharp and round-edged geometries and in both the low and high free stream turbulent flows is shown in Fig. 5. This figure indicates that for a sharp-edged cylinder, in the range Re≈750-7500, the Strouhal number is rather insensitive to a change in Reynolds number and turbulence characteristics and has a value of approximately 0.137. It is noteworthy to mention that, for this cross-section, the measurement of Bearman and Trueman (Ref. 14) in a wind tunnel with a low turbulence level (intensity of the order of 0.3%) and in the Reynolds number range Re≈2×10^4-7×10^4, gives S=0.131. For round-edged cylinders with r/h equal to 0.141 and 0.187, in the range of Re≈10^3-7500, the Strouhal numbers are independent of change in Reynolds number and turbulence characteristics and are roughly equal to 0.169 and 0.183 respectively. The increase in the Strouhal number with the increase in r/h is because the rounding corners of the cylinder helps the coupling together of the shear layers.

5.2   Lateral force measurements.

To use the quasi-steady approach, it is important to obtain the lateral force coefficients for Reynolds numbers close to that of the oscillating cylinder whose response is to be investigated. Due to the symmetry of the cylinder cross section at zero angle of flow attack, no lateral force could exist and this hydrodynamic characteristic was used to establish the turntable position which gave zero angle of attack. Fig. 6 represents the lateral force coefficients for sharp and round-edged rectangular cylinders in both low and high free stream turbulent flows. The lateral force measurements for the sharp-edged cylinder was performed at two Reynolds numbers 5000 and 8600. They were not carried out for any lower Reynolds number because of inaccuracy in the measurement of low forces involved. Table 2 represents the coefficients of polynomial approximation to the measured lateral force coefficients using the least squares approach. On Fig. 6, is also plotted the experimental results of Novak (Ref. 6) obtained in a wind tunnel, presumably at Reynolds numbers one order of magnitude greater than those of present investigation. Novak's curve of 5% turbulence intensity was measured in a turbulent boundary layer, while his 11% curve was established in a turbulent flow behind a grid. Some discrepencies can be seen between the results of the present investigation and the corresponding results of Novak. Similar discrepencies exist between the results of Novak and those reported by Laneville and Parkinson (Ref. 15), even between the corresponding curves of very smooth flow. Whilst experimental errors could in part be responsible for these differences, it is important to note that parameters such as aspect ratio of the models tested and Reynolds number may, collectively or individually, be significant factors.

Fig. 6 indicates that turbulence has a profound effect on the behaviour of lateral force coefficient. The effects of increasing turbulence intensity on $C(\alpha)$ for sharp-edged cylinder are seen to be a reduction in the maximum value of $C(\alpha)$ and its associated value of α and a contraction of the range of α in which $C(\alpha)$ is positive; maximum lateral force occurs when reattachment at the trailing edge of the relative flow-ward face becomes permanent. The reason behind this can be found in the work of Gartshore (Ref. 16). He suggested that the effect of increasing free stream turbulence is to increase the turbulent mixing in the shear layers and this increases the rate of entrainment of fluid from the wake and decreases the radius of curvature of shear layers. Hence, the maximum lateral force coefficient occurs at an angle of incidence which is smaller in the high free stream turbulent flow than in low free stream turbulent flow. Moreover, Fig. 6 denotes that by rounding the corners of the sharp edged-rectangular cylinder by $r=0.141h$ and in high free stream turbulence, the maximum lateral force coefficient drops by 30% and occurs at an angle α which is 67% lower than the corresponding value of the sharp-edged cylinder. The value of $C(\alpha)$ for this cylinder in low free stream turbulent flow is entirely negative. With the increase in the corner radius $r=0.187h$, the lateral force coefficient becomes negative in both the low and high free stream turbulent flows. This is probably because the separation points are mobile on the rounded edges of the cylinder as opposed to the case of the sharp-edged cylinder where the separation points are fixed.

5.3   Still water added mass of the cylinders.

In each test, the net mass of the system in still water was measured by direct weighing. The natural circular frequency of the system, ω_n was obtained by plucking excitation in still water and analysing its damped free oscillation. The initial displacement to the system relative to the zero position was roughly $Y_0 \approx 50mm$. The half-cycle peak amplitudes Y_i and the half-cycle durations were measured from the stripchart recording of the oscillatory response. To within the accuracy of the measurements (errors of the order of 1%), the half-cycle duration was found to be constant in each test. That is, the response frequency in still water and hence the total vibrating mass (i.e., the mass of the system in still water plus the added mass of the cylinder and the wetted parts of the arms) were constant and independent of the amplitude of oscillation from $Y_0 \approx 50mm$ down to measurable limits of amplitude in all tests. The added mass of the cylinder and wetted parts of the arms were found by subtracting the measured mass of the system in still water from the total vibrating mass. This value however, could be considered as the added mass of the cylinder itself, since some checks (by removing the cylinder from the arms and repeating the above procedure) proved that the added mass of the wetted parts of the arms was negligible compared to that of the cylinder. In short, it can be concluded that, in each test, if the dimensionless vibrational amplitude in still water Y_o/h is less than η (see Table 1), then the still water added mass of the cylinder and hence the vibratory Reynolds number γ are constant and independent of the amplitude of oscillation. Using the technique of damped free oscillation in still water, Skop et al. (Ref. 17) arrived at the same conclusion for a circular cylinder in the range $\eta=Y_o/h<2$ and $920<\gamma=\omega_n h^2/\nu<20880$, where h in this case is the cylinder diameter.

The values of the added mass coefficient C_M and vibratory Reynolds number γ associated with each test is shown in Table 1. The added mass coefficient was calculated in a conventional manner by dividing the value of added mass by $\rho\pi L(b/2)^2$ (Ref.18). As can be seen in this table, the values of C_M and γ for sharp-edged geometries vary between 2.9-3.4 and 775-1495 respectively. The theoretical value of C_M for this cross-sectional geometry, which has been evaluated through the mathematical application of two-dimensional potential theory is 1.70 (Ref. 18). The observed range is roughly 70%-100% greater than the theoretical value. These differences mainly arise from low values of γ in the tests, However, the low aspect ratios (departure from two-dimensionality) in some tests as well as the proximity of the cylinder to the side walls could have also contributed to these differences; the proximity effects of outside boundaries on the added mass of a vibrating body has been observed by many investigators (Refs. 19, 20). Table 1 clearly indicates that within the group of cylinders A, there is a decrease in observed values of C_M with the increase in γ. This behaviour has also been noted by Miller (Ref. 21).

In each test, once the natural circular frequency of the system in still water ω_n was found, the system structural damping could be obtained by removing the cylinder from the arms and placing enough additive mass on the cross bar, so that the circular natural frequency of the system in still water would have a value equal to ω_n as found above. The structural damping of this system could be calculated from the free damped oscillation in stagnant water. The damping thus obtained was found to be viscous down to measurable limits of amplitude in all tests.

5.4   Dynamic test results

Table 1 indicates the important details of the dynamic tests which were performed for both sharp and round-edged cylinders and in both the low and high free stream turbulent flows. In each test, a gradual but continual increase of flow velocity in the flume, together with a simultaneous recording of the vibration amplitude Y, the response frequency ω_c and the flow velocity V was made. Indicated in Columns (16) and (11) of Table 1, are the values of the vortex resonance speed $U_r=1/(2\pi S)$ where S is the Strouhal number for a stationary cylinder and the response parameter $K_s=\beta/(nU_r^2)$. The response parameter has widely been used in previous work concerned with flow-induced vibration in both air and water to corrolate mass parameter and reduced damping with vibration amplitudes and to determine instability regions. However, it should be noted that since U_r is constant for each group of experiments, the response parameter is therefore directly proportional to β/n within each group.

Figs. 7, 8, 9,10 and 11 summarize the experimental results of the dynamic tests performed in this investigation. The first four figures show the results in a non-dimensionalized form with twice reduced amplitude $\bar{a}=na/(2\beta)$ and response frequency ratio $\Omega=\omega_c/\omega_n$ (where ω_c is the system response frequency), plotted against twice reduced velocity $\bar{U}=nU/(2\beta)$. The results of Novak (Ref. 22) is plotted on Fig. 8. The small arrows on $\bar{U}$ axes indicate the twice reduced vortex resonance speed $\bar{U}_r=nU_r/(2\beta)$. Adopting the quasi-steady approach, the two theoretical galloping curves for sharp-edged cylinder in both the low and high free stream turbulent flows (turbulence types 1 and 2) are shown on Figs. 7, 8 and 9. It should be noted that these curves are all identical but drawn on different scales. Similarly, the theoretical galloping curve for round-edged cylinder with r/h=0.141 in turbulence type 2 is indicated on Fig. 10. These curves were computed from Eq. (7) using the coefficients A_i as indicated in Table 2; the full lines represent the stable galloping amplitudes while the unstable amplitudes are indicated by dashed lines. The dimensionless variables $\bar{a}$ and $\bar{U}$ make it possible to represent the galloping response independently of mass and damping.

The theoretical galloping curve for sharp-edged cylinder in the low free stream turbulent flow indicates that the zero position always remains stable. This can also be concluded from the corresponding curve of the variation of $C(\alpha)$ versus α in Fig. 6, since $dC(\alpha)/d\alpha|_{\alpha=0}=A_1=-0.001$. However, in this case, in order to excite the cylinder to gallop, for a given velocity, an initial displacement greater than the corresponding unstable amplitude is required. In contrast, for the sharp-edged cylinder in the high free stream turbulent flow, small disturbances which are always present in the flow can set the cylinder into galloping vibration, provided that $\overline{U} \geq 1/A_1$ or equivalently $U \geq U_0 = 2\beta/(nA_1)$. This is also true for the round-edged cylinder with $r/h = 0.141$ in turbulence type 2.

As is clearly indicated in Figs. 7, 8 and 9, two distinct and well established types of response, namely, the vortex resonance and galloping can be seen in all the experiments with sharp-edged cylinders (group A). However, as Figs. 10 and 11 show, while the round-edged cylinders did exhibit vortex resonance in both the low and high free stream turbulent flows, galloping oscillation was only observed for the round-edged cylinder with $r/h = 0.141$ in turbulence type 2. This indicates that rounding the edges of a sharp-edged cylinder would help to suppress the galloping vibration. This conclusion is also supported by the measurements of the lateral force coefficients of both sharp and round-edged cylinders in both types of turbulence (Fig. 6). A smooth and thin curve was drawn through the experimental points of the vortex resonance response of the aforementioned figures. On Figs. 7, 8, 9 and 10, the vortex resonance regions are shown magnified in the upper left parts of these figures but drawn in a different non-dimensionalized form with the reduced amplitude, a plotted versus the reduced velocity $U = V/(\omega_n h)$. On U axis of these figures and also on that of Fig. 11, the corresponding values of the vortex resonance speed U_r are indicated. Note that the magnified resonance curves in Figs. 7, 8 and 9 are in an increasing order of the response parameter.

As the magnified vortex resonance curves and Fig. 11 clearly indicate, the oscillation in all runs started from zero position with the amplitude growing rather rapidly with an increase in the flow velocity. In the vortex resonance regions, no appreciable change can be seen in the vibration amplitude and the frequency response with a change in turbulence. Fig. 12 summarizes the influence of mass and damping on vortex resonance in the form of stability diagrams. The graphs show the variation of the peak amplitude and the corresponding flow speed as well as the upper and lower velocity bounds for some limiting permissible displacement, with the response parameter. As Fig. 12 shows, within each group of experiments, there appears to be a slight increase in the observed vortex resonance speed where the resonance amplitude peaks, with an increase in the response parameter. This pattern can also be seen in the work of Novak (Ref. 22) on the model of tall building of a sharp-edged rectangular cross-section with side ratio $b/h = 0.5$ and an aspect ratio 3.48 in a smooth flow at Reynolds number range $Re \approx 3.5 \times 10^4 - 6 \times 10^4$. However, although this increase is rather distinct for round-edged cylinders, for sharp-edged cylinders the increase can be considered to be negligible and it can be concluded that the vortex resonance responses peak at approximately $U = 1.22$; the resonance speed $U_r = 1/(2\pi S)$ being 1.16. This observed value of the resonance speed is in a very close agreement with the approximate value 1.19 obtained by Novak (Ref. 22). Similar to sharp-edged cylinders, the measured resonance speeds for round-edged cylinders can be seen to be generally slightly greater than the corresponding values based on the measurements of the Strouhal number. Moreover, comparison between Runs A3 and C3 and similarly between Runs A6, B2 and C2 reveals that with a fixed magnitude of the response parameter, the vortex resonance peak amplitude increases with an increase in the value of r/h.

The response frequency ratio Ω associated with the peak in the vortex resonance responses, for both the sharp and round-edged cylinders, occurred in the close vicinity of 1. This could mean that at the peak of vortex resonance, the still water added mass was unaffected by streaming flow; this behaviour has also been noted by King (Ref. 23) for a cantilevered circular cylinder. As was expected, within each group of experiments, the peak amplitude decreased with an increase in the response parameter. A close examination of the aforementioned figures reveals that with the exception of group C, where no galloping vibration was observed at any flow velocity, there exists two patterns of response depending on the magnitude of response parameter. The first pattern is associated with Runs A1, A2 and A5 of group A and Runs B1 of group B, having low values of K_s. In these runs, after the peak vortex resonance was reached, an increase in the flow velocity resulted in a continual decrease in amplitude with a minimum at $U = 1.5$ for the aforementioned runs of group A and at $U = 1.7$ for Run B1, while the response frequency ratio remained at roughly $\Omega = 1$. With further increase in flow velocity, the galloping vibrations started, having an approximately linear relationship with the flow velocity. However, in each test, while the cylinder was galloping, some checks were made on the cylinder's stability at zero position. This was done by stopping the vibration manually and setting the cylinder at its zero position; the experimental points on the $\bar{U}$ axes denote these checks. It was found that the zero position was stable in turbulence type 1 while unstable in turbulence type 2. Nevertheless, in the former case, for a given flow velocity, an initial displacement greater than the corresponding unstable amplitude was found to set the cylinder into galloping vibration again. This could be the reason that in turbulence type 1, since the amplitude associated with the peak at vortex resonance was greater than the corresponding unstable amplitude, the galloping vibration follwed the vortex resonance almost immediately. However, in turbulence type 2, although the minimum reduced velocity for galloping can be seen to be less than the observed velocity at vortex resonance peak, yet, the galloping mechanism was completely suppressed by the vortex resonance mechanism. In this case, the galloping oscillation occurred at higher flow velocity.

The second pattern is associated with Runs A3, A4, A6 and A7 of group A and B2 and B3 of group B, having higher values of K_s. In these runs, after the peak vortex resonance was reached, an increase in the flow velocity resulted in a continual decrease in amplitude from peak to zero with the response frequency exceeding somewhat beyond 1. With still further increase in flow velocity, two different behaviours were observed, depending on the type of turbulence. In turbulence type 2, the galloping oscillation for both the sharp and round-edged cylinders started at a reduced velocity very close to U_0 predicted by galloping theory (Column 15 of Table 1). The results of Novak (Ref. 22) for 11% turbulence intensity and a turbulence length scale ratio of 6.33 with minimum reduced velocity for galloping $\bar{U}=1/A_1=2.34$ at two levels of response parameter 0.79 and 1.59 fall into this category of behaviour (Fig. 8). Novak's results show a reasonable agreement with the corresponding results of the present investigation beyond $\bar{U}>3.4$. The second behaviour was observed in turbulence type 1, where the zero position remained stable. Moreover, with the sharp-edged cylinders in this type of turbulence and at a given flow velocity, an initial displacement greater than the corresponding unstable amplitude was necessary to trigger the galloping vibration. It should be noted that in spite of artificial disturbances, the round-edged cylinder with r/h=0.141 in turbulence type 1 (Runs B1, B2 and B3), did not show any galloping instability and the zero position remained always stable; the experimental points on $\bar{U}$ axis of Fig. 10 indicate this. This point can also be inferred from its corresponding lateral force measurements (Fig. 6).

The above aspects clearly indicate that turbulence has a strong effect on galloping response. A cylinder typically stable in turbulence type 1 becomes unstable in turbulence type 2. The dynamic response figures show that for a given turbulence type, the experimental points of $\bar{a}$ versus $\bar{U}$ associated with the galloping vibrations fall into a band with experimental points of turbulence type 2 lying generally below those of turbulence type 1. However, although the bands associated with Runs A1,A2,A5 and A6 having low values of response parameter K_s are rather thick, they become narrower for Runs A3 and A7 of group A having higher values of K_s (note that the scale of Figs. 7, 8 and 9 are different) and also for group B. From these results, it appears that the turbulence scale has very little effect on cylinder amplitude response. Moreover, although the agreement between the experimental points and the corresponding theoretical curves for Runs A3 and A7 of group A and for group B are reasonably good, for other runs of group A having lower values of response parameter, the theoretical galloping curves give conservative estimation of galloping response. The response frequency ratio Ω associated with galloping vibration for both sharp and round-edged cylinders occurred in the close vicinity of 1, irrespective of flow velocity and vibration amplitude. This very probably means that the still water added mass was unaffected by streaming flow.

6. CONCLUDING REMARKS

The vortex-induced oscillation and hydrodynamic galloping response of rectangular rigid cylinders, mounted elastically and restricting oscillations only to a plane normal to the incident water flow was studied both experimentally and analytically. Wake observations of stationary cylinders indicated that in the range of the Reynolds numbers tested, the Strouhal number was roughly constant and insensitive to turbulence characteristics. This number was estimated to be 0.137, 0.169 and 0.183 for a sharp-edged cylinder and round-edged cylinders with r/h=0.141 and 0.187 respectively. The still water added mass of both sharp and round-edged cylinders was found to be independent of the dimensionless vibrational amplitude. The still water added mass of sharp-edged cylinders was found to be considerably higher than the value predicted through the application of the potential theory.

As opposed to vortex resonance, turbulence was found to have profound effects on galloping vibrations. The variation of the lateral force coefficients with the angle of flow attack as well as the dynamic tests revealed that tendency to galloping instability increased with an increase in turbulence intensity. The response frequency at peak vortex resonance and during galloping oscillations was found to be equal to the natural frequency of the system in still water.

The response parameter was found to have a marked effect on the cylinder response. For values of the response parameter K_s less than roughly 0.3 and 0.2, for a sharp-edged cylinder and a round-edged cylinder with r/h=0.141 respectively, it was observed that the galloping was suppressed by the vortex resonance in the vicinity of U_r. In this case, a quasi-steady approximation of the hydrodynamic force derived from the transverse force characteristics was found to lead to a conservative estimation of the galloping amplitudes. For high values of the response parameter, the galloping was separated from the vortex resonance and the quasi-steady approach was found to predict the galloping response adequately. With rounding the edges of a sharp-edged cylinder to r=0.187h, while the vortex resonance peak amplitude increased, the galloping instability disappeared in both the low and high free stream turbulent flows.

7. ACKNOWLEDGEMENTS

The authors wish to express their sincere gratitude to Professor J.M.T. Thompson of the Department of Civil Engineering, University College London for his continuous encouragement and unfailing support throughout the course of this project. Thanks are extended to Mr. D. Vale , Mr. L. Morris and Mr. M. Gregory for their technical assistance. The authors are indebted to Dr. J. D. Hardwick and especially to Mr. M.J. Kenn of Imperial College London for their kind assistance.

8. REFERENCES

1. Den Hartog, J.P.: "Mechanical vibrations". New York, McGraw-Hill, New York, 1956.

2. Parkinson, G.V. and Brooks, N.P.H.: "On the aeroelastic instability of bluff cylinders". Transactions of the ASME, Journal of Applied Mechanics, 28, June 1961, pp. 252-258.

3. Parkinson, G.V.: "Aeroelastic galloping in one degree of freedom". Symposium on Wind Effects on Buildings and Structures, National Physical Laboratory, Teddington, U.K., 2, June 1963, pp. 581-609.

4. Parkinson, G.V. and Smith, J.D.: "The square prism as an aeroelastic non-linear oscillator". Quarterly Journal of Mechanics and Applied Mathematics, 17, Part 2, 1964, pp. 225-239.

5. Novak, M. and Davenport, A.G.: "Aeroelastic instability of prisms in turbulent flow". Journal of the Engineering Mechanics Division, ASCE, 96, No. EM1, Proc. Paper 7076, Feb., 1970, pp. 17-39.

6. Novak, M.: "Galloping oscillation of prismatic structures". Journal of the Engineering Mechanics Division, ASCE, 98, No. EM1, Proc. Paper 8692, Feb. 1972, pp. 27-46.

7. Novak, M. and Tanaka, H.: "Effects of turbulence on galloping instability". Journal of the Engineering Mechanics Division, ASCE, 100, No. EM1, Proc. Paper 10338, Feb. 1974, pp. 26-46.

8. Kwok, K.C.S. and Melbourne, W.H.:"Free stream turbulence effects on galloping". Journal of the Engineering Mechanics Division, ASCE, 106, No. EM2, Proc. Paper 15356, Apr. 1980, pp. 273-288.

9. Parkinson, G.V. and Wawzonek, M.A.: "Combined effects of galloping instability and vortex resonance". Wind Engineering, Proceedings of the Fifth International Conference, Fort Collins, Colorado, U.S.A., 2, July 1979, pp. 673-684.

10. Hallam, M.G., Heaf, N.J. and Wootton, L.R.: "Dynamics of marine structures". London, Ciria Underwater Engineering Group, Report UR8, 1977, 198-200.

11. King, R. and Every, M.: "Design for underwater service". Engineering Materials and Design, Mar. 1980, pp. 31-34.

12. Krylov, N. and Bogoliulov, N.: "Introduction to non-linear mechanics". Princeton, N.J., Princeton University Press, 1949.

13. Resch, F.J. and Abel, R.: "Spectral analysis using Fourier Transform Techniques". International Journal of Numerical Methods in Engineering, 1975, 9, 869-902.

14. Bearman, P.W. and Trueman, D.M.: "An investigation of the flow around rectangular cylinders". Aeronautical Quarterly, 23, Aug. 1972, pp. 229-237.

15. Laneville, A. and Parkinson, G.V.: "Effects of turbulence on galloping of bluff cylinders". Proceedings of the Third International Conference on Wind Effects on Buildings and Structures, Tokyo, Japan, Sept. 1971, pp. 787-797.

16. Gartshore, I.S.: "The effects of free stream turbulence on the drag of rectangular two-dimensional prisms". Research Report BLWT-4-73, Boundary Layer Wind Tunnel Laboratory, Faculty of Engineering Science, University of Western Ontario, London, Ontario, Canada, 1973.

17. Skop, R.A., Ramberg, S.E. and Ferer, K.M.: "Added mass and damping forces on circular cylinders". Paper 76-Pet-3, American Society of Mechanical Engineers, 1976.

18. Blevins, R.D.: "Formulas for natural frequency and mode shape". New York, Van Nostrand Reinhold Co., 1979, pp. 391-392.

19. Chen, S.S., Wambsganss, M.W., and Jendrzejczyk, J.A.: "Added mass and damping of a vibrating rod in confined viscous fluids". Transactions of the ASME, Journal of Applied Mechanics, 98, No. 2, June 1976, pp. 325-329.

20. Yang, C.I. and Moran, T.J.: "Calculations of added mass and damping coefficients for hexagonal cylinders in a confined viscous fluid". Flow-Induced Vibrations, Chen, S.S. and Bernstein, M.D. (eds.), 1979, ASME, New York, pp. 97-103.

21. Miller, R.R.: "The effects of frequency and amplitude of oscillation on the hydrodynamic masses of irregular shaped bodies". MSc Thesis, University of Rhode Island, Kingston, R.I., 1965.

22. Novak, M.: "Galloping and vortex-induced oscillations of structures". Proceedings of the Third International Conference on Wind Effects on Buildings and Structures, Tokyo, Japan, Sept. 1971, pp. 799-809.

23. King, R.: "The added mass of cylinders". The British Hydromechanics Research Association, BHRA, Report TN-1100, 1971.

Table 1 Experimental programme

(1) Run number	(2) Cylinder dimensions (b×h×L) in (mm×mm×mm)	(3) Corner radius ratio $p=r/b$	(4) Aspect ratio L/h	(5) Initial displacement ratio $\eta=Y_0/h$	(6) Natural circular frequency ω_n in (Hz)	(7)[*] Vibratory Reynolds number $\gamma=\dfrac{\omega_n h^2}{\nu}$	(8) Coefficient of added mass C_M	(9) Reduced damping β	(10) Mass parameter n	(11) Response parameter $K_S=\dfrac{\beta}{nU_r^2}$	(12)[**] Turbulence length scale ratio L_x/h — Turbulence type 1	(13)[***] Turbulence length scale ratio L_x/h — Turbulence type 2	(14)[**] Minimum reduced velocity for galloping $U_0=\dfrac{2\beta}{nA_1}$ — Turbulence type 1	(15)[***] Minimum reduced velocity for galloping — Turbulence type 2	(16) Vortex resonance speed $U_r=1/(2\pi S)$	(17)[*] Reynolds number $Re=Vh/\nu$
A1[+]	9.4×18.8×298	–	15.85	2.66	4.23	1495	3.0	0.0204	0.1104	0.137	0.85	0.42	–	0.54	1.16	800–6000
A2[+]	9.4×18.8×298	–	15.85	2.66	4.14	1463	3.0	0.0432	0.1057	0.304	0.85	0.42	–	1.20	1.16	700–6000
A3[+]	9.4×18.8×298	–	15.85	2.66	4.12	1456	2.9	0.1347	0.1047	0.956	0.85	0.42	–	3.79	1.16	1000–8000
A4[+]	9.4×18.8×298	–	15.85	2.66	4.17	1474	3.0	0.1770	0.1070	1.228	0.85	0.42	–	4.87	1.16	1500–2500
A5[+]	6.65×13.3×298	–	22.41	3.76	4.38	775	3.3	0.0190	0.0592	0.239	1.20	0.60	–	0.95	1.16	700–5500
A6[+]	6.65×13.3×298	–	22.41	3.76	4.39	777	3.4	0.0444	0.0595	0.555	1.20	0.60	–	2.20	1.16	500–5000
A7[+]	6.65×13.3×298	–	22.41	3.76	4.44	785	3.3	0.0732	0.061	0.892	1.20	0.60	–	3.53	1.16	700–5500
B1[×]	8.5×17×297	0.141	17.47	2.94	4.34	1254	3.4	0.0183	0.0947	0.218	0.94	0.47	–	0.99	0.94	900–7000
B2[×]	8.5×17×297	0.141	17.47	2.94	4.30	1243	3.4	0.0432	0.0929	0.526	0.94	0.47	–	2.39	0.94	700–8000
B3[×]	8.5×17×297	0.141	17.47	2.94	4.35	1257	3.4	0.0572	0.0952	0.680	0.94	0.47	–	3.09	0.94	900–8000
C1[×]	8.5×17×297	0.187	17.47	2.94	4.38	1266	3.7	0.0196	0.0964	0.269	0.94	0.47	–	–	0.87	700–7000
C2[×]	8.5×17×297	0.187	17.47	2.94	4.35	1257	3.6	0.0459	0.0954	0.635	0.94	0.47	–	–	0.87	1000–6500
C3[×]	8.5×17×297	0.187	17.47	2.94	4.38	1266	3.6	0.0695	0.0964	0.953	0.94	0.47	–	–	0.87	900–6500

[+] Cylinders made of P.V.C. and filled with Araldite.
[×] Cylinder made of hard wood.
[*] $\nu=1\times10^6\ \mathrm{m^2/s}$.
[**] $(v^2)^{\frac{1}{2}}/V\approx6.5\%$, $L_x\approx16\mathrm{mm}$.
[***] $(v^2)^{\frac{1}{2}}/V\approx11.9\%$, $L_x\approx8\mathrm{mm}$.

Table 2   Polynomial coefficients

| Polynomial coefficients | Sharp-edged cylinder | | Round-edged cylinder,r/h=0.141 |
| | Turbulence type 1 | Turbulence type 2 | Turbulence type 2 |
(1)	(2)	(3)	(4)
A_1	-0.001	0.679	0.389
A_3	-0.003	5.189	0.271
A_5	77.910	-54.793	-46.938
A_7	-382.269	87.347	137.435
A_9	169.370	-	-
A_{11}	734.037	-	-

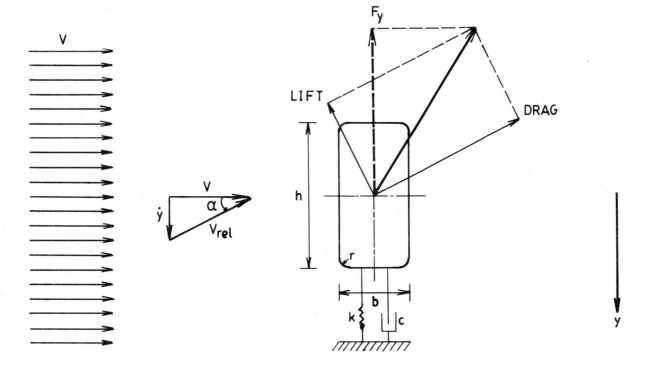

Fig. 1   Vibrating cross section in a steady flow.

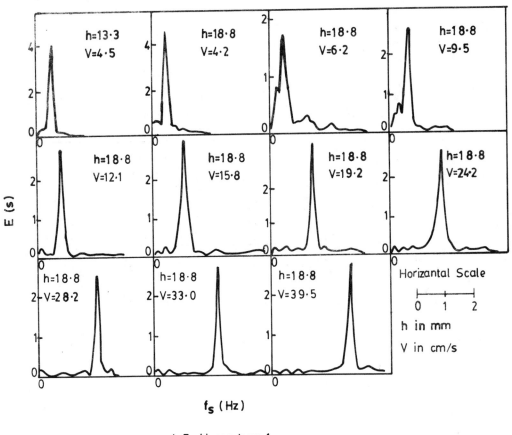

a) Turblence type 1

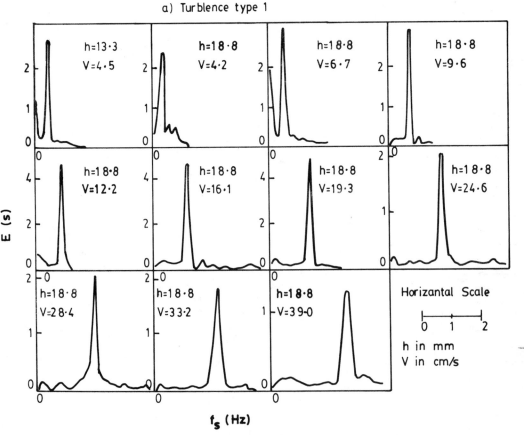

b) Turblence type 2

Fig. 2 Normalized spectra of wake fluctuations past stationary sharp-edged
rectangular cylinders.

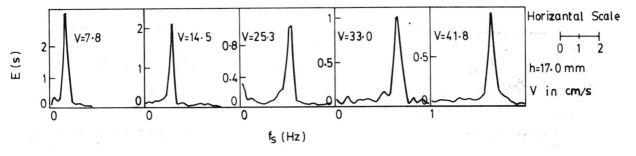

a) Turblence type 1

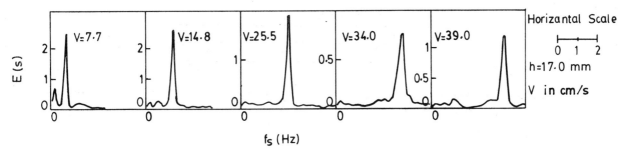

b) Turblence type 2

Fig. 3   Normalized spectra of wake fluctuations past stationary round-edged
rectangular cylinder with r/h=0.141.

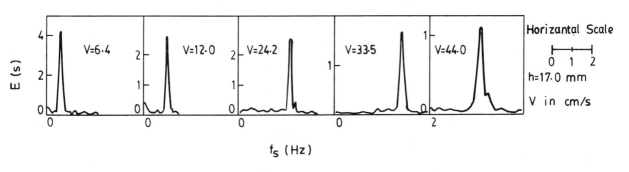

a) Turblence type 1

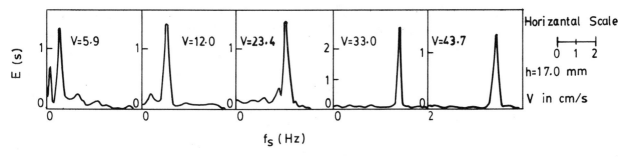

b) Turblence type 2

Fig. 4   Normalized spectra of wake fluctuations past stationary round-edged
rectangular cylinders with r/h=0.187.

123

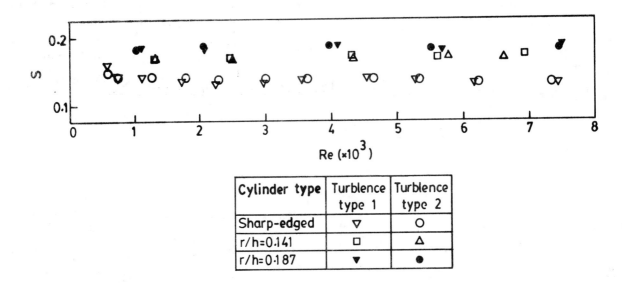

Cylinder type	Turblence type 1	Turblence type 2
Sharp-edged	▽	○
r/h=0.141	□	△
r/h=0.187	▼	●

Fig. 5   The Strouhal-Reynolds number relationship for sharp and round-edged rectangular cylinders.

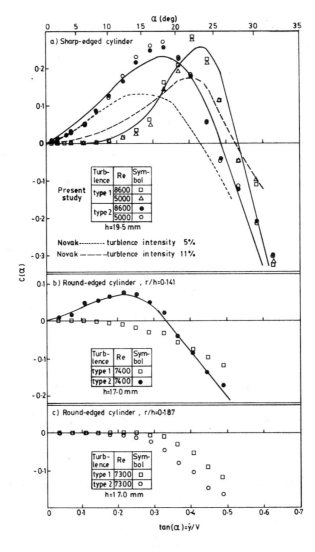

Fig. 6   Lateral force coefficients of sharp and round-edged rectangular cylinders in both the low and high free stream turbulent flows.

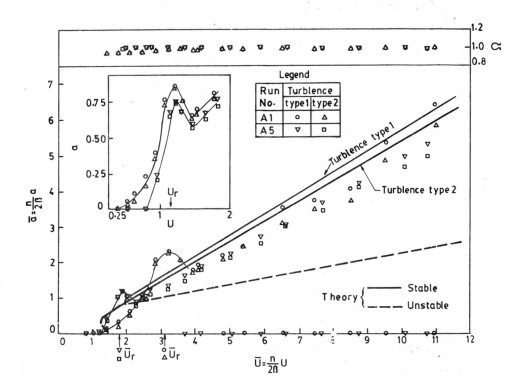

Fig. 7 Vortex-excited oscillation and hydrodynamic galloping response of sharp-edged rectangular cylinders in both the low and high free stream turbulent flows.

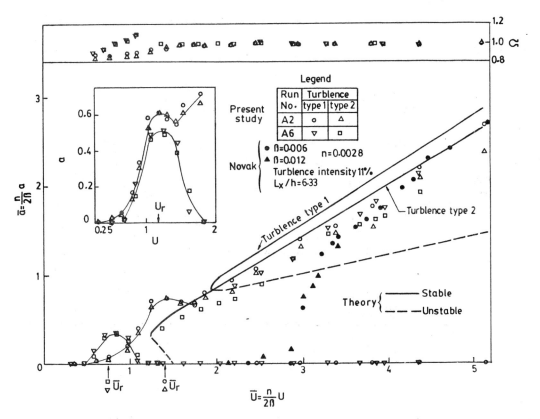

Fig. 8 Vortex-excited oscillation and hydrodynamic galloping response of sharp-edged rectangular cylinders in both the low and high free stream turbulent flows.

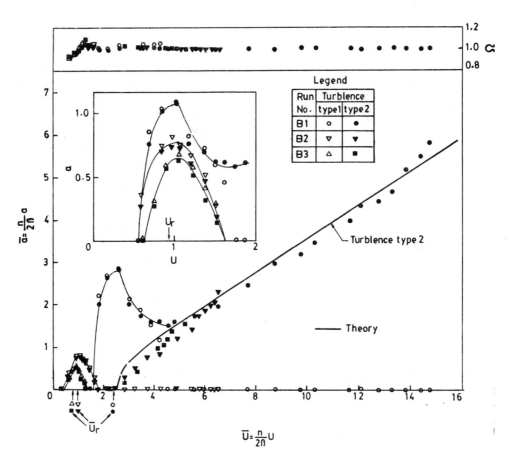

Fig. 9   Vortex-excited oscillation and hydrodynamic galloping response of sharp-edged rectangular cylinders in both the low and high free stream turbulent flows.

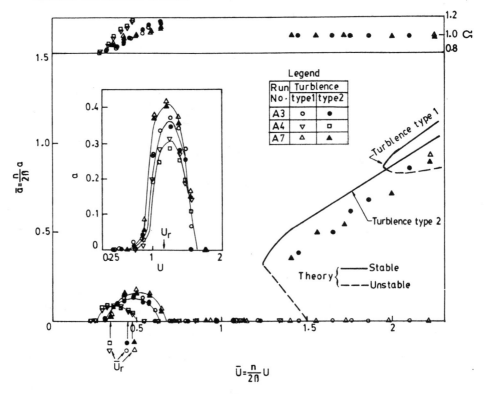

Fig. 10   Vortex-excited oscillation and hydrodynamic galloping response of the round-edged rectangular cylinder with r/h=0.141 in both the low and high free stream turbulent flows.

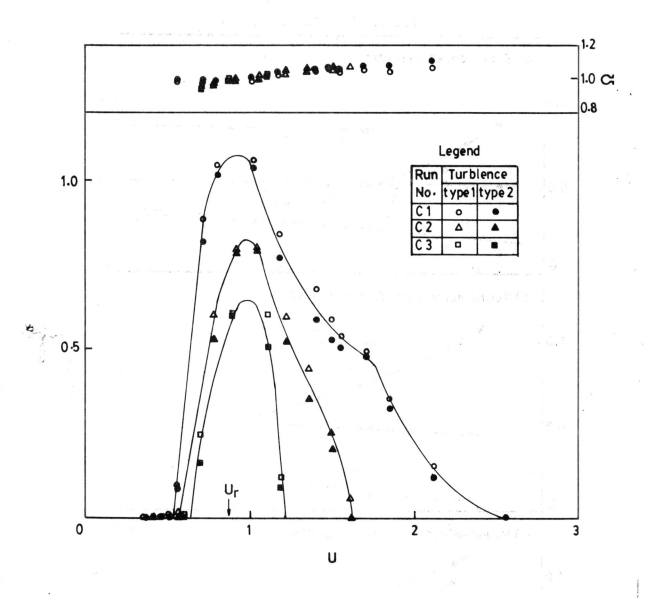

Fig. 11 Vortex-excited oscillation of the round-edged rectangular cylinder with r/h=0.187 in both the low and high free stream turbulent flows.

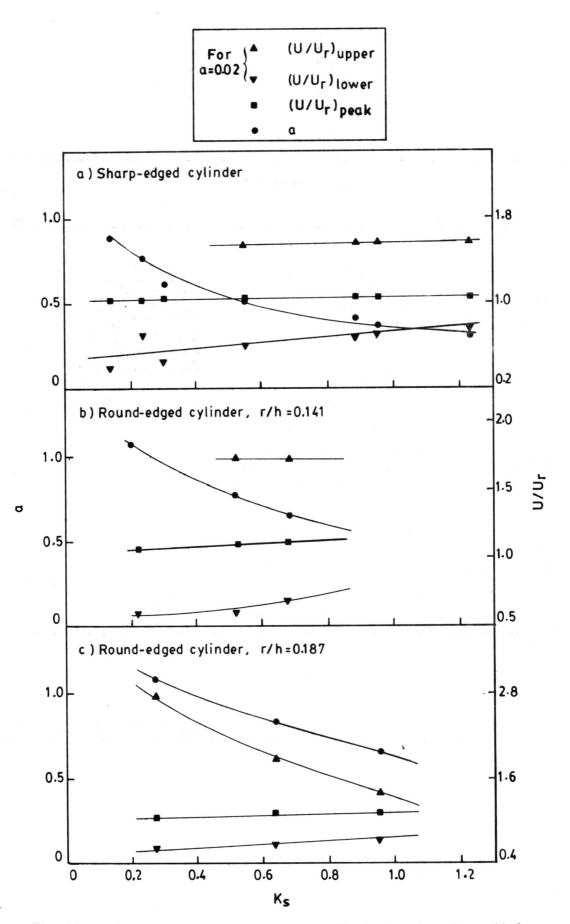

Fig. 12 Stability diagrams of sharp and round-edged rectangular cylinders experiencing vortex-excited oscillation.

128

A: PLATE
B: BASEPLATE ADJUSTABLE SCREWS
C: AIRBEARING HOUSING
D: AIRBEARING
E: MANIFOLDS
F: AIRBEARING ALIGNMENT SCREW
G: AIRFILTER
H: AIRBEARING SHAFT
I: ARM
J: CYLINDER
K: CROSS BAR
L: SPRING
M: SHIELDED RECEIVING TRANSDUCER
N: SHIELDED TRANSMITTING TRANSDUCER
O: RIGID FRAME
P: DAMPER CONNECTING SHAFT
Q: OIL POT
R: ADJUSTABLE HOT FILM HOLDER

Plate 1   Dynamic test apparatus.

**FLOW INDUCED VIBRATIONS
IN FLUID ENGINEERING**

Reading, England: September 14-16, 1982

FLOW INDUCED VIBRATION OF ROUGHENED CYLINDERS

T. Sarpkaya

Naval Postgraduate School, U.S.A.

Summary

Measurements are presented of the forces acting on sand-roughened circular cylinders forced to oscillate in the transverse direction to a uniform flow with amplitude ratios from 0.25 to 1.00 and reduced velocities in the range 3 to 8, which includes the vortex lock-in regime. The in-phase component of the force for a rough cylinder was found to be nearly identical with that for a smooth cylinder. However, substantial differences were noted between the out-of-phase components. The difference was largest at perfect synchronization and increased with increasing roughness Reynolds number. For the largest amplitude ratio, both the in-phase and the out-of-phase component of the force exhibited hysteresis. The phenomenon was attributed to two modes of vortex shedding.

Organised and sponsored by
BHRA Fluid Engineering, Cranfield, Bedford MK43 0AJ, England

NOMENCLATURE

A = amplitude of cylinder oscillation

$C_{d\ell}$ = out-of-phase force coefficient, [see Eq. (3)]

C_L = total lift coefficient, [see Eq. (1)]

$C_{m\ell}$ = in-phase force coefficient, [see Eq. (4)]

D = cylinder diameter

F(t) = transverse fluid force

f_c = frequency of cylinder oscillation, $f_c = 1/T$

K = Keulegan-Carpenter number, $K = 2\pi A/D = U_m T/D$

k_s = mean roughness height

L = cylinder length

Re = Reynolds number, $Re = VD/\nu$

Re_k = roughness Reynolds number, $Re_k = Vk_s/\nu$

T = period of cylinder oscillation

t = time

U_m = amplitude of cylinder velocity

V = velocity of the uniform flow

V_r = reduced velocity, $V_r = VT/D$

ζ = logarithmic damping ratio

θ = ωt

ν = kinematic viscosity of fluid

ρ = density of fluid

ω = $2\pi f_c$

1. INTRODUCTION

Numerous experiments have shown that when the vortex shedding frequency brackets the natural frequency of an elastic or elastically-mounted rigid bluff body with a suitable afterbody, the body takes control of the shedding frequency in apparent violation of the Strouhal relationship. This phenomenon, called lock-in or synchronization, increases spanwise correlation, vortex strength, and the in-line drag force. The subject has been reviewed most recently by Sarpkaya (Ref. 1) where reference to numerous previous investigations can be found. This and other reviews have shown that many fundamental questions remain unresolved regarding the effect of body motion on the formation, growth and motion of vortices. The previous investigations dealt mostly with the response of smooth circular cylinders and cables, primarily because of their practical significance. A few investigations were devoted to the flow-structure interaction of rectangular bodies partly for the purpose of research into the phenomena involved and partly for the practical purpose of understanding the dynamic response of bridge-like structures. Clearly, there is much to be done aside from reducing the insufficiently-substantiated experimental data into a set of design criteria.

In this paper we examine the flow around a sand-roughened circular cylinder forced to oscillate at frequencies at and away from its natural vortex shedding frequency. It has been known for quite sometime that roughness plays a very important role in the determination of the hydrodynamic forces acting on a body [see e.g., Sarpkaya and Isaacson (Ref. 2)]. A close examination of the flow about a rough body shows that roughness induces earlier transition in the boundary layer, increases the thickness of the boundary layer, improves the spanwise correlation, and increases the strength of the vortices shed from the body.

In steady flow, the effect of roughness is expressed in terms of a relative roughness parameter k_s/D, where k_s is the mean roughness height and D, the characteristic dimension of the body. The one-parameter characterization of the effect of roughness needs some justification. It is of course recognized that not only the relative size of the roughness elements but also their shape and distribution may be quite important. It is in fact partly for the difficulty of uniquely specifying the 'roughness' and partly for the differences in other test conditions that there are considerable differences between the steady-flow data reported by various workers, particularly in the drag crisis region. It is because of this reason that the experiments reported herein were carried out with cylinders roughened with sand of uniform size and packing rather than with various types of roughness elements.

The effect of roughness on hydroelastic oscillations has not been studied in detail. In general, cables and structures subjected to the ocean environment are seldom smooth. Thus, it was deemed necessary to examine the problem in some detail. Nakamura (Ref. 3) has measured the steady drag forces and Strouhal frequencies on rough circular cylinders at supercritical Reynolds numbers, and has observed strong regular vortex shedding. In this Reynolds number range the vortex-excited cross-flow displacement of a rough cylinder increased substantially from the corresponding smooth cylinder experiment. Szechenyi (Ref. 4) carried out a series of tests with rough cylinders in steady flow and determined the drag and lift coefficients as a function of the Reynolds number and roughness Reynolds number, defined as $Re_k = Vk_s/\nu$. The results have shown that the lift coefficient increases rapidly as Re_k increases from about 300 to 700 and then remains nearly constant at a value about three times larger than that for the smooth cylinder. Similar conclusions have been reached by Sarpkaya (Ref. 5) in connection with harmonically oscillating flow about rough cylinders undergoing transverse oscillations. The significance of these conclusions is that the lift coefficient which is directly related to the hydroelastic oscillations increases dramatically in a range of roughness Reynolds numbers where the steady drag coefficient shows no abnormalities. In other words, there is a roughness Reynolds number in the range of subcritical Reynolds numbers above which the transverse oscillations of a rough cylinder can be significantly affected by the surface roughness.

It is on the basis of the foregoing that a series of experiments were carried out with sand-roughened cylinders forced to oscillate in the transverse direction in a uniform flow at frequencies at and away from the natural vortex shedding frequency.

2. TEST EQUIPMENT AND PROCEDURES

The experiments were performed in an open-return water tunnel with a working section 0.51 m high and 0.36 m wide. It is an improved version of that described in (Refs. 6 and 7). The end effects caused by tunnel-wall boundary layers are thought to be eliminated by using two end plates. The plates were attached to the vertical walls and not to the cylinder. The circular cylinder models, placed centrally and horizontally in the test section, spanned the 0.33 m distance between the two plates. A gap of 0.08 cm was provided between the plate and the end of the cylinder.

Oscillations were obtained by means of a small, variable speed-drive motor, a flywheel, and a yoke assembly. The motor, mounted so as to isolate its vibration from the remainder of the test apparatus, transmitted the rotary motion through a flexible belt to a flywheel. A long rigid bar transmitted the vertical component of motion of a bearing, attached to the flywheel, to a shaft connected to the yoke assembly and the cylinder. The resulting motion of the cylinder was not distinguishable from a purely sinusoidal motion. This was verified by comparing the cylinder acceleration signal with a sine wave on a dual-beam oscilloscope.

Two streamlined leaf arms connected the ends of the cylinder to the yoke assembly outside the test section. Small rectangular holes at the ceiling of the test section accommodated the connecting arms and their vertical oscillation. Water level was allowed to rise about 5 cm above the ceiling of the test section into two tubes for that purpose. Thus, the connecting arms did not contact the tunnel.

The strain gages, from which the unsteady forces were sensed, were cemented to the flattened portions of the surfaces on the yoke assembly. Also mounted on the yoke was a beam-type accelerometer whose signal was electronically subtracted from the total force signal by means of a summing-differencing unit. The mass of the accelerometer beam and the sensitivity of the signal amplifier were properly adjusted so as to obtain a zero net signal when the cylinder assembly was oscillated in air at desired amplitudes and frequencies. Since the experiments were conducted in water, the contribution of air to the added mass and force signal was ignored.

The natural frequency of the accelerometer was about 80 times the largest vortex-shedding frequency. The accelerometer had a logarithmic damping ratio of $\zeta = 0.006$. Thus, it has correctly detected the inertial force acting on the system without any amplification.

Dynamic calibration tests were made by attaching known masses to the cylinder and oscillating them at known fixed amplitudes and frequencies. It was found that the magnification factor of the force-transducer-amplifier-recorder system was nearly unity. The natural frequency of the entire system was 45 Hz in air and 40 Hz in water.

The velocity of the fluid in the test section was regulated by adjusting the variable speed pump. Extensive velocity surveys with a hot film anemometer were made to determine the velocity and turbulence distribution. The velocity was found to be uniform in both the vertical and horizontal directions within 2 percent of the mean velocity. The turbulence level was about 0.3 percent for a mean velocity of 1.20 m/s.

Circular cylinders of 19 mm, 25.4 mm, 38 mm, and 51 mm diameter with a relative roughness of $k_s/D = 1/100$ were used. Cylinders were oscillated at amplitude ratios of $A/D = 0.25$, 0.50, 0.75 and 1.00. The reduced velocity, VT/D, was varied between 3 and 8 (up and down), usually changing the cylinder oscillation frequency f_c and holding V constant. The cylinder Reynolds number varied over the range 20,000 to 50,000; and the roughness Reynolds number, $Re_k = Vk_s/\nu$, over the range 200 to 500.

Fifty cycles of fluid force, $F(t)$, in the direction of cylinder oscillation, together with the corresponding cylinder acceleration, were recorded in digital form with a sampling rate of $6.94 \times T$ ms. The averaged-force cycle and the corresponding averaged-acceleration cycle were used to determine the in-phase and the out-of-phase components of the force. The results have not been corrected for blockage effects. The simple geometric blockage for the largest cylinder was 10 percent. No information is available regarding the correction of forces, pressures, etc., acting on bodies undergoing in-line and/or transverse oscillations in steady or harmonic flows. Thus, no attempt was made to apply the stationary cylinder correction methods to the results presented herein.

3. DATA ANALYSIS

The force acting on the cylinder in the direction of cylinder oscillation is expressed in terms of Morison's equation as (Ref. 6)

$$C_L = \frac{2F(t)}{\rho LDV^2} = C_{m\ell}\pi^2 (U_m T/D)(D/VT)^2 \sin \omega t - C_{d\ell}(U_m T/D)^2(D/VT)^2 |\cos \omega t|\cos \omega t \qquad (1)$$

in which $C_{m\ell}$ and $C_{d\ell}$ are the inertia and drag coefficients, respectively; and $U_m = 2\pi A/T$. Evidently, the normalized force as well as the drag and inertia coefficients depend on the parameters $K = U_m T/D = 2\pi A/D$, $V_r = VT/D$, $Re = VD/\nu$, and $Re_k = Vk_s/\nu$. The establishment of the required relationships between the said parameters is an extremely difficult task. Here attention will be devoted to the range of synchronization and to an assessment of the effect of roughness on the force-transfer coefficients.

For the purpose under consideration, it is more advantageous to expand the $|\cos \omega t|\cos \omega t$ term in series and to retain only the first term. This procedure yields

$$C_L = C_{m\ell}\pi^2 (K/V_r^2) \sin \omega t - \frac{8}{3\pi} C_{d\ell} (K/V_r)^2 \cos \omega t \qquad (2)$$

in which $C_{d\ell}$ and $C_{m\ell}$ are given by their Fourier averages as (Ref. 6)

$$C_{d\ell} = -\frac{3}{4} \int_0^{2\pi} \frac{F \cos\theta \, d\theta}{\rho U_m^2 DL} \qquad (3)$$

and

$$C_{m\ell} = \frac{2K}{3} \int_0^{2\pi} \frac{F \sin\theta \, d\theta}{\rho U_m^2 DL} \qquad (4)$$

where $\theta = \omega t = 2\pi t/T = 2\pi f_c t$.

The drag and inertia coefficients obtained as described above are obviously a consequence of the use of Eq. (2) and the method of Fourier averaging. It was, therefore, necessary to compare the predictions of Eq. (2) through the use of the calculated coefficients with those measured directly. Only such a comparison could show whether the decomposition of the instantaneous force into an in-phase and an out-of-phase component is justifiable. The calculations performed along these lines have shown that the measured force is fairly well represented by Eq. (2) for V_r values in the neighborhood of perfect synchronization where $-C_{d\ell}$ reaches its maximum value. The representation of the measured force by Eq. (2) becomes progressively poor as V_r acquires values smaller than about 4 and larger than about 6. Then it becomes necessary to consider the higher harmonics of the force for a correct representation of the measured force. These results will not be presented here for sake of brevity. Suffice it to note that it is not a meaningful exercise to discuss the variation of the two force coefficients ($C_{d\ell}$ and $C_{m\ell}$) in terms of the governing parameters for $6 < V_r < 4$.

4. DISCUSSION OF RESULTS

Figures 1 and 2 show $C_{m\ell}$ and $C_{d\ell}$ as a function of V_r for $A/D = 0.25$, 0.50, 0.75, and 1.00 for $Re = 51,200$ and $Re_k = 512$. Additional data are given in (Ref. 8). It is clear from these figures that important variations occur in both $C_{m\ell}$ and $C_{d\ell}$, particularly in the vicinity of the Strouhal frequency where the natural eddy shedding is both enhanced and correlated by the oscillations. The inertia coefficient or the normalized in-phase component of the transverse force undergoes a rapid drop as the frequency of oscillations approaches the Strouhal frequency. In other words, synchronization is manifested by a rapid decrease in inertial force and a rapid increase in the absolute value of the drag force.

$C_{m\ell}$ reaches a value of about 2 at a V_r value slightly under that corresponding to perfect synchronization. $C_{m\ell}$ becomes nearly equal to unity when $C_{d\ell}$ reaches its maximum negative value at perfect synchronization. Evidently, the use of $C_{m\ell} = 1$ (for a circular cylinder) in off-resonance conditions is not justified. King (Ref. 9) conducted a series of experiments in still water (by plucking a cantilevered beam) and also in flowing water at the fundamental mode of the flow-excited vibrations. From a comparison of the measured frequencies he calculated that the still-water and flow-excited frequencies are virtually equal, i.e., the added mass is unaffected by streaming flow. Apparently, this conclusion is valid only at the resonance conditions.

For larger values of V_r, $C_{m\ell}$ becomes negative. Sarpkaya (Ref. 10) has shown that the averaged negative added mass is a consequence of the averaging process and does not actually contradict reality. It simply shows that when the variable mass is growing, the Fourier-averaged $C_{m\ell}$ is much larger than that required to define the quantity of fluid being accelerated. The shedding of a vortex brings about an abrupt reduction in the fluid mass which can and does reduce the average added mass coefficient even to negative values. One must, however, bear in mind the fact that the use of Morison' equation in regions away from perfect synchronization is not justified and requires the inclusion of additional harmonics.

The two most interesting features of the results shown in Figs. 1 and 2 are that (i) roughness does not change $C_{m\ell}$ relative to that for a smooth cylinder (see Refs. 6 and 7), and (ii) roughness does significantly increase $C_{d\ell}$. In other words, it is not the in-phase component but rather the out-of-phase component of the force that is affected by roughness.

The reasons for the insensitivity of $C_{m\ell}$ to roughness, within the range of Re, Re_k, and V_r values encountered, are not entirely clear. The Keulegan-Carpenter number for the transverse motion varied from 1.57 to 6.28 (corresponding to A/D = 0.25 and A/D = 1). In the absence of uniform flow, the transverse force is said to be inertia dominated for K smaller than about 8 (Ref. 2). In this regime, the inertia coefficient is not affected by roughness (see e.g., Ref. 2). It does not follow directly that $C_{m\ell}$ should not vary with k/D or Re_k either, at least within the range of the variables cited. Further evidence is needed to ascertain the role of roughness on $C_{m\ell}$ as well as on $C_{d\ell}$.

Additional comparisons of the smooth and rough cylinder results have shown that the effect of roughness is extremely complex, particularly for $C_{d\ell}$. For $Re_k \simeq$ 200, the smooth and rough cylinder $C_{m\ell}$ and $C_{d\ell}$ values are nearly identical. The deviation of $C_{d\ell}$ for rough cylinder from that for the smooth cylinder increases with increasing Re_k, at least for Re_k values up to about 500. Furthermore, for a given Re and Re_k, the effect of roughness on $C_{d\ell}$ depends on V_r. The results have shown that, at perfect synchronization, $C_{d\ell}$ is about 20 percent larger than that for the smooth cylinder for $Re_k = 300$; about 35 percent larger for $Re_k = 400$; and about 50 percent larger for $Re_k = 512$. On either side of the perfect synchronization, however, the effect of roughness is somewhat smaller.

Figure 2 also shows that $C_{d\ell}$ becomes negative for V_r values in the neighborhood of 5. Outside this range the drag is mostly positive, thus in the opposite direction to the motion of the cylinder. Within the range of $V_r = 5$, the drag force is in phase with the direction of motion of the cylinder and helps to magnify the oscillations rather than damp them out. For this reason, the range in which $C_{d\ell}$ is negative is sometimes referred to as the negative damping region. The fact of the matter is that this is not damping in the proper use of the word but rather an energy transfer from the fluid to the cylinder via the mechanism of synchronization. The maximum absolute value of $C_{d\ell}$ in the synchronization range decreases rapidly as A/D increases. For A/D = 1 both $C_{d\ell}$ and $C_{m\ell}$ exhibited hysteresis as shown in Figs. 1 and 2. For increasing V_r, $C_{d\ell}$ is negative near $V_r = 5$, indicating that the limiting amplitude has not yet been reached. For decreasing V_r, $C_{d\ell}$ remained positive, indicating that no more energy was being transferred from the fluid to the cylinder. This behavior has not been observed in experiments with smooth cylinders (Refs. 6 and 7). The reasons for the hysteresis are not yet clear. Zdravkovich (Ref. 11) attributed the hysteresis effect to the two modes of timing in synchronized vortex shedding. According to him, the first mode (in the lower region of the synchronization range) corresponds to the shedding of the vortex formed on one side of the cylinder when the cylinder is near to the maximum amplitude on the opposite side. Likewise, the second mode (in the upper region of synchronization) corresponds to the shedding of the vortex formed on one side of the cylinder when the cylinder is near its maximum displacement on the same side. These observations appear to be plausible. Detailed measurements of the spanwise pressure distribution and flow visualization are currently underway for a much closer examination of the consequences of roughness in general and of the hysteresis in particular. It is already clear that the experimental facts regarding the effect of roughness will need to be accounted for in any design and in any application of the mathematical models devised to predict the dynamic response of cylinders.

5. CONCLUSIONS

Sand-roughened circular cylinders were forced to oscillate in the transverse direction to a uniform flow with amplitude ratios from 0.25 up to 1.00 and reduced velocities in the range 3 to 8. The Reynolds number ranged from about 20,000 to 50,000; and the roughness Reynolds number, from about 200 to 512. Vortex lock-in was observed in the neighborhood of the Strouhal frequency. The Fourier-averaged transverse force coefficients exhibited significant variations in the vicinity of the resonant velocity. The inertia coefficient is larger than unity for oscillation frequencies larger than the Strouhal frequency and may be negative for oscillation frequencies lower than the Strouhal frequency.

Substantial differences were noted between the out-of-phase component of the force for a rough cylinder and that for a smooth cylinder. The difference was largest at perfect synchronization and increased with increasing roughness Reynolds number. The in-phase component of the force for a rough cylinder was nearly identical with that for a smooth cylinder. At perfect synchronization, the in-phase component was nearly equal to unity and the out-of-phase component was about 20 percent larger than that for a smooth cylinder for $Re_k = 200$; about 35 percent larger for $Re_k = 400$; and about 50 percent larger for $Re_k = 512$.

For the largest amplitude ratio (A/D = 1), both the in-phase and out-of-phase components of the force exhibited hysteresis. For increasing reduced velocities, the lock-in near the Strouhal frequency indicated that the limiting amplitude is larger than the cylinder diameter. For decreasing reduced velocities, the out-of-phase component of the force remained positive and there was no lock-in. The phenomenon was attributed to two modes of vortex shedding.

6. ACKNOWLEDGEMENTS

This work was sponsored by the National Science Foundation. The author wishes to thank Dr. George K. Lea for his continued support. The experimental system was constructed by Mr. Jack McKay whose help and ingenuity are always appreciated.

7. REFERENCES

1. Sarpkaya, T.: "Vortex-induced oscillations - a selective review". Journal of Applied Mechanics, ASME, 46, June 1979, pp. 241-258.

2. Sarpkaya, T. and Isaacson, M.: "Mechanics of Wave Forces on Offshore Structures". New York, Van Nostrand Reinhold, 1981, pp. 108-136.

3. Nakamura, Y.: "Some research on aeroelastic instabilities of bluff structural elements". Proceedings of the Fourth International Conference on Wind Effects on Buildings and Structures, London, Cambridge University Press, 1976, pp. 359-368.

4. Szechenyi, E.: "Supercritical Reynolds number simulation for two-dimensional flow over circular cylinders". Journal of Fluid Mechanics, 1975, 70, 3, pp. 529-542.

5. Sarpkaya, T.: " Dynamic response of piles to vortex shedding in oscillating flows" Offshore Technology Conference Paper No. 3647, May 1979, Houston, TX.

6. Sarpkaya, T.: "Fluid forces on oscillating cylinders". Journal of Waterways, etc. Div., ASCE, 104, WW4, 1978, pp. 275-290.

7. Sarpkaya, T.: "Transverse oscillations of a circular cylinder in uniform flow". Naval Postgraduate School Technical Report No: NPS-69SL77071-R, Dec. 1977.

8. Sarpkaya, T.: "Fluid forces on oscillating rough cylinders". Naval Postgraduate School Technical Report (under preparation), 1982.

9. King, R.: "The added mass of cylinders". The British Hydromechanics Research Association, BHRA, Report TN-1100, 1971.

10. Sarpkaya, T.: "A critical assessment of Morison's equation". Proceedings of the International Symposium on Hydrodynamics in Ocean Engineering, The Norwegian Institute of Technology, Trondheim, 1981, pp. 447-467.

11. Zdravkovich, M. M.: "Modification of vortex shedding in the synchronization range". ASME Paper No. 81-WA/FE-25, November 1981.

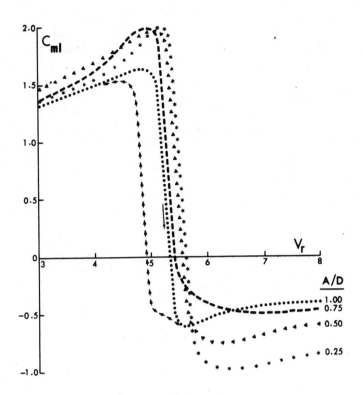

Fig. 1 The inertia coefficient versus reduced velocity
for various amplitude ratios. Only the mean
lines through the data points are shown.
$Re = 51,200$ and $Re_k = 512$.

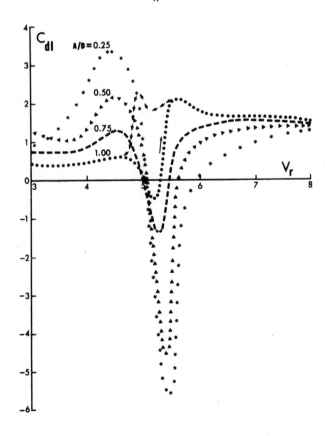

Fig. 2 The drag coefficient versus reduced velocity
for various amplitude ratios. Only the mean
lines through the data points are shown.
$Re = 51,200$ and $Re_k = 512$.

FLOW INDUCED OSCILLATIONS OF TWO
INTERFERING CIRCULAR CYLINDERS

M.M. Zdravkovich

University of Salford, U.K.

Summary

Flow interference between two circular cylinders in various arrangements imposes continuous and discontinuous changes in vortex shedding. The resulting oscillations induced by the vortex shedding are considerably modified by and strongly depend on the arrangement of the two cylinders. A systematic classification of flow interference regimes is linked to the observed vortex shedding responses for a wide range of arrangements.

The discontinuous change of flow regimes in some arrangements can excite and maintain large amplitude oscillations beyond a certain critical high velocity. The critical velocity and magnitude of oscillation strongly depends on the initial location and subsequent displacement of the cylinder by the fluid elastic forces. The wide variety of fluid elastic responses is categorised into three characteristic types:

1.   Flow instability rapidly builds up to extremely large amplitude predominantly in the streamwise direction.

2.   Flow instability slowly builds up amplitude to a certain level, the oscillations are mostly in the streamwise direction.

3.   Flow instability gradually builds up to large amplitude predominantly in the transverse direction.

It has been found that all types of the fluid elastic oscillations were confined within the regions of interference.

Organised and sponsored by
BHRA Fluid Engineering, Cranfield, Bedford MK43 0AJ, England

NOMENCLATURE

A_L = amplitude of oscillation in streamwise direction

A_T = amplitude of oscillation in transverse direction

D = diameter of cylinders

L = streamwise spacing between axes of cylinders

M = Mass per unit length of cylinder

$R_e = \dfrac{VD}{\nu}$ = Reynolds number

$S_c = \dfrac{2M\delta}{\rho D^2}$ = Scruton number

$S_t = \dfrac{n\,D}{V}$ = Strouhal number

T = transverse spacing between axes of cylinders

V = free stream velocity

$W = \dfrac{V}{n_c D}$ = reduced velocity

W_{cr} = critical reduced velocity corresponds to:

    1) Maximum amplitude for vortex induced oscillation

    2) Initial increase of amplitude for fluid elastic oscillation

n_c = natural frequency of cylinders

n_v = frequency of vortex shedding

$n_v = n_c$ = only in synchronization range

δ = logarithmic decrement of successive amplitudes

ν = kinematic viscosity of fluid

ρ = density of fluid

1. INTRODUCTION

Two parallel circular cylinders have been often used in various engineering structures which are exposed to wind, current, tide or waves. There are infinite numbers of possible arrangements of two parallel cylinders relative to the cross flow. When the two cylinders are far apart and the rear cylinder is well outside the wake of the front one there is no interference between them. The flows around both cylinders are the same as that around a single cylinder.

The interference between the two cylinders will start either when they are sufficiently close to each other or when the rear cylinder is adjacent to or within the wake of the front one. The first category will be called the proximity-interference and the second one the wake-interference. There is another distinction between the two categories:

(i) The formation of both vortex streets is affected by the interference when the cylinders are in close proximity.

(ii) When the wake interference takes place only the formation of the vortex street behind the rear cylinder is affected for sufficiently large spacing between the cylinders.

The two cylinders can be positioned side by side relative to the approaching flow direction, one behind the other in tandem arrangement or staggered relative to the velocity vector. The proximity - interference encompasses all three arrangements up to a certain spacing between the axes of the cylinders. The wake-interference extends very far downstream but it is confined to the tandem and slightly staggered arrangements. Fig. 1 schematically shows the regions of no-interference, proximity and wake-interference.

The flow interference between two circular cylinders has been reviewed for all three arrangements (Ref. 1). The proximity effects for two cylinders in side by side arrangement may be conveniently divided into three flow regimes as depicted in Fig. 2:

a) When the spacing between the cylinders is small, $1 \leqslant T/D < 1.2$, a single vortex street is formed downstream. Two cylinders behave as a single bluff body with base bleed in the gap between them.

b) When the spacing between the cylinders is in the range $1.2 < T/D < 2.2$ narrow and wide wakes are formed divided by a biased flow through the gap. The biased flow is bistable and the narrow and wide wake can intermittently interchange between the two cylinders. The frequency of vortex shedding is different in the two wakes.

c) When the spacing is further increased both vortex streets have the same frequency but they are coupled in an out-of-phase mode. The vortices are simultaneously formed and shed on the gap side and then simultaneously on the outer sides. The coupling gradually decreases and finally disappears beyond T/D of about 4.

The proximity effects for two cylinders in tandem arrangement may be separated into two distinct regimes:

1) For spacings up to the some critical range of L/D in the downstream direction, the vortex street is suppressed behind the front cylinder.

2) Beyond the critical spacing range both cylinders form vortex streets.

The first regime may be further subdivided (Ref. 2) according to the flow changes in the gap between the cylinders see Fig. 2. The subdivisions are as follows:

a)   When the spacing is very small, say less than 1.1 the two cylinders behave as a single slender body with high Strouhal number. The region between the cylinders is stagnant and the shear layers separated from the front cylinder do not reattach onto the downstream cylinder.

b)   When the spacing is between $1.1 < L/D < 1.6$ an alternate reattachment of the shear layers takes place on the front side of the rear cylinder in the rhythm of vortex shedding of the latter.

c)   When the spacing is $1.6 < L/D < 2.4$ quasi-steady reattachment of separated shear layers is observed on the rear cylinder.

d)   Beyond $L/D = 2.5$ occasionally one of the reattachments is disrupted but still no regular vortex shedding behind the **front** cylinder.

e)   Bistable range exists around the critical spacing where the vortex shedding behind the front cylinder persists for some time and then it is intermittently suppressed and replaced by the reattachment flow regime.

These subtle changes of the gap flow will have a profound effect on the oscillations of the front cylinder as will be shown later.

Finally, the least known and explored, are the proximity effects in the staggered arrangements bounded by the side and tandem arrangements. Two flow regimes can exist:

1)   The vortex streets are formed in the narrow wake behind the **front** cylinder and in the wide wake behind the rear cylinder with high and low frequency of vortex shedding respectively.

2)   When the transverse spacing is sufficiently small the vortex shedding behind the front cylinder is suppressed and a strong gap flow induces large transverse component of force on both cylinders (Ref. 3).

The main object of the present paper is to link the observed vortex shedding induced oscillations to the interference flow regimes. The fluid-elastic oscillations will be related to the discontinuous change of flow regimes as affected by the fluid elastic displacement of the cylinders.

2.   EXPERIMENTAL   ARRANGEMENTS

The vortex shedding excitation occurs in the reduced velocity range which may be one order of magnitude below the critical reduced velocity for the fluid-elastic excitation. These features required different models and wind tunnels for the tests.

The oscillations induced by the vortex shedding were investigated in the 0.3m x 0.6m wind tunnel over the velocity range from 6 to 36 m/s. The cylinders were made of aluminium tubing of 25.3mm outer diameter and were 296mm long. Both ends were closed with bungs and on one side 9mm threaded rod was screwed into the bung. The other side of the threaded rod protruded outside the test section and was fixed to a mild steel plate 0.6m x 0.6m x 0.1m. The frequency of both cylinders was tuned to 71 Hz ($\pm$ 0.2 Hz). The damping expressed as a logarithmic decrement was 0.013. The Scruton number was 23 and the Reynolds number range was from 1×10^4 up to 8×10^4.

The oscillations induced by the fluid-elastic forces were investigated in the 0.45m x 0.45m wind tunnel. Two sets of cylinders were made of aluminium tubing and were of two sizes: 38.1mm and 50.8mm outer diameter; both were 406mm long and 1.45mm thick. Both ends of the cylinders were sealed. The cylinders were suspended by means of threaded rods 7.5mm in diameter. The rods were attached to a mild steel plate 0.45m x 0.51m x 0.025m which was clamped on the top and outside the test section. Air leakage was minimised by sealing the plate into a box.

The rods acted as a flexible cantilever and by altering their length the frequency was tuned to 10.5 Hz. The logarithmic decrement was 0.07 and the Scruton number 80. The vortex shedding could not excite oscillations and both cylinders remained stable at low values of reduced velocity. The Reynolds number range was from 1.5×10^4 to 9.5×10^4.

The amplitude and frequency of oscillations were monitored in both wind tunnels by means of four capacitance probes. The probes were attached to the steel plate and arranged to monitor streamwise and transverse displacements of both cylinders. Small cubes made of aluminium were attached to the threaded rods and two pairs of capacitance probes at right angles to each other were positioned to face the aluminium blocks. This non-contact system provided an accurate and continuous record of oscillation on the oscilloscope screen and UV-recorder. It has been described in more detail elsewhere (Ref. 4).

3. VORTEX SHEDDING EXCITATION

The vortex shedding has been the most common form of fluid dynamic excitation for the single cylinder. The oscillations are always confined within the synchronization range. The frequency of vortex shedding of two cylinders depends on their arrangement and the Strouhal number can vary from 0.1 up to 0.38 (Ref. 5). This means that there are two synchronization ranges which can initiate oscillations at the reduced velocities from 2.6 up to 10. The cylinder having high Strouhal number starts to oscillate at low reduced velocity, while the other one may remain stable despite the fact that both cylinders are of the same size and have the same natural frequency. Another possibility, which is non-existent for the single cylinder, is that the oscillations of one cylinder can strongly affect the vortex shedding and subsequent synchronization of the second cylinder.

The present experiments covered the proximity-interference, wake-interference and no-interference regions as depicted in Fig. 3. The extent of the synchronization range varied considerably within all three regions as shown in Fig. 3. The reduced velocities for the threshold and end of synchronization are written within each circle in Fig. 3 while the critical reduced velocity, where maximum amplitude developed, is given in each circle in brackets.

The tandem arrangement corresponded to the flow regime with suppressed vortex shedding behind the front stationary cylinder. The flexible cylinders remained stable up to $W = 6.5$. The maximum amplitude of oscillation developed at $W_{cr} = 8$ and slowly died down at $W = 16$. The typical maximum response showed that the front cylinder oscillated with larger amplitude than the rear one. The initial lateral displacement of the rear cylinder triggered the regular synchronized vortex shedding behind the front cylinder. The latter developed the dominant oscillation and disrupted the vortex shedding behind the rear cylinder. A similar response was found in water for tandem cylinders arranged at $L/D = 1.5$ and 1.75 (Ref. 6) and $L/D = 2.5$ (Ref. 7). The quasi-steady reattachment regime was disrupted behind the front cylinder and synchronized vortex shedding was established instead.

The tandem arrangements chosen at the beginning of the two-vortex-street regime produced the most vigorous response of the rear cylinder. The large transverse oscillations disrupted the vortex shedding behind the front cylinder which oscillated irregularly. The tandem arrangement 3 showed a considerable reduction in amplitude. Stable oscillations were observed for the front cylinder while the rear cylinder was occasionally disrupted and then gradually built up the amplitude to a maximum shown by, the dotted line in Fig. 4. A similar response was found in water for 4 < L/D < 7 (Ref. 7). Further increase in the streamwise spacing gradually diminshed the interference effect.

The side by side arrangements have produced three different flow regimes behind two stationary cylinders in the interference region. The regime of coupled vortex streets excited an out-of-phase mode of oscillation of flexible cylinders in the synchronization range (Ref. 8) and the single-vortex-street regime imposed an in-phase oscillation of both cylinders. The biased gap-flow regime exhibited both modes: the low synchronization range excited by the vortex shedding in the narrow wake was in the out-of-phase mode while the second synchronization range at high reduced velocity was in the in-phase mode of oscillation (Ref. 8).

The arrangement 4 was chosen as representative of the biased gap-flow regime. The first synchronization regime was not excited probably due to weak vortex shedding in the narrow wake. The second synchronization range slowly developed maximum amplitude at W_{cr} = 15. The oscillation was small even in comparison with the no-interference arrangement 14.

The out-of-phase mode excited by the coupled vortex shedding was found in the arrangement 5. The oscillation was more vigorous than in the arrangement 4. The maximum response in the out-of-phase mode was found at T/D = 2 (Ref. 8) just before the break-down of the coupled vortex streets regime.

The front cylinder remained very stable in the staggered arrangements, 6-9, all in the proximity region. The narrow wake behind the front cylinder could not excite oscillation and the wide wake behind the rear cylinders produced significant amplitudes. As the cylinder approached the wake boundary of the front cylinder, the oscillations became more vigorous (compare arrangements 6 and 7 in Fig. 4). When the rear cylinder was submerged into the wake of the front cylinder the oscillations were reduced, particularly, when the rear cylinder was located nearer to the front one (see arrangements 8 and 9 in Fig. 4).

The second group of staggered arrangements 10-13 was in the wake-interference region. In contrast to the previous group, the vortex shedding behind the front cylinder was not suppressed. The stable oscillations of the front cylinder were recorded at W_{cr} for the arrangements 11 and 13, while the rear cylinder displayed a build up and decay of amplitude. The dotted line showed the maximum loop for these time-varying oscillations.

Another notable feature was the significant increase of the streamwise component and tilting of the oscillation loops. This was caused by the higher drag force when the cylinder was less submerged in the wake and lower drag when it was deeper in the wake. This mechanism alone is called wake galloping and produce large amplitude oscillations of twin-transmission lines at high reduced velocities (ref. 9). The maximum response was found when the cylinder was fully submerged in the wake (see arrangement 10 in fig. 4).

The arrangement 14 was chosen to represent the no-interference region. Both cylinders exhibited very stable oscillations and the amplitude was considerably less than that found in the wake interference region. The rear cylinder oscillated with higher amplitude which indicated that the genuine no-interference region was not reached for that transverse spacing.

4. FLUID-ELASTIC EXCITATION

The fluid-elastic forces result from the interaction between the motion of the tubes and the flow around them. If the tubes are arranged within the regions where the discontinuous change of flow regime takes place then the oscillating tube can trigger and control the change of flow regimes. The discontinuous change of the fluid-elastic force coupled with some hysteretic delay can build up and maintain a large amplitude of oscillation. The fluid-elastic oscillations are not confined within the finite range of reduced velocities. Once initiated they continue unabated by clashing of tubes up to the highest attainable velocity.

When the tubes were sufficiently far apart and outside the wake-interference region the fluid-elastic excitation was not found. The typical response denoted by RO is shown in Fig. 5 where up to the highest reduced velocity the amplitude of oscillation in both directions remains unchanged.

An entirely opposite response can be seen in Fig. 6 for side-by-side arrangement. Initially both cylinders are subjected to the same drag force with a stable bleed flow through the gap. When the reduced velocity reaches 45 the irregular displacement of the tubes brings them into the region of the biased-gap-flow with narrow and wide wakes and different drag forces. The latter are in phase with the displacement leading to a steep rise in amplitude in both directions. The large component of oscillation in the transverse direction indicates that the displaced tubes are subjected to large transverse force as well. The jet-switch mechanism of excitation which was found (Ref. 10) for a single-row of alternately fixed and flexible tubes is coupled with the significant transverse displacement for the two flexible cylinders.

As the transverse spacing between the axes of the cylinders is increased the fluid-elastic instability becomes less and less steep and occurs at a progressively higher critical reduced velocity as shown for T/D = 3 in Fig. 6. This, type R1, response is also found in some staggered arrangements as will be discussed later.

Another type of response, R2, with a predominant oscillation in the streamwise direction is shown in Fig. 7. The moderate build-up of amplitude ends with a large amplitude which does not change with further increase of reduced velocity. Note that this type does not indicate a more stable arrangement because the instability initiates at a lower reduced velocity than for the type R1.

Finally, there is the type R3 response with a predominant oscillation in the transverse direction as shown in Fig. 8. It is characterised by a gradual build up of oscillations either of both cylinders or only of the rear cylinder. This type of fluid-elastic oscillation was particularly vigorous when the front cylinder was rigid and the rear cylinder could oscillate only in the transverse direction (Ref. 11). The mechanism of excitation was attributed to the alternation of strong gap flow and its full suppression by the transverse oscillation of the rear cylinder.

All tested arrangements are shown in Fig. 9. The circle denotes the position of the rear cylinder. The number written inside the circle designates the type of response and the number next to it, the critical reduced velocity at which a considerable increase in amplitude commences. Two concentric circles designate that the maximum peak to peak amplitude exceeded two diameters and three circles indicate steep increase in amplitude above three diameters in magnitude. The maximum amplitude is not reached in tests with the type R1 response.

The side-by-side arrangements produced always the R1 response. The reduction of the transverse spacing increased the amplitude of oscillation and decreased the critical reduced velocity. The only exception was the arrangement T/D = 1.125 where instability shown in Fig. 6 occurred at higher W_{cr} than for T/D = 1.25. The type R1 response was also found in the neighbouring staggered arrangements and at L/D = 1.3 and T/D = 0.75. The latter was tested twice and both tests showed consistently the type R1 response.

The tandem arrangements displayed a variety of responses. The smallest spacing L/D = 1.125, produced a random response with irregular fluctuations of amplitude for both cylinders. Next at, L/D = 1.25, response was similar to that at L/D = 4, Fig. 8, but both cylinders oscillated with large amplitude in the transverse direction. Further increase in spacing stabilized the cylinders and the amplitude of oscillations remained small up to W = 90. The same was observed for T/D = 2.0 except that at W = 87 a sudden jump to large amplitude occurred for the rear cylinder. A similar trend was found for L/D = 2.5 and 3 where two types of responses were possible beyond W_{cr}.

The type R3 response was also found in an isolated region of staggered arrangements as shown in Fig. 9. Both cylinders developed transverse amplitude of oscillations and in that respect were similar to the staggered arrangement L/D = 1.25.

All the other staggered arrangements were characterised by the type R2 response. The large amplitude response occurred for larger transverse spacing at higher W_{cr}. The arrangements close to the front cylinder became unstable at very low W_{cr}.

The staggered arrangements above the dotted line remained stable up to W = 90. This trend supported the argument that the origin of the fluid-elastic forces was related to the different flow regimes within the interference regions. Once the cylinders were outside the interference region the fluid elastic oscillations could not be excited.

5. CONCLUSIONS

Flow induced oscillations of two flexibly mounted cylinders were excited either by vortex shedding or by fluid elastic displacement of the cylinders within the flow interference regions.

The vortex induced oscillations were strongly dependent upon the arragement of cylinders. The oscillations in the proximity-interference region were significantly magnified in comparison with the no-interference region only when the rear cylinder was in the vicinity of the wake boundary of the front cylinder. The rear cylinder oscillated with large amplitude while the front cylinder displayed an order of magnitude less amplitude. Both cylinders oscillated with large amplitudes further downstream in the wake-interference region.

The fluid-elastic oscillations were characterised by very large amplitudes at high values of the reduced velocity. Three different types of response were found; two with the dominant component in the streamwise direction and one with the dominant component in the transverse direction. The largest amplitude of oscillation was excited in the streamwise direction and all severe oscillations occurred in the proximity interference region. An extremely complex relationship was found between the type of oscillation, critical reduced velocity and the arrangement of the two cylinders. A far more exhaustive set of experiments would have to be undertaken before any definitive conclusions could be drawn about the fluid elastic oscillations of two circular cylinders.

6. ACKNOWLEDGEMENTS

The author would like to give credit to his former students who carried out experiments: Mr D L Pridden in 1974 (fluid-elastic oscillations) and Mr K Mojtahedi in 1978 (vortex induced oscillations).

7. REFERENCES

1. Zdravkovich, M M "Review of flow interference between two circular cylinders in various arrangements", Journal Fluids Engineering, Trans. ASME, 99, 1977, pp 618-633.

2. Igarashi, T "Characteristics of the flow around two circular cylinders arranged in tandem", Bull. Japan, Society Mechanical Engineers, 24, 1981, pp 323-331.

3. Zdravkovich, M M and Pridden, D L "Interference between two circular cylinders, series of unexpected discontinuities", Journal of Industrial Aerodynamics, 2, 1977, pp 255-270.

4. Southworth, P J and Zdravkovich, M M, "Cross-flow-induced vibrations of finite tube banks in in-line arrangements", Journal of Mechanical Engineering Science, 17, No 4, 1975, pp 109-148.

5. Kiya, M., Arie, M., Tamura, H and Mori, H "Vortex shedding from two circular cylinders in staggered arrangement", Journal Fluids Engineering, Trans. ASME, 102, 1980, pp 166-173.

6. Jendrzejczyk, J A., Chen, S S and Wambsganss, M W "Dynamic response of a pair of circular tubes subjected to liquid cross flow", Journal of Sound and Vibration, 67 (2), 1979, pp 263-273.

7. King, R and Johns, D J "Wake interaction experiments with two flexible circular cylinders in flowing water", Journal of Sound and Vibration, 45 (2), 1976, pp 259-283.

8. Quadflieg, H "Vortex induced load on pair of cylinders in incompressible flow at high Reynolds number", (in German) Forschung Ingenieurwesen. 43, 1977, pp 9-18.

9. Cooper, K R and Wardlaw, R L "Aeroelastic instabilities in wakes", Proc. Int. Symp. Wind Effects on Buildings and Structures, Tokyo, 1971, paper IV.1.

10. Roberts, B W "Low frequency aeroelastic vibrations in a cascade of circular cylinders", Inst. Mech. Eng. (London), Mechanical Engineering Science Monograph, No. 4, 1966.

11. Zdravkovich, M M "Flow induced vibrations of two cylinders in tandem and their suppression" in "Flow Induced Structural Vibrations" Ed. E. Naudascher, Springer, Berlin, 1974, pp 631-639.

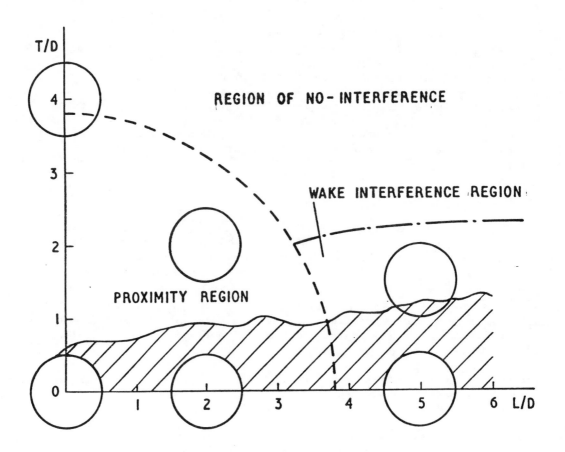

Fig. 1   Classification of interference regions.

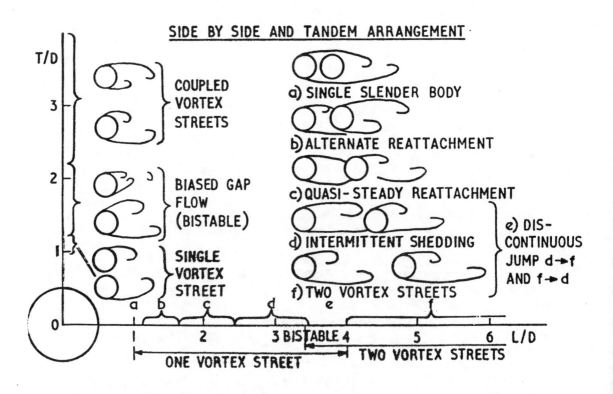

Fig. 2   Classification of flow regimes in side-by-side and tandem
arrangements for stationary cylinders.

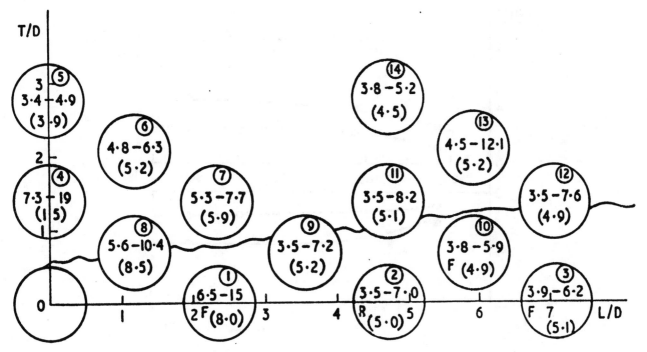

Fig. 3 Tested arrangements for the vortex shedding excited oscillation.

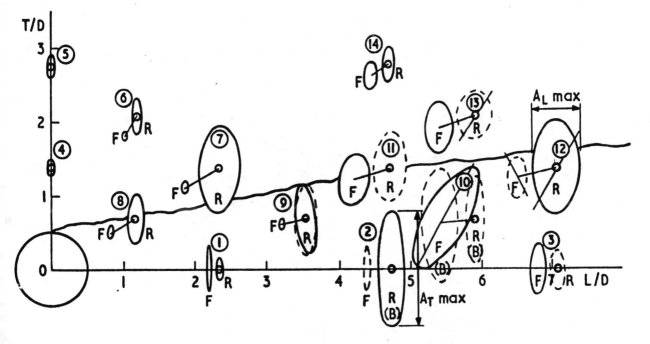

Fig. 4 Typical oscillation loops at maximum amplitude for vortex-
shedding excitation. The response of all front cylinders,
F, are shown next to rear ones, R. The amplitudes of
cylinder tips are magnified 10 times.

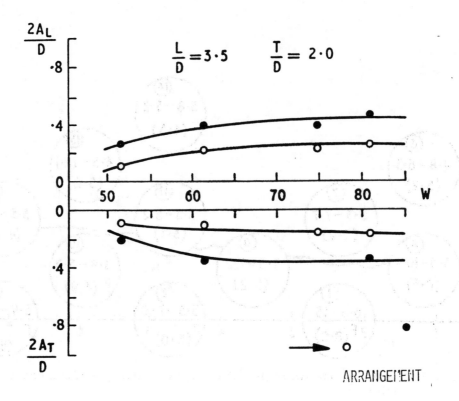

Fig. 5 Type RO response to fluid-elastic excitation.

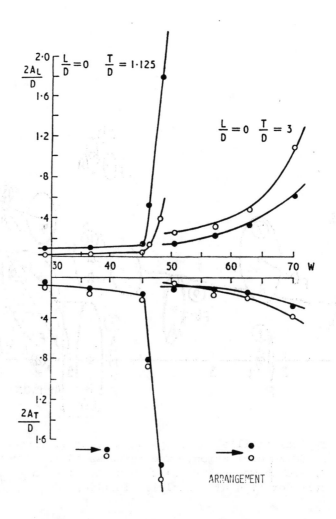

Fig. 6 Type R1 response to fluid-elastic excitation.

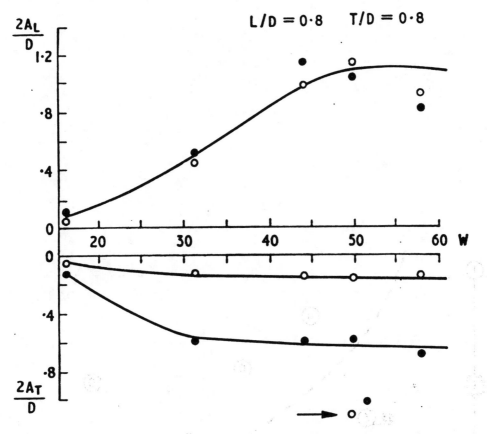

Fig. 7 Type R2 response to fluid-elastic excitation.

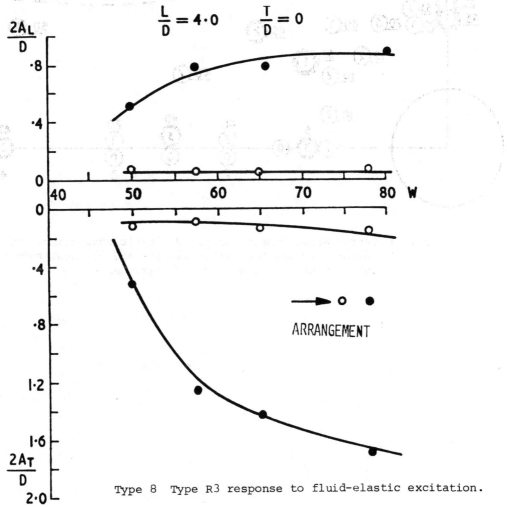

Type 8 Type R3 response to fluid-elastic excitation.

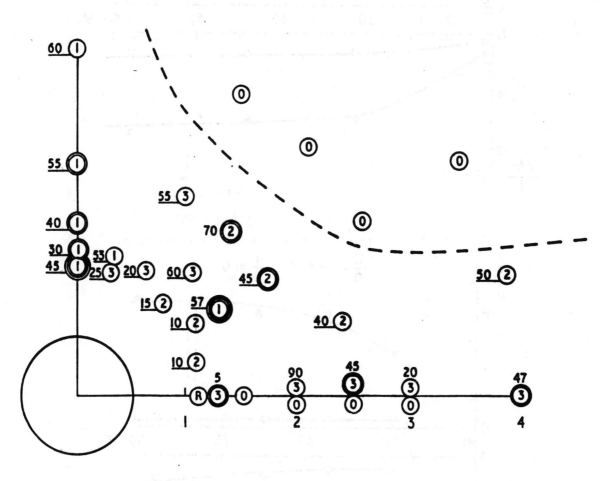

Fig. 9 Tested arrangements for fluid elastic excitation. Type of response is written within circles and critical reduced velocity next to them for each arrangement.

FLOW INDUCED VIBRATIONS
IN FLUID ENGINEERING

Reading, England: September 14-16, 1982

A MODELLING TECHNIQUE FOR THE DETERMINATION
OF DYNAMIC WIND LOADS ON BUILDINGS

R. A. Evans and B. E. Lee

University of Sheffield, U. K.

Summary

At the previous Symposium on Practical Experience of Flow Induced Vibrations held at Karlsruhe in 1979 a number of papers on critical unresolved problems in the field of wind engineering were presented. Outstanding amongst the problems discussed was that of determining the dynamic wind input forces which are responsible for the cross wind motion and torsional motion of buildings and structures. Additionally whilst it was recognized that combinations of theoretical and empirical methods of solution were useful for predicting the input wind forces responsible for in-wind motion, there remained a shortage of hard information either from full scale tests or model tests with which to compare such prediction methods. Finally it was acknowledged that for wind directions oblique to a building, external surface wind tunnel tests could provide the only means of determining the wind inputs.

The present paper describes a wind tunnel modelling technique for determining the wind input forces, usable for all wind directions, applicable either to in-wind or cross-wind modes of response. The technique, which utilizes rigid models mounted on one component and two component balance transducers can be used to operate either as a resonant or as a non-resonant measurement system.

Organised and sponsored by
BHRA Fluid Engineering, Cranfield, Bedford MK43 0AJ, England

1. INTRODUCTION

Steady progress has been made in the description of the response of taller buildings to wind where the main components of the response come from each of the fundamental modes, excited by the modal forces. The mechanical transfer function, relating the load function to the response, is straightforward though the aerodynamic transfer function, relating the gust structure to the forces, has to be determined experimentally in most cases. A further complication exists if body motion effects interact with the load function producing aerodynamic damping terms. Multi-degree-of-freedom aeroelastic wind tunnel models are believed to provide the most truthful results, but they are expensive and time consuming and extrapolations for different building properties than those modelled are difficult. Rigid models pivoting at the base are less expensive, but they do not provide information on torsional response and are normally modelled for particular structural properties as well. If aerodynamic damping effects are negligible, which is usual for most buildings at practical wind speeds and practical structural damping values, and if a high frequency model can be made, then it is possible to measure the load function directly since the modal loads are a function only of the wind structure and aerodynamic shape of the building.

Tschanz (Ref. 1) describes the design and operation of such a high frequency balance system capable of resolving five components of the total dynamic force acting on model buildings.

2. DIRECT MEASUREMENT OF TOTAL DYNAMIC LOADS

The resonant test method described here, for the measurement of total dynamic loads, is a particular case of the more general high frequency method described by Tschanz. In this method the frequency of the wind tunnel model - balance transducer combination is deliberately chosen to have a low value at frequencies excitable by the turbulence field of the simulated boundary layer. The frequency of the model, f_m may be related to the frequency of a particular building f_f by the relationship

$$\frac{f_m}{f_f} = \frac{V_m}{V_f} \cdot \frac{D_f}{D_m} \tag{1}$$

where D represents linear dimension and V represents velocity. The operation of the model - balance transducer combination at resonance instead of below resonance only represents a full aeroelastic test where the corresponding values of frequency and damping have been deliberately chosen to coincide with a particular full scale building. Elsewhere the resonance method simply reflects the utility of the high frequency method but with the advantage of the resonance phenomena, provided that scaling laws can be invoked to permit the normalization of data with respect to the values of damping and frequency at which the model - balance transducer operates. The advantages of resonance testing are that the model - balance transducer acts as a well characterized mechanical amplifier in which the useful signal is amplified by the mechanical resonance factor but in which the system noise remains unaffected, thereby increasing the signal to noise ratio by up to 2 orders of magnitude. This signal magnification is most significant if the order of size of the test results is considered. Since the purpose of tests such as these is to produce the mode generalized power spectral density function, Sn, and

$$\sigma^2 = Sn \frac{\pi f}{4\zeta} \tag{2}$$

we can write

$$\frac{Sn_m}{Sn_f} = \frac{\sigma^2_m}{\sigma^2_f} \cdot \frac{f_f}{f_m} \tag{3}$$

156

in order to relate model to full scale values. Since σ^2 represents a mean square response term in units of force squared, and force scaling can be represented by

$$\frac{\text{Force m}}{\text{Force f}} = (\frac{V_m}{V_f})^2 \cdot (\frac{D_m}{D_f})^2 \qquad (4)$$

the modal P.S.D. scaling relationship becomes

$$\frac{Sn_m}{Sn_f} = (\frac{V_m}{V_f})^3 (\frac{D_m}{D_f})^5 \qquad (5)$$

Using typical values of wind tunnel to full scale velocity scaling ratios and length scaling ratios of

$$\frac{V_f}{V_m} = 3 \text{ and } \frac{D_f}{D_m} = 500$$

the PSD scaling relationship is

$$Sn_f = Sn_m \times 8 \times 10^{14}$$

Hence, if a typical full scale value of Sn_f for a tall building is of the order of 10^8 N^2/Hz then the equivalent model value is 10^{-7} N^2/Hz. This may serve to illustrate the cause of the measurement difficulties referred to previously in the literature as well as to indicate the advantages of a 10^2 amplifier with low noise characteristics. Additionally the resonance test method ensures that the model balance transducer acts as a narrow band filter restricting the information signal to frequencies of interest.

The resonance test method for the determination of total dynamic loads may be applied in two ways. The first method in its simplest form involves determining the model response from an r.m.s. meter reading. The r.m.s. level of the resonance peak, suitably filtered to exclude the background response contribution, can then be converted to a modal force PSD using equation 2, assuming the value of the damping, ζ, to be known from an impulse decay test. This method has been shown to be a most precise way of determining the mode generalized PSD function but is available for only one value of $V_{/nD}$ at a time, tests at different tunnel speeds being necessary to establish a useful range of $V_{/nD}$. The second method of using the resonance test method is to employ a digital spectrum analyser. Here the full response spectrum is measured and subsequently divided by the mechanical transfer functions of the model - balance transducer system in order to yield the full spectrum of the total dynamic load. These two methods have been employed on the same model and have been shown to produce identical results. The two methods could be regarded as being complementary where the r.m.s. level is used as a precise spot check on the full spectrum method.

The model - balance transducer system itself consists of a series of strain gauged balance transducers, each one of which, when fitted with the appropriate model building, resonates at a predetermined frequency. Additionally the system contains an integral variable damping mechanism, thus enabling both model frequency and model damping to be varied independently. Figure 1 illustrates the model - balance transducer system and Figure 2 its location in relation to the wind tunnel turntable. The response characteristics of modelled buildings have been investigated at specific frequencies within the range 60-125 Hz and for damping values in the range $0.005 < \zeta < 0.05$ in atmospheric boundary layer flow. The fundamental translational modes have been examined for both translational modes and both resonant and non-resonant dynamic response components have been evaluated. A typical total system response is shown in Figure 3 where the different contributions from the background and resonant dynamic response components can be seen. It should be noted that the resonant response, used to evaluate the modal loads, is always made with high pass and low pass filters set at 60 Hz and 149 Hz respectively, in operation. This approximation may result in an underestimate in the modal forces of the order of 5%.

The resonance conditions of the model - balance configuration are dependant on the geometry and material of the balance and the mass of the model. For the single component balance shown in Figure 1 the thickness of the balance web may be determined by an equation which relates it to the desired resonance frequency and the effective mass of the model, treated as a lumped mass cantilever system. Thus, balance transducers may be designed to give a particular resonance frequency condition once the size and mass of the model are known. However the model mass is only important in determining the operational frequency of any model - balance transducer combination, and is not otherwise, significant in determining the balance transducer output, except as a second order effect on sensitivity, or in the scaling up procedure to full scale values. This may be illustrated by Figure 4 in which a light model, 600 grammes, has been mounted on a nominal 100 Hz balance and the combination found to have an exact frequency of 97 Hz. The output of the light model is contrasted with that of a heavy model, 1050 grammes, mounted on a nominal 125 Hz balance where the combination has an exact frequency of 95 Hz. The nominal balance frequencies have assumed the use of a light balance. Figure 4 demonstrates that the r.m.s. resonant response as a function of wind tunnel speed is unaffected by model mass.

Multiple graphs such as Figure 4 may be obtained for given conditions of building shape and wind regime, for all values of model - balance transducer frequency and damping level. The frequency dependant data may be collapsed by expressing the r.m.s. response as a function of $V_{/nD}$. The damping level, independently controllable by a viscoelastic compression ring, controls the dynamic resonance magnification which in turn governs the conversion of r.m.s. peak level resonance responses to modal P.S.D.'s. An extensive series of tests has been carried out, using the resonance test method in both modes, for a wide range of values of frequency, damping and bandwidth which indicate methods of non-dimensional data collapse. It is the intention of the authors Evans and Lee, that the results of these tests be full published in due course, 6th International Conference on Wind Engineering, Gold Coast, Australia, 1983.

The number of force and moment components to be resolved by high frequency dynamic force balances may be worth comment. The full six components of the traditional aeronautical balance may appear at first sight to be a laudable aim and the attainment of 5 components, as described by Tschanz, and also obtained by Saunders (Ref. 2) for work on vehicle aerodynamics, a creditable achievement. However the production of such a balance system may produce quantity of results at the expense of the simplicity and signal to noise ratio of the one or two component balances as used by Evans and Lee (Ref. 3) and Cermak et-al (Ref. 4). Greater detail of a variable frequency one component balance is shown in Figure 5 and a two component balance is shown in Figure 6 whose uncoupled outputs can be seen in Figure 7. Such considerations prompt the question, how many components does the structural engineer need for design purposes. If the design problem can be solved by the specification of a maximum tip deflection in two dimensions for structural purposes and a maximum resultant acceleration for human comfort purposes, then as few as three components may be adequate. Since it seems that little success is to be had currently with the problem of modelling the fundamental torsional mode of oscillation, Tschanz reporting the same difficulties as those described by Ruscheweyh (Ref. 5) at the Colorado Wind Engineering Conference in 1979, then it seems that only the two common translational modes about the vertical major and minor axes are required to produce large amounts of useful design data. Such a philosophy may lead to the design and operation of considerably less complex balances than that described by Tschanz, which in turn may have the benefit of encouraging a wider sphere of potential users.

3. COMPARISON OF MODEL RESULTS WITH FULL SCALE DATA

The comparison of model results with full scale data has been extensively discussed by Lee (Ref. 6). In this comparison full scale data derived from wind response recording, carried out on the University of Sheffield Arts Tower building have been used.

Examples of the available full scale modal load data are given in Table 1 where the modal loads are expressed as root mean square values. The corresponding mode generalized power spectral density function may be obtained using equation (2) where $f = 0.68$ Hz for the N-S mode and $f = 0.86$ Hz for the E-W mode and the value of ζ in both cases is 0.0086.

The corresponding data scaled up from the wind tunnel tests, derived using the model - balance transducer system described in this paper, are shown in Table 2. The agreement for corresponding groups of wind speed and wind direction is considered to be good.

ACKNOWLEDGEMENT

This work has been performed under contract from the U.K. Building Research Establishment and is reproduced by their kind permission.

REFERENCES

1. Tschanz, A. "Measurement of total dynamic loads using elastic models with high natural frequencies." International Workshop on Wind Tunnel Modelling Criteria and Effects. N.B.S. Gaithersburg, Md., 1982.

2. Saunders, J. Private communication 1982.

3. Evans, R.A. and Lee, B.E. "The assessment of dynamic wind loads on a tall building : a comparison of model and full scale results." Proc. 4th US National Conf. Wind Engineering Research, Seattle, 1981.

4. Cermak, J., Sadeh, W. and His, G. "Fluctuating moments on tall buildings produced by wind loading." Proc. Meeting on Wind Loads on Buildings and Structures. N.B.S. Building Science Series 30, 1970.

5. Rusheweyh, H. "Dynamic response of high rise buildings under wind action." Proc. 5th Int. Conf. on Wind Engineering, Colorado, 1979.

6. Lee, B.E. "Model and full scale tests on the Arts Tower at Sheffield University." International Workshop on Wind Tunnel Modelling Criteria and Effects. N.B.S. Gaithersburg, Md., 1982.

TABLE 1

Full Scale - Modal Wind Loads

Mean Wind Direction 0	Mean Wind Speed (84m) m/s	Equivalent r.m.s. Modal Force $\sigma_{r_{r.m.s.}}$ KN	
		N-S mode	E-W mode
254	20	2.71	2.03
254	24.1	3.13	1.59
305	18.3	2.01	1.68
305	25	4.25	2.86
305	29.7	5.23	3.40
305	34.6	5.50	3.50
338	22.9	1.95	0.81
338	27.2	2.77	1.47
338	30.4	3.12	1.86

TABLE 2

Model Data - Modal Wind Loads (Scaled Up)

Mean Wind Direction	Mean Wind Speed (84m) m/s	Equivalent r.m.s. Modal Force $\sigma_{r_{r.m.s.}}$ KN	
		N-S mode (site model)	E-W mode (no site model)
254	28.2	2.07	1.21
305	28.2	1.68	1.00
338	28.2	2.12	1.26

160

Figure 1.   Model - Balance Transducer

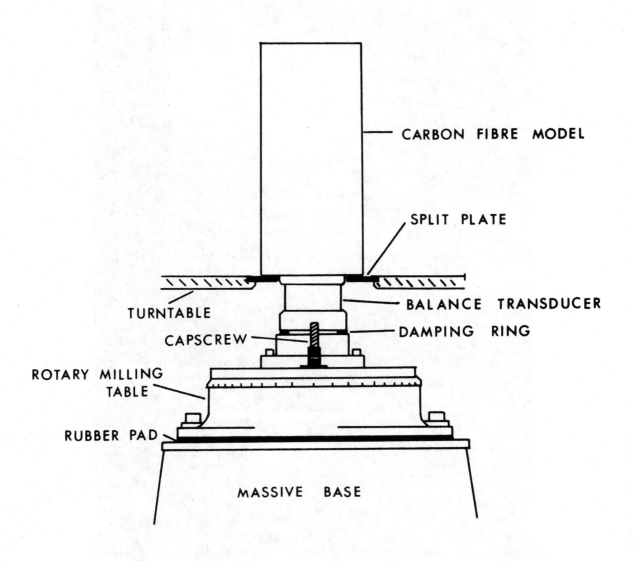

CARBON FIBRE MODEL

SPLIT PLATE

BALANCE TRANSDUCER

DAMPING RING

TURNTABLE

CAPSCREW

ROTARY MILLING
TABLE

RUBBER PAD

MASSIVE BASE

Figure 2.   Experimental Arrangement

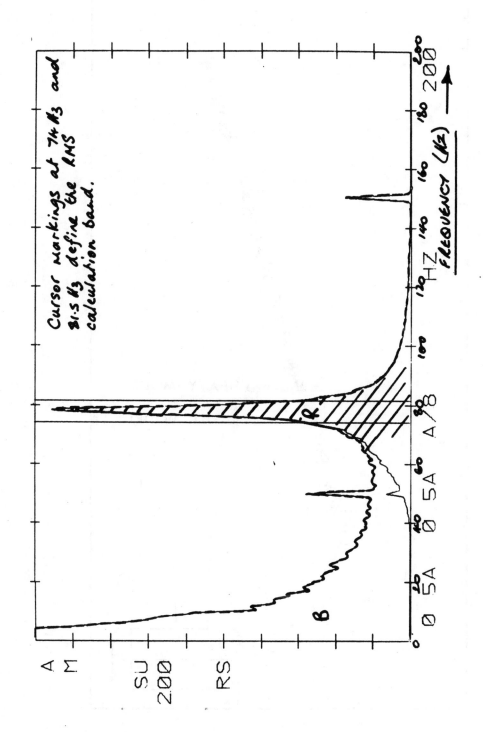

Figure 3.   RMS Spectrum of Balance Output

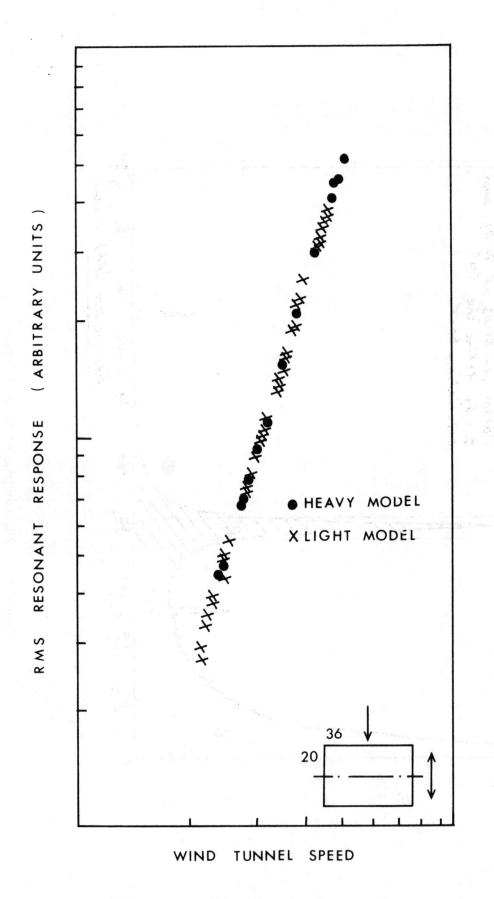

Figure 4.   Effect of Model Mass on Output

Figure 5.   Variable Frequency Single Component Balance

Figure 6.   Two Component Balance

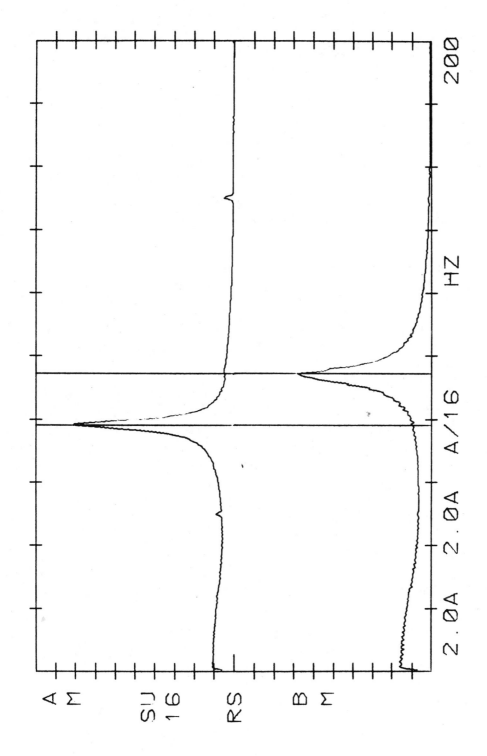

Figure 7.   Output Spectra from Two Component Balance

FLOW INDUCED VIBRATIONS
IN FLUID ENGINEERING

Reading, England: September 14-16, 1982

THE AERODYNAMIC STABILITY OF THE PROPOSED
WESTERN-SCHELDT SUSPENSION BRIDGE

A. J. Persoon and C. M. Siebert

National Aerospace Laboratory (NLR), The Netherlands

Summary

A windtunnel study was performed on the aerodynamic stability of the proposed Western-Scheldt suspension bridge. The experiments aimed to predict the flutter stability of the bridge deck and the wind-induced response of the pylons. For the flutter predictions aerodynamic derivatives were measured using an oscillating bridge deck model. Flutter calculations based on these derivatives resulted in an aerodynamically unstable bridge deck. The predicted response of the pylon appeared to be acceptable.

Organised and sponsored by
BHRA Fluid Engineering, Cranfield, Bedford MK43 0AJ, England

NOMENCLATURE

b	damping ($b = {}^c/cr$)	–
C	chord of the bridge deck (Ref. 1)	m
d	height of the bridge deck	m
f	vibrational frequency	Hz
I	mass moment of inertia of the model	kgm^2/m
k	reduced frequency ($k = \frac{\omega \ell}{V}$)	–
k_A	unsteady derivative of the lift (heave)	–
k_B	unsteady derivative of the lift (pitch)	–
2ℓ	chord of the bridge deck (Ref. 3)	m
L	unsteady lift force	kg/m
M	unsteady moment	kgm/m
m	mass of the bridge deck model	kg/m
m_A	unsteady derivative of the moment (heave)	–
m_B	unsteady derivative of the moment (pitch)	–
R_e	Reynolds number	–
S	Strouhal number: $S = \frac{f.d}{V}$	–
V	wind velocity	m/s
z	amplitude of vibration	m
Θ	angle of rotation	(rad)
ω	circular frequency	(1/s)

1. INTRODUCTION

In the Netherlands a suspension bridge (Fig. 1) is being designed by the government (Rijkswaterstaat Directie Bruggen). The bridge, crossing partly the Western-Scheldt estuary, is situated about 20 m above sea-level. This type of structures is generally sensitive to fluctuating wind loads. A number of aerodynamic phenomena may cause unwanted vibrations of the main span or pylons. These phenomena have been well described in literature (for example Ref. 1 and 2).

The reason for the present study was twofold:

a. The cross-section of the bridge deck was not aerodynamically shaped so vibrations might pose a stability problem. Grills were installed in the bridge deck to enhance the stability. To the authors knowledge no reliable aerodynamic data existed in literature. The stability of this bridge deck was therefore open to question and had to be investigated.

b. The wind-induced response of a pylon was investigated to establish the need of guying the pylons during an erection stage when they are standing free.

The study consisted of wind tunnel measurements on an oscillating bridge deck model from which the unsteady aerodynamic derivatives were obtained. Then a flutter calculation was performed on the basis of these aerodynamic derivatives together with computed vibration modes of the full-scale structure.

The response of the pylon was investigated on an elastically scaled model. Applying similarity rules, the vibrational behaviour of the concrete full-scale structure was predicted.

This paper deals with the experiments and presents the results.

2. SHORT DESCRIPTION OF THE BRIDGE

The main span of the suspension bridge amounts 918 m (2755 ft). With both side spans the complete bridge has a length of 1450 m (4350 ft). The two pre-stressed concrete pylons have a height of 119 m (357 ft). This bridge may be considered as a large one in Europe. The main span consists of two steel box girders along the two edges of a slotted road deck. The dimensions of the cross-section as proposed during an early design stage are presented in figure 2.

3. VIBRATIONAL CHARACTERISTICS OF THE BRIDGE

As part of the study the natural frequencies and vibration modes of the bridge structure (including pylons and cables) were calculated with a finite-element method. Some results of frequencies and vibration modes are given in figure 3.

A comparison in the table below of natural frequencies with those of existing bridges of more or less similar length shows a good agreement.

BRIDGE	MAIN SPAN (m)	VERTICAL BENDING (Hz) SYMM.	ANTI SYMM.	TORSION (Hz) SYMM.	ANTI SYMM.
Western-Scheldt	918	0.154	0.135	0.299	0.310
Tacoma	855	0.133	0.145	0.166	0.173
Bosporus	1090	0.156	0.117	0.315	0.430
Triborough	701	0.187	0.201	0.274	0.362

4. THE UNSTEADY AERODYNAMIC FORCES ON A BRIDGE DECK

When the oncoming wind passes the bridge deck, the vibration modes generate aerodynamic forces which may alter the stability of the bridge structure. Because the bridge deck may be considered as a very slender structure, the aerodynamic loading is thought to be distributed over a number of crosswise strips which do not interfere with each other. The oscillatory motion of each strip can always be approximated by superimposing a heaving and a pitching motion. Therefore, to measure the aerodynamic forces by a wind tunnel model it is sufficient to consider only one such a strip, vibrating in heave or in pitch. Chordwise deformations of the bridge deck are neglected.

The unsteady aerodynamic forces are formulated in the present wind tunnel test using definitions usual in analyzing aircraft flutter stability. The definitions correspond to the AGARD notation (Ref. 3). To facilitate comparisons with existing data in literature for different bridge deck sections, also the definitions used in reference 1 are given.

The unsteady lift per unit length can be written as (see also Fig. 4):

$$\text{AGARD: } L = \pi \rho V^2 \ell [(k_A' + ik_A'') A + (k_B' + ik_B'') B]$$

$$\text{Ref.1: } L_h = \tfrac{1}{2}\rho V^2 (2C) (KH_1^* \frac{\dot{h}}{V} + KH_2^* \frac{C\dot{\alpha}}{V} + K^2 H_3^* \alpha),$$

and the moment per unit length (about midchord point) as:

$$\text{AGARD: } M = \pi \rho V^2 \ell^2 [(m_A' + im_A'') A + (m_B' + im_B'') B]$$

$$\text{Ref.1: } M_\alpha = \tfrac{1}{2}\rho V^2 (2C)^2 [KA_1^* \frac{\dot{h}}{V} + KA_2^* \frac{C\dot{\alpha}}{V} + K^2 A_3^* \alpha].$$

Both sets of expressions can be related to each other.
Putting $k_A' = 0$ and $m_A' = 0$:

$$k_A'' = - 4/\pi \; k^2 \; H_1^*$$

$$k_B' = - 8/\pi \; k^2 \; H_3^*$$

$$k_B'' = - 8/\pi \; k^2 \; H_2^*$$

and

$$m_A'' = - 8/\pi \; k^2 \; A_1^*$$

$$m_B' = - 16/\pi \; k^2 \; A_3^*$$

$$m_B'' = - 16/\pi \; k^2 \; A_2^*.$$

where k denotes the reduced frequency $k = \frac{\omega \ell}{V}$ (AGARD).

5. THE BRIDGE DECK MODEL AND TEST SET-UP

A wooden sectional model of the bridge deck of 1100 mm length was built at scale 1:60. A vibration test which preceeded the wind tunnel test did not reveal significant deformations at the oscillation frequencies used during the wind tunnel measurements. So, the model could be considered as sufficiently rigid.

The model oscillated in the wind tunnel between two end plates (Fig. 5). The distance to the ground floor of the test section represented the height above sea-level.

The model was supported by three electrodynamic shakers (Fig. 6). The amplitude and phase of each shaker could be adjusted in such a way that a pure heaving or pitching motion (about midchord point) was obtained. These motions could be measured by accelerometers. Between model and each shaker an one-component strain gauge balance was installed, which measured the local unsteady lift force.

The bridge deck model was investigated in three different configurations (Fig. 7):

. Configuration 1: Bridge deck closed
. Configuration 2: Bridge deck with grills
. Configuration 3: Bridge deck with grills and side walks.

6. MEASURING PROCEDURE AND EQUIPMENT

The measuring procedure to obtain the unsteady derivatives (k_A, k_B, m_A and m_B) can be explained by considering the equations of motion for the model (Fig. 8). Besides the unsteady aerodynamic loads also forces due to vibrations of the model mass, and mass moment of inertia are acting in terms of $m\ddot{z}$ and $I\ddot{\theta}$. During the wind tunnel tests the inertia forces were compensated electrically by a procedure which is described here briefly (also Ref. 4).

First the model was vibrated in still air in one of the two modes at a certain frequency and amplitude. The balance forces which were measured then, resulted each from inertia forces generated by the oscillating model and for a small part by the vibrating mass of surrounding air. Next, an analog summation of the output of the three balances was made of which the summed signal was compensated electrically by one of the accelerometer signals, after adjusting proper phase and amplification of this signal. Then the air flow was started and care was taken that the previous vibration level was adjusted again. The sum of the balance signals, now being measured, was directly proportional to the local unsteady aerodynamic load. The measuring equipment used for this procedure is (schematically) presented in figure 9.

The determination of the magnitude of the unsteady aerodynamic load and its phase lag with respect to the model motion was performed with a Fourier analyzer. The cross-spectrum between the summed balance signals and the model acceleration was determined and the result, a complex vector, presented in a Nyquist diagram (Fig. 10). This vector was multiplied with the oscillation frequency squared. Finally, using the formulas in figure 8 (without inertia terms) the result was scaled to the required derivatives, both real and imaginary.

7. PRESENTATION OF RESULTS

7.1 Derivatives of the flat plate model

In figure 11 the measured unsteady aerodynamic derivatives for a streamlined flat plate model are presented in the AGARD notation, as a function of the reduced frequency k up to values of about k = 1.5.

Comparison with theoretical data for an infinitely thin aerofoil shows an satisfactory agreement, which justifies the measuring procedure as already described.

7.2 Derivatives of the proposed bridge deck

Results of the proposed bridge deck cross-section (Conf. 2) are shown in figure 12. The results of the configurations 1 and 3 will be briefly discussed in the next section.

The unsteady derivatives of conf. 2 are presented in a reduced frequency range of k = 0.4 to k = 1.5, corresponding to actual wind speeds of about 50 m/s down to 15 m/s, related to a torsional frequency of the bridge of f = 0.30 Hz and a bending frequency of f = 0.15 Hz respectively. At higher values of k (< 15 m/s) the presence of vortex shedding was dominating, an unwanted effect which should not interfere with measuring the derivatives. A Strouhal number of S = 0.14 (corresponding to k = 2.5) was established.

Comparison with theoretical data for the infinitely thin aerofoil learns that significant differences occur especially at the derivatives of the pitching motion. For example, both coefficients m_B' and m_B'' have opposite signs. The latter, representing the torsional aerodynamic damping, points to unstable flutter characteristics of the bridge deck. This characteristic of m_B is most probably due to flow separation from the sharp corners of the main girders.

A further observation shows that the results as measured in heaving (k_A) and pitching (k_B) for different amplitudes and frequencies (Fig. 12) behave in a linear way.

The derivatives were only measured in a smooth oncoming flow because reference 5 shows that the disparity between results obtained under turbulent and smooth flow conditions for $k > 0.3$ is of the order of 15 % or less. Therefore no significant differences were expected in the predicted flutter characteristics by applying turbulent flow conditions.

7.3   Correlation of measured data with literature

In reference 1 (page 301 and 302) unsteady derivatives are presented for other bridge deck cross-sections. Using the conversion rules given in section 4, a correlation was made. The results for the pitching motion, being the most important one, are presented in figure 13.

Of special interest is the derivative A_2^*, representing the aerodynamic damping in the torsional mode of the bridge deck. The results of conf. 1 as well as conf. 2 have a positive sign, which points to a flutter instability of the bridge deck. They are pretty much the same as those of the original Tacoma Narrouws bridge, being another indication that the present bridge deck is flutter sensitive. As the differences between the results of conf. 1 and conf. 2 are small, it is obvious that installing grills in the bridge deck like proposed has a negligible stabilizing effect. Probably, the dimensions of the grills had to be larger to improve the stability. Only from adding side walks (Conf. 3) a stabilizing effect may be expected at higher reduced velocities.

7.4   Flutter predictions

To illustrate the aerodynamic instability, flutter calculations were performed on the basis of the measured derivatives and computed vibration modes. The results (Fig. 14), showing the aerodynamic damping and frequency of each vibration mode as functions of the wind speed learn indeed that flutter instability occurs over the entire speed range. The type of flutter is a typical single degree-of-freedom instability of the bridge deck torsional motion, the bending modes are all stable. From literature it is known that this phenomenon often occurs at H-shaped cross-sections. The behaviour of the aerodynamic loads around this particular cross-section seems to be similar. The results also show that this bridge should possess unrealistic high structural damping to prevent aerodynamic instability.

8.   THE WIND-INDUCED RESPONSE OF THE PYLON

Another part of the study concerned the wind-induced response of the pylon standing free during a part of the erection stage of the bridge. The wind tunnel test was performed on an elastic model of which the vibration modes in bending and torsion agreed with computed modes of the full-scale structure. By measuring the modal properties of the pylon at wind-off (mode shapes, generalized masses and structural damping values) and the model responses during vortex excitation, a prediction could be made of the wind-induced response of the full-scale structure. The procedure has been described in reference 1, page 298.

174

The pylon model and test set-up

        The model of the pylon (Fig. 15) was made of steel and built at a scale
1:60. With a number of accelerometers mounted inside, the required vibrational
characteristics were obtained. The model was clamped on a rigid support of the turntable in the test section, so that the wind direction could be varied.

        At wind-on, the response was measured by the two accelerometers at the top
of the model, of which the signals were reduced by a Fourier analyzer into root mean
square (RMS) bending and torsional amplitudes. The measurements were performed in a
smooth oncoming flow.

        A matter of some concern were the rounded edges of the both pylon columns
which could induce an unwanted sensitiviness of the vortex excitation mechanism to
Reynolds number at the higher wind speeds. This problem could arise when the Reynolds
number would exceed the subcritical range (Re < 3.5×10^5, based on an average
distance between the columns). The nature of the wake within that range agrees roughly
with the wake in the supercritical range (Re > 3.5×10^6), occurring at the full-scale
structure. Thus, it was necessary to keep the occurrance of vortex excitation of the
model within the subcritical Reynolds number range. This could be achieved by adding
masses inside the model in order to make the resonance frequencies of the bending and
the torsion modes low enough. The vibration modes were hardly affected.

8.2   Some experimental results

        At various wind directions (β) the model response was measured. At $\beta = 0$
(Fig. 16),when the wind was parallel to the plane of the pylon, the response appeared
to be the largest. At a wind speed of V = 9.5 m/s a definite bending response was
observed and at V $\approx$ 32 m/s a torsional response. The corresponding Strouhal numbers S
were 0.142 and 0.126, respectively. These values agree well with data of similar
structures, e.g. the Severn bridge and the Forth Road bridge, of which it is known
that vibrations due to vortex-shedding occur at Strouhal numbers of S = 0.11 to 0.15.
(Ref. 6).

        A demonstration of vortex-induced responses of the pylon, when the correspond-
ing Strouhal numbers agree to one of the above-mentioned values, is given in
figure 17. Power spectra of one of the accelerometer signals are shown at increasing
wind speed. When the frequency of the wake vortices coincides with the bending
frequency (V = 9.5 m/s) or the torsion frequency (V = 32 m/s) a nearly harmonic
response occurs.

        The responses of the full-scale concrete pylon were predicted after proper
scaling the measured model response. It was necessary to assume a structural damping
of the full-scale pylon. According to reference 6, values of b (= c/c_r) = 0.015 to
0.03 were chosen. The corresponding full-scale response appeared to be negligible, as
is shown for both wind speeds in figure 18. Thus, guying the pylons during the
construction of the bridge is not necessary.

        The smallness of the pylon response looked somewhat surprising. For that
reason the pylon wake was investigated separately by flow visualization, using a
smoke trail that was ejected into the flow. Further, hot-wire measurements were made.

        A picture of the pylon wake is shown in figure 19 for a wind speed at which
vortex shedding occurred. Distinct vortices are shown between the pylon columns, but
not behind the downstream column. This result was confirmed by the results of the
hot-wire measurements (Fig. 20). A distinct peak in the spectrum was only detected
between the columns. The absence of clear vortices in the wake behind the downstream
column may explain the moderate pylon response.

9.   CONCLUSIONS

The wind tunnel study of the proposed Western-Scheldt suspension bridge reveals:

a. The particular shape of the cross section is aerodynamically unstable in spite of the presence of grills in the roadway surface.
b. The measured unsteady derivatives due to torsion, are similar to those of the original Tacoma Narrows bridge, or more generally to bridge sections of the H-shaped type.
c. Flutter calculations with measured unsteady derivatives show a predominantly single-degree-of-freedom instability in the fundamental torsion mode of the bridge deck, this instability is probably due to flow separation.
d. The wind-induced response of the free-standing pylon is small and does not necessitate additional guying.

10.   ACKNOWLEDGEMENT

The authors are indebted to "N.V. Westerschelde-Oeververbinding" and "Rijkswaterstaat, Directie Bruggen" to present the results of this part of the wind tunnel study of the Western-Scheldt suspension bridge. They also acknowledge the contribution of Mr. H.H. Ottens, structures department at NLR, of calculating the vibration modes of the bridge, and the support of Mr. A. Steiginga, Department of Aëroelasticity of NLR, in performing the flutter calculations.

11.   REFERENCES

1.   Simiu, E and Scanlan, R.H.: "Wind effects on structures; an introduction to wind engineering", A Wiley-Interscience publication, 1978.

2.   Scanlan, R.H. and Wardlaw, R.L: "Aerodynamic stability of bridge decks and structural members". Proceedings of the symposium on cabled stayed bridges, Washington D.C., 1978.

3.   AGARD: "Manual on aeroelasticity", Volume VI, 1968.

4.   v. Nunen, J.W.G.; Persoon, A.J. and Tijdeman, H: "Windtunnel test to determine the instationary aerodynamic derivatives on a model of a twin-bridge. Deutsche Gesellschaft für Luft und Raumfahrt; Mitteilung 72-06, Köln, Germany (1972).

5.   Scanlan, R.H. and Wen-Huang Lin: "Effects of turbulence on bridge flutter derivatives". Journal of the engineering mechanics division, august 1978.

6.   Sachs, P: "Windforces in engineering"; Pergamon Press, Oxford 2nd edition, 1978.

Fig. 1 Artists impression of the bridge

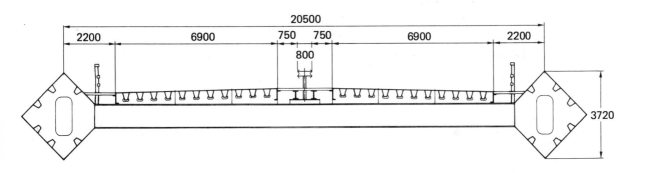

DIMENSIONS IN mm

Fig. 2 Dimensions of the cross-section

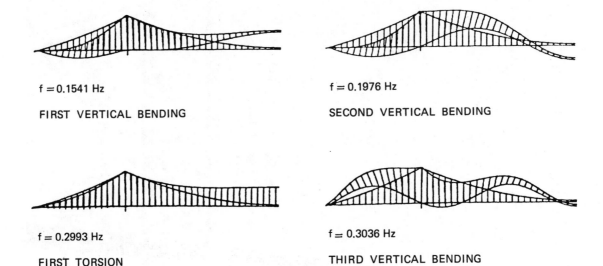

f = 0.1541 Hz

FIRST VERTICAL BENDING

f = 0.1976 Hz

SECOND VERTICAL BENDING

f = 0.2993 Hz

FIRST TORSION

f = 0.3036 Hz

THIRD VERTICAL BENDING

Fig. 3   Calculated vibration modes and natural frequencies

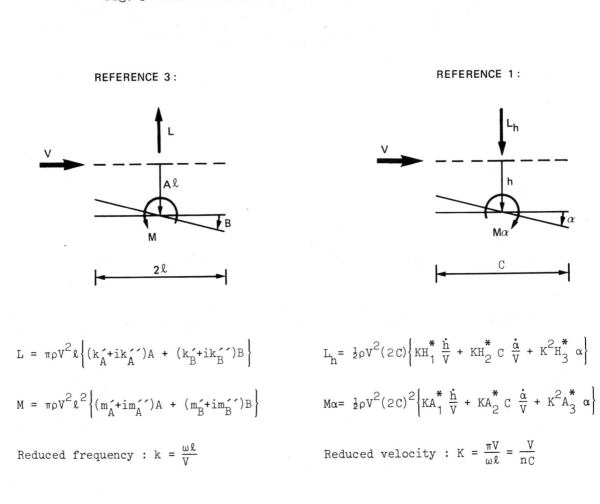

REFERENCE 3:

REFERENCE 1:

$$L = \pi\rho V^2\ell\left\{(k_A' + ik_A'')A + (k_B' + ik_B'')B\right\}$$

$$L_h = \tfrac{1}{2}\rho V^2(2C)\left\{KH_1^*\frac{\dot{h}}{V} + KH_2^* C\frac{\dot{\alpha}}{V} + K^2H_3^*\alpha\right\}$$

$$M = \pi\rho V^2\ell^2\left\{(m_A' + im_A'')A + (m_B' + im_B'')B\right\}$$

$$M\alpha = \tfrac{1}{2}\rho V^2(2C)^2\left\{KA_1^*\frac{\dot{h}}{V} + KA_2^* C\frac{\dot{\alpha}}{V} + K^2A_3^*\alpha\right\}$$

Reduced frequency : $k = \dfrac{\omega\ell}{V}$

Reduced velocity : $K = \dfrac{\pi V}{\omega\ell} = \dfrac{V}{nC}$

Fig. 4   The unsteady forces on a bridge deck with sign convention

Fig. 5   The model mounted in the windtunnel

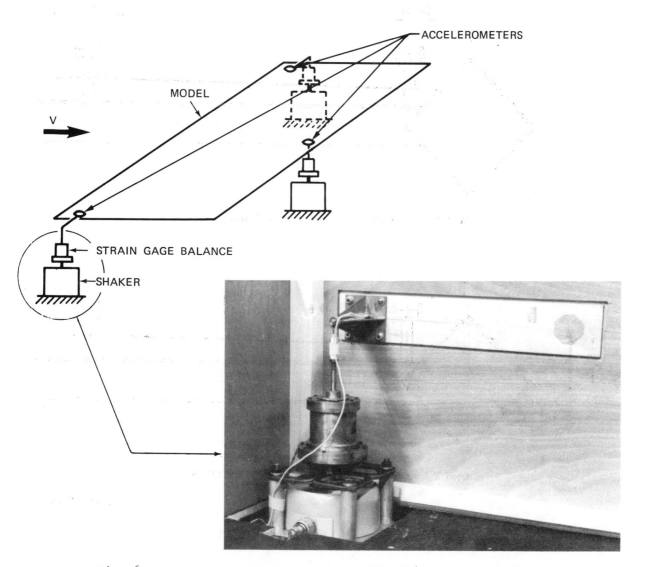

Fig. 6   The model supported by the shakers and strain gage balances

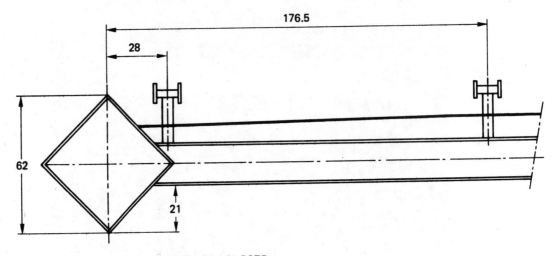

Conf. 1 BRIDGE DECK CLOSED

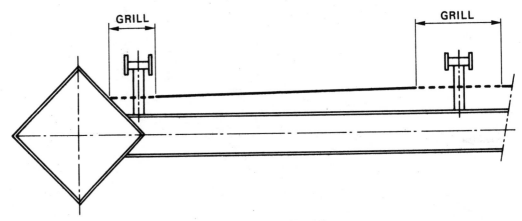

Conf. 2 BRIDGE DECK PROVIDED WITH GRILLS

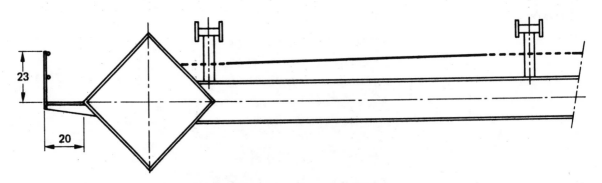

Conf. 3 BRIDGE DECK WITH GRILLS AND SIDE WALKS

DIMENSIONS IN mm

Fig. 7 The present model configurations

HEAVING

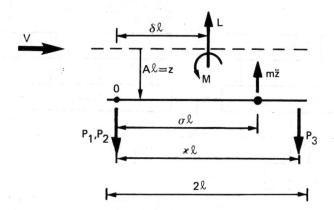

1) Balance of forces :

$$\Sigma P_i = m\ddot{z} + \pi\rho V^2 \, b \, k_A \, A\ell$$

2) Balance of moments about 0 :

$$P_3 x\ell = m\ddot{z} \, \sigma\ell + \pi\rho V^2 \, b \, k_A \, A\ell \, \delta\ell + \pi\rho V^2 \, b\ell \, m_A \, A\ell$$

PITCHING

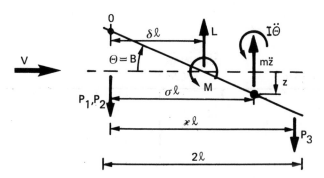

3) Balance of forces :

$$\Sigma P_i = m\ddot{z} + \pi\rho V^2 \, b\ell \, k_B \, B$$

4) Balance of moments about 0 :

$$P_3 x\ell = I\ddot{\Theta} + m\ddot{z} \, \sigma\ell + \pi\rho V^2 \, b\ell \, k_B \, B \, \delta\ell + \pi\rho V^2 \, b\ell^2 \, m_B \, B$$

Fig. 8 Equations of motion of the model with sign convention

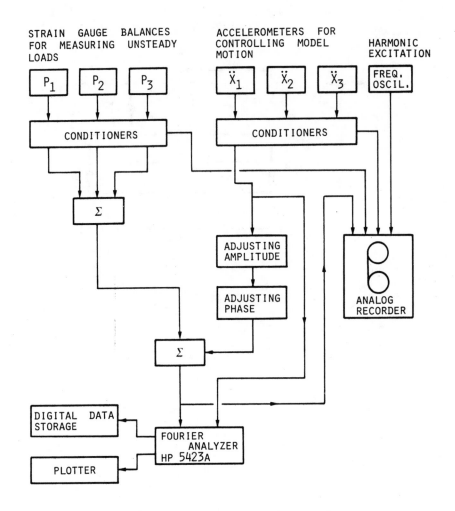

Fig. 9   Diagram of the measuring equipment

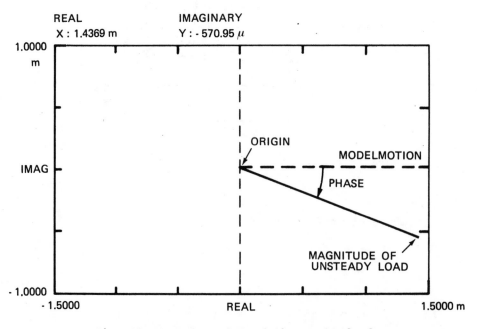

Fig. 10   A vector plot of the unsteady force

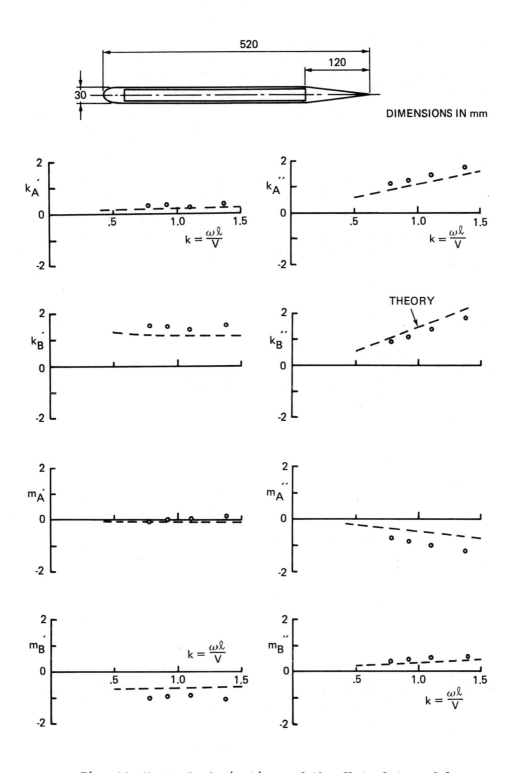

Fig. 11 Unsteady derivatives of the flat plate model

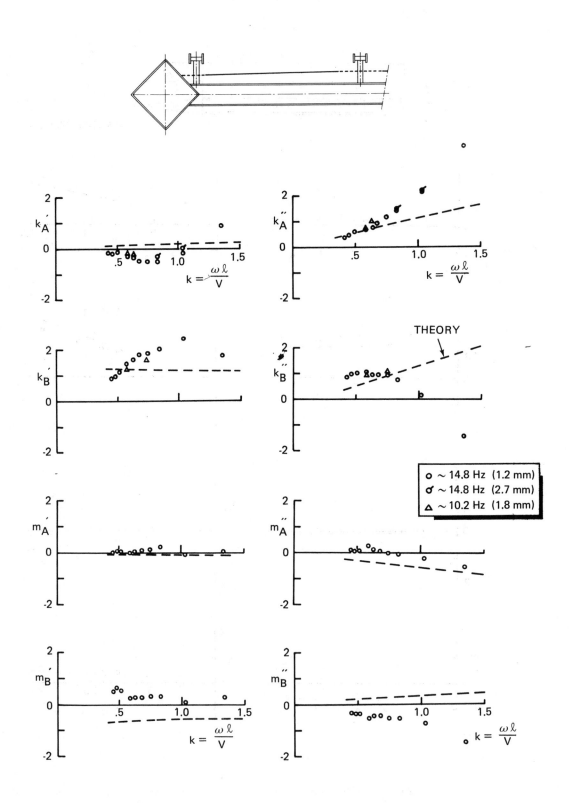

Fig. 12 Unsteady derivatives of the proposed cross-section with grills (conf. 2)

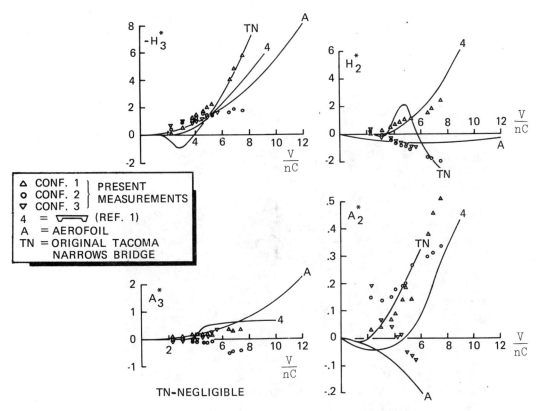

Fig. 13 Comparison of present data with literature

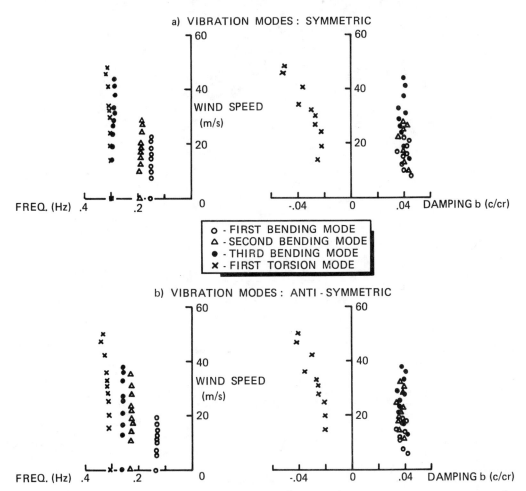

Fig. 14 Flutter diagram of configuration 2 (bridge deck with grills)

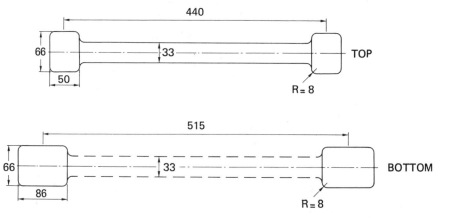

DIMENSIONS IN mm

Fig. 15 Model of the pylon in the windtunnel

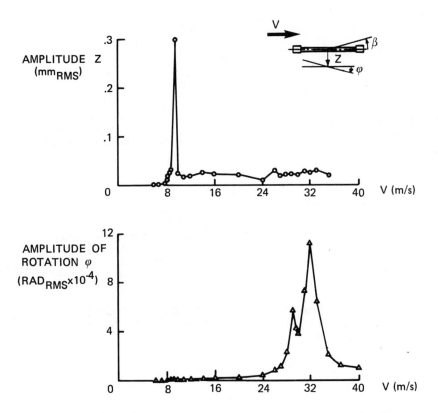

Fig. 16　The response of the pylon at $\beta = 0°$

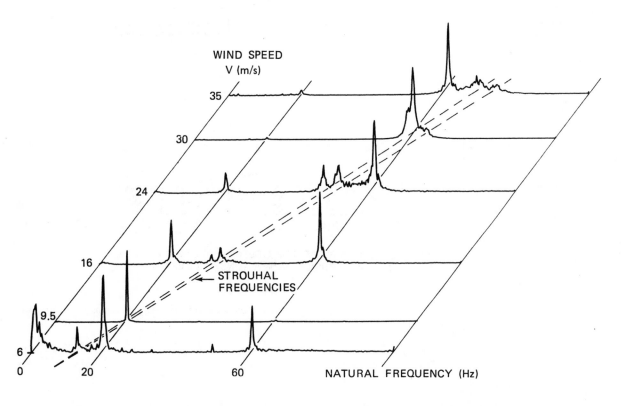

Fig. 17　The Strouhal frequency as a function of the windspeed

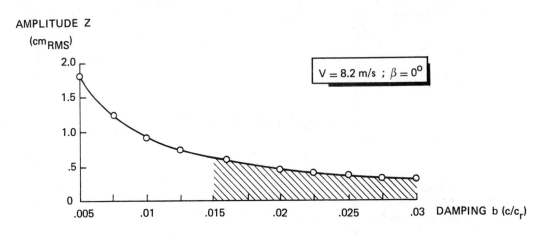

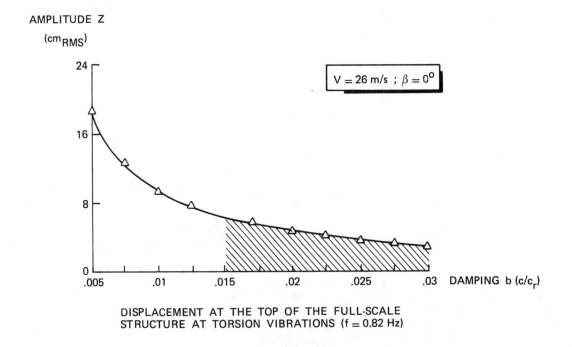

Fig. 18   The response of the full-scale structure as predicted from measured data

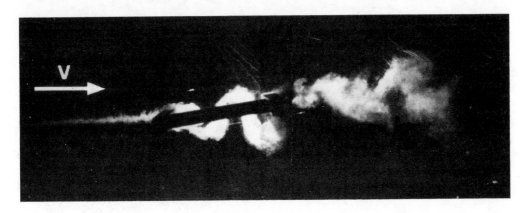

Fig. 19 Photographs of the vortex eddies around the model of the pylon

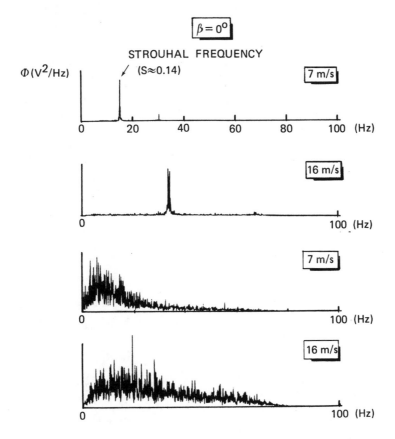

Fig. 20 Power spectral density plots measured between the
columns of the pylon and in the wake behind the pylon

WIND TUNNEL TESTS ON A TWIN-DECK BRIDGE MODEL

D.J. Johns and F.G. Maccabee

Loughborough University of Technology, U.K.

Summary

Wind tunnel tests have been conducted on a sectional model (1/40 scale) of a twin-deck arrangement of a proposed road bridge over the River Tay at Friarton, Scotland. The tests covered the speed range corresponding to a full scale 13 - 73 m/s (28 - 140 m.p.h.).

In the twin-deck arrangement the variables studied included individual deck flexural frequencies; added mass; simulated "snow"; relative vertical stagger; angle of incidence; structural damping. The basic case studied corresponded to the full scale structure in terms of geometry and predicted value of non-dimensional mass-damping parameter.

Two critical speeds were found in almost all the cases studied corresponding to non-dimensional reduced velocities of approximately 10 and 30. The corresponding non-dimensional maximum vibration amplitudes were 25×10^{-3} and 50×10^{-3} respectively of the overall deck depth.

In general the front deck had a greater response than the aft deck, particularly at high wind speeds.

For the single deck (erection phase) condition there was only one pronounced resonance – at a reduced velocity of about 8.

Organised and sponsored by
BHRA Fluid Engineering, Cranfield, Bedford MK43 0AJ, England

NOMENCLATURE

D   Depth of model from deck upper surface to bottom of box girder (metres). (Suffix m for model; p for full scale).

k_s Non-dimensional mass-damping parameter $= 2m\delta / \rho D^2$

m   Mass per unit spanwise length (kg/metre).

n   Natural frequency in still air in vertical flexure (Hz) ($= \omega_F$ or ω_A at zero V).

V   Wind velocity (metres/sec.).

$\overline{V}$ Reduced velocity $= V/nD$

z   Vertical relative separation (+ve forward deck higher).

α Angle of approach wind to bridge horizontal datum.

δ Logarithmic decrement.

η Non-dimensional vertical oscillatory amplitudes (peak to peak) with respect to model depth, D.

ρ Air density (Kg/metre^3).

ω_F, ω_A Measured natural frequencies of forward and aft decks (Hz).

1. INTRODUCTION

The Friarton Bridge crosses the River Tay, approximately 2 km east of Perth, where it flows through a wide valley. The south side of the valley slopes moderately steeply but the north side rises almost vertically in an impressive 200m escarpment. Thus a high level bridge design was called for which resulted in an overall structure length of 831 m and a main span over the river of 174 m. Overall there are nine spans and these are likely to experience significant winds from the prevailing westerly (or eastern) directions. Fig. 1 shows the geographical location.

The bridge design was undertaken by Freeman, Fox and Partners in accordance with the Merrison Box Girder Design Rules and it was decided that the deck would be of composite, lightweight, concrete construction over the whole length of the structure. Another fundamental decision was to divide the deck along its centre-line so that the structure became essentially two separate bridge decks, side by side. This change considerably simplified problems concerning inequality of web and bearing loading which are inherent in wide, cross-connected box girders carrying in-situ poured deck concrete. Fig. 2 shows typical cross-sections.

Since existing published data was inadequate for a split design of Friarton proportions it was considered essential that the aerodynamic characteristics of the proposed design should be determined by a wind tunnel study. These tests were conducted at Loughborough University of Technology in 1973 and covered both static and dynamic characteristics. Only the latter will be considered in depth here, though flow visualisation and drag studies were conducted and gave some insight into the dynamic behaviour.

Because of the considerable difference in natural frequency in vertical flexure (0.453 H_z) and torsion (6.2 H_z) it was decided that only the former needed to be modelled. No attempt was to be made to model the turbulence of the natural wind.

2. OUTLINE OF TEST PROGRAMME

The major emphasis was to be on the twin-deck arrangement and the suscept-ability to wind excited vibrations was to be explored over a range of wind tunnel speeds corresponding to full scales speeds of 0 - 50 m/sec, and with small variations in the absolute (and relative) frequencies in vertical flexure of the two individual deck structures.

Other effects to be considered included relative deck vertical stagger; incidence changes of the configuration; simulation of snow build-up and variations in damping.

In addition the single deck condition - corresponding to the erection stage - was to be studied but without any frequency variations or other parameter changes.

Table 1 lists the final agreed programme of tests.

3. DESIGN OF TEST MODEL

The test programme was required to be performed extremely quickly and the design of model, mountings, springs and dampers, and choice of isntrumentation, reflected the need for haste.

In practice the maximum size of model which could be fitted into the wind tunnel was 1/40 scale, and a sectional model was made, representative of the mid-section of the river crossing span (between piers 6 and 7 of Fig. 2), as shown in Fig. 3. The model was of constant section and constructed of wood and aluminium alloy with a span of 1.07 m (42 in.).

The dynamic characteristics of the (rigid) model were scaled to be close to the anticipated full scale behaviour in vertical flexure only, by the use of a spring-damper combination of a very simple type. The model decks were each supported by rigid, horizontal arms to which the spring attachments were connected. The springs formed part of a separate superstructure attached to the roof of the wind-tunnel room and comprised cantilevered flexible flat plates of spring steel whose bending stiffness could be varied by adjusting their lengths. Oil-hydraulic dampers of long stroke, small bore were found to give reasonable results. Figs. 4 and 5 are photographs showing the overall, and damper, layout.

The full-scale structure had a quoted mean weight of 85 kN/m for a complete single deck. The corersponding mean weight of a model single deck would, therefore, be

$$85 \times (D_m/D_p)^2 \; \triangleq \; 0.055 \text{ kN/m}$$

The required overall model weight for a span of 1.07 m was 59 N but, in fact, it was 67 N, i.e. 13.6% larger. It was hoped, therefore, that variations in δ, the logarithmic decrement, would enable correct scaling to be achieved in the non-dimensional mass-damping parameter $k_s = 2m\delta / \varrho D^2$, despite the fact that inertial scaling was not achieved.

From the definition of $\overline{V}$ (= V/nD), the reduced velocity, the actual velocity and frequency scales for sectional models are interdependent and, therefore, interchangeable. With a wind tunnel maximum speed of approximately 35 m/sec and a model scale of 1 : 40 a basic natural frequency for vertical flexure of approximately 8H$_z$ was required for the model.

4. THE WIND TUNNEL

The Loughborugh University of Technology Wind Tunnel used is of open jet, closed return type, and, to achieve two-dimensionality of flow end plates were fitted to the tunnel contraction so as effectively to "close" the tunnel sides. To allow for mvoement of the model decks under wind action it was necessary to have narrow gaps between the end plates and model sections. These did not (despite the absence of any labyrinth or other seal) affect the flow two-dimensionality as leakage through the gaps was minimal and the possible interference to motion was thereby avoided. Drag wires werefitted to each deck and these had to be adjusted at each wind speed to ensure that the horizontal gap between each deck was maintained at its correct value.

5. INSTRUMENTATION

The information to be obtained included:

(i)   the frequencies of any wind excited oscillations in flexure or torsion
(ii) the phase relationships of different parts of the structure when vibrating
(iii) the amplitude of these oscillations
(iv) the structural damping (logarithmic decrement) for different conditions.

Low mass, piezoelectric accelerometers were used being positioned at the ends of the model out of the air stream. These were connected as required to four channels of source-following high impedance amplifier, with switched gain settings of X1, X10, X100 and X1000. The outputs of the four channels were taken through low-pass filters to remove any unwanted high frequency signals. Two types of filter were used: 12 Hz low-pass for measurements taken from transducers normally on a centre line of each bridge section (so as to correctly identify flexure motions without other spurious signals causing confusion), and 35 Hz low-pass for the transducers normally positioned at the rear of each bridge section used to detect the higher (torsion) frequencies.

The four channels of data were displayed continuously on a four beam storage oscilloscope. A permanent record was also taken on ultra-violet film, these records being used subsequently for the detailed analysis.

The accelerometers were calibrated at 1g with the transuucer amplifiers set at a gain of X1. A fine external gain control was then used to set a constant deflection on both oscilloscope and u.v. recorder for 1g on all channels. For msot of the test, the amplifier gain was set to X10, this therefore representing 0.1g for the calibration levels obtained on X1 setting.

The positions eventually selected for the four accelerometers, based on initial observed responses, were one at each end of the centre-line of the forward deck (to measure flexure only) and two at one end of the aft deck - centre line and rear - (to measure flexure and torsion). For the single deck tests there were two at each end as for the aft deck above.

6. FLOW VISUALISATION

The results obtained from smoke and wool tuft investigations are shown schematically in Fig. 6.

All of these tests, except for high positive incidence ((a) in Fig. 6) showed that the flow coming over the top of the bridge deck re-attached to the surface of the bridge near the middle and near to the gap between decks, or even halfway across the downstream (aft) deck; with some minor secondary flow effects off the crash barriers between the two decks.

The simulated snow ((b) in Fig. 6) was added in a position at the side of the carraigeway where compacting due to the passage of vehicles was considered to be less likely and where it was likely to be built up by snow-ploughs on the carriageway. In the event the snow was effectively within the re-circulation zone behind the front separation point.

7. DYNAMIC TESTS

Initial tests were concerned with proving the rig, checking that frequency variations could be accomplished easily, and learning how to solve such problems as relative streamwise movement of the two decks under wind-on-conditions and the similar effects of relative incidence and vertical separation of the two decks. It was accepted that adjustments in the geometry during a particular series of tests would be required should such effects be important.

Such variations as those in flexure frequency, mass and addition of snow could easily be accomplished. Conversely vertical stagger between the decks and angle of incidence were tedious variables to change and it was not easy to vary in a controlled manner the damping of each deck. However it is believed that sufficient results have been obtained to indicate the significance of the damping.

Figs. 7 - 19 show the results of some of the dynamic tests as plots of amplitude against $\overline{V}$ (= V/nD), with values of the appropriate natural frequencies and logarithmic decrements.

It should be noted that the non-dimensional amplitudes plotted on the figures represent the maximum values occurring in the flexure mode during a given length of recorded measurements. Often these amplitude variations were such that the peak values were quite sparse (usually when the amplitudes were small in any case); in other cases they were intermittent and at the "resonance" points they were usually almost continuous. Thus if r.m.s. values had been plotted instead of maximum values, the resonances at V/nD = 10 and 30 would have appeared far more "peaky"

It was also noted that in almost all of the cases studied the flexure motions of the two decks in the region of the "resonances" were in phase. Thus one suggestion which had been made, to fit special damping devices across the gap between the decks to help limit the amplitudes of vibration would appear to have little merit.

As it was not possible to "tune" the damping as required, in some cases the two decks had different logarithmic decrements. This and slight differences in frequencies are believed to be contributory factors to some of the differences seen in the figures between otherwise similar cases.

At the higher wind speeds in almost every case the torsional motion of the aft deck was excited to very large amplitude.

However, since the torsional modes were deliberately not adequately simulated these effects were neglected.

It was also noted that at the higher wind speed there was a tendency for the flexure mode to be forced at a higher frequency than its still air value. This was especially true of the aft deck motions. Typically the 8 Hz frequency was raised to 10 Hz.

Comparing Figures 7. and 8. it is seen that the addition of mass to the front deck to lower its frequency has lowered its amplitude response in the region of V/nD = 10. This is possibly due to the effective increase in $m\delta/\rho D^2$.

Comparing Figures 7, 9 and 10 it is seen that the addition of simulated snow has considerably increased the dynamic response particularly at the resonances at V/nD

= 10 and 30.

The effect of vertical separation between the two decks (Figures 11 and 12) appears to indicate that larger amplitudes occur on the rear deck at V/nD = 10 when the rear deck is lower than the front deck, and on the front deck at all other values of V/nD.

Figures 13 - 16 show the effects of incidence. Clearly, positive incidence is more critical than negative incidence, though the differences between Figures 13 and 14 when the incidence changes from $+1\frac{1}{4}^{\circ}$ to $+2\frac{1}{4}^{\circ}$ are not great.

Apart from small changes in damping and freqeuncy Figure 17 should be similar to Figure 7. Over much of the V/nD range the differences are small but at V/nD $\simeq$ 28 these are large.

Figures 17 and 18 show the pronounced effect of a large increase in logarithmic decrement.

Figure 19 for the single deck (α = 0) case shows that the principal resonance is at V/nD $\simeq$ 8 and that the growth in response as V/nD $\rightarrow$ 34 is progressive. The single deck case is probably less critical than the twin-deck case (compare Figures 7, 17, 19).

8.   CONCLUSIONS

The test programme has shown that the twin-deck arrangement is probably more critical than the single deck case. Both arrangements show a resonance in the region V/nD = 8 - 10 and the twin-deck shows a more pronounced resonance at V/nD $\simeq$ 30. In general the front deck experiences larger amplitudes of vibration than the aft deck, particularly at high values of V/nD.

The experimental difficulties associated with testing twin-deck sectional models - particularly in maintaining correct relative vertical, horizontal and angular positions for all wind speeds and incidences were considerable and hence the results should perhaps be mainly interpreted as qualitative and in quantitative terms the accuracy of the data is probably no better than $\pm$ 30%.

9.   ACKNOWLEDGEMENTS

The authors wish to acknowledge the assistance of Dr. R. Nataraja in the conduct of the original work on which this paper is based, and of Freeman, Fox and Partners in allowing the work to be published.

TABLE 1

PROGRAMME OF TESTS

CASE	REMARKS	FIG. NO.
	TWIN DECK CONFIGURATION:	
1	Original Configuration	7
2	As Fig. 7 but m_F increased by 16% $\omega_F = 7.2$Hz; $\omega_A = 7.75$Hz	8
3	As Fig. 7. "Snow" added at front of front deck to 8% D	9
4	As Fig. 7. "Snow" added at front of front deck to 18% D	10
5	As Fig. 7. but $\omega_F = \omega_A = 8$Hz. $\delta_F = .02$; $\delta_A = .05$. $z = -8\%$ D	11
6	As Fig. 11 but $\delta_F = \delta_A = .05$. $z = +8\%$ D	12
7	As Fig. 11 but $\delta_F = \delta_F = .03$ $\alpha = -1\frac{1}{4}^O$ $z = 0$	13
8	As Fig. 13. but $\delta_F = \delta_A = .04$. $\alpha = +1\frac{1}{4}^O$	14
9	As Fig. 14. but $\alpha = +2\frac{1}{4}^O$	15
10	As Fig. 15. but $\alpha = -2\frac{1}{4}^O$	16
11	As Fig. 7 but $\omega_F = 8$Hz; $\omega_A = 7.5$Hz. $\delta_F = .02$; $\delta_A = .04$	17
12	As Fig. 7. but $\omega_F = \omega_A = 10.5$Hz $\delta_F = \delta_A = .09$	18
13	SINGLE DECK CONFIGURATRION: $\omega = 8$Hz $\delta = .03 \longrightarrow .01$ as amplitude decreases. $\alpha = 0$	19

N.B.   In Cases 1 - 12 both decks of equal mass except case 2.

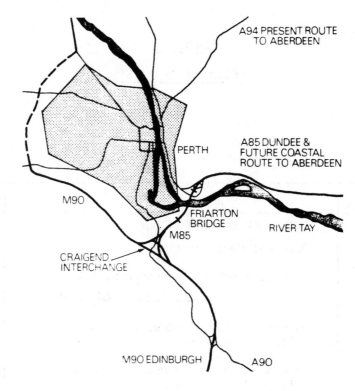

Fig. 1   GEOGRAPHICAL LOCATION

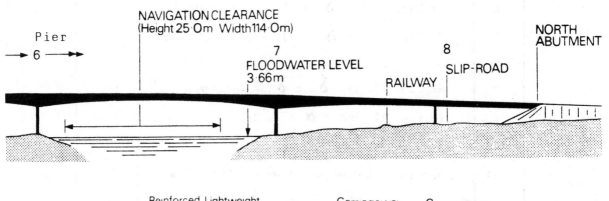

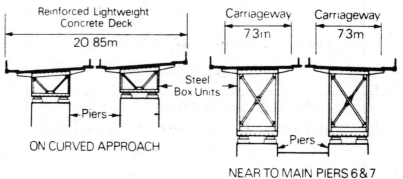

TYPICAL CROSS SECTIONS

Fig. 2   TYPICAL CROSS SECTIONS

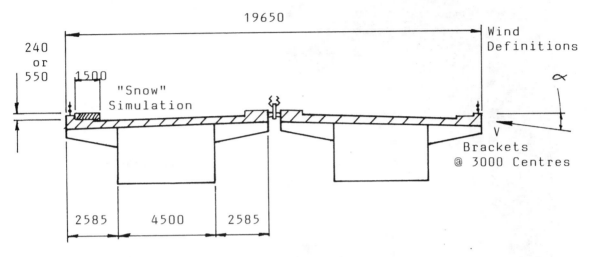

All Dimensions in mm

FIG. 3   FULL-SCALE BRIDGE CROSS-SECTION AT MID-SPAN

Fig. 4   OVERALL LAYOUT

Fig. 5   DAMPER LAYOUT

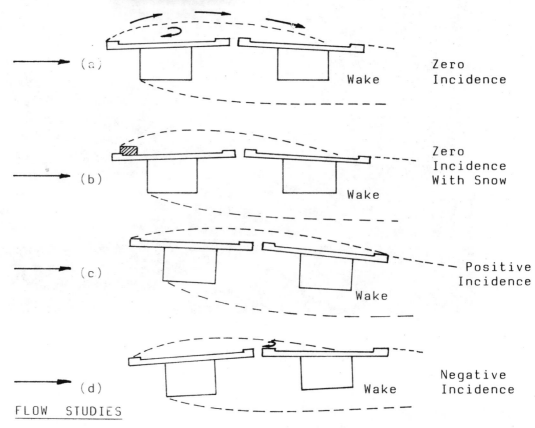

(a)   Zero Incidence
Wake

(b)   Zero Incidence With Snow
Wake

(c)   Positive Incidence
Wake

(d)   Negative Incidence
Wake

Fig. 6   FLOW STUDIES

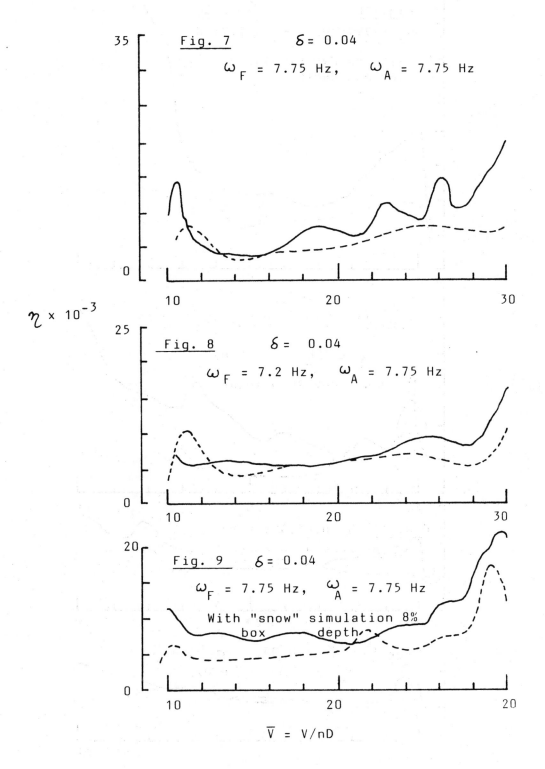

$\eta \times 10^{-3}$

Fig. 7 $\delta = 0.04$

$\omega_F = 7.75$ Hz, $\omega_A = 7.75$ Hz

Fig. 8 $\delta = 0.04$

$\omega_F = 7.2$ Hz, $\omega_A = 7.75$ Hz

Fig. 9 $\delta = 0.04$

$\omega_F = 7.75$ Hz, $\omega_A = 7.75$ Hz

With "snow" simulation 8%
box   depth

$\overline{V} = V/nD$

THE FIGURES ON THIS AND THE FOLLOWING TWO PAGES
SHOW THE DYNAMIC BEHAVIOUR FOR VARIOUS CASES

_____ Front Deck

--------- Aft Deck

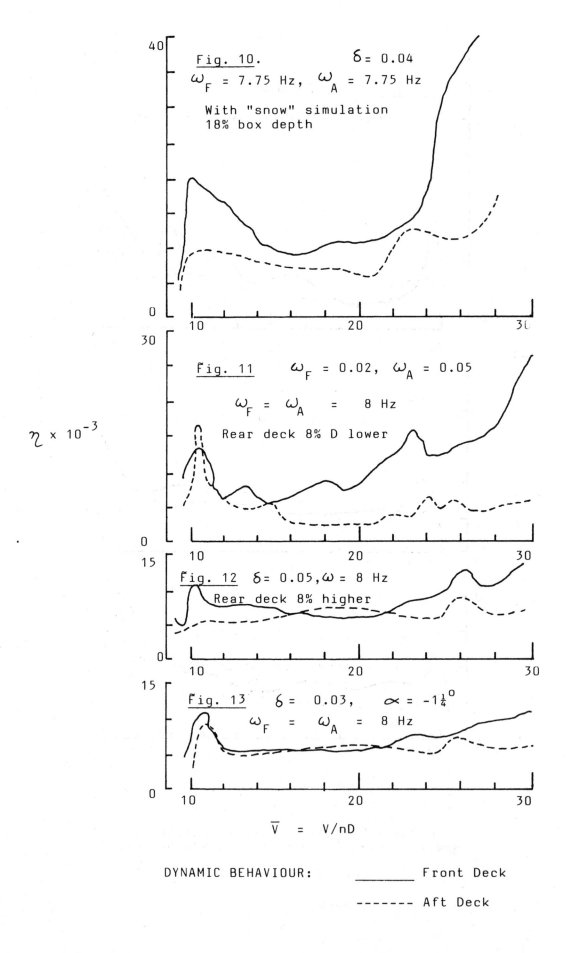

Fig. 10. $\delta = 0.04$

$\omega_F = 7.75$ Hz, $\omega_A = 7.75$ Hz

With "snow" simulation
18% box depth

$\eta \times 10^{-3}$

Fig. 11 $\omega_F = 0.02$, $\omega_A = 0.05$

$\omega_F = \omega_A = 8$ Hz

Rear deck 8% D lower

Fig. 12 $\delta = 0.05$, $\omega = 8$ Hz
Rear deck 8% higher

Fig. 13 $\delta = 0.03$, $\alpha = -1\frac{1}{4}^0$
$\omega_F = \omega_A = 8$ Hz

$\overline{V} = V/nD$

DYNAMIC BEHAVIOUR: ———— Front Deck

                      ------- Aft Deck

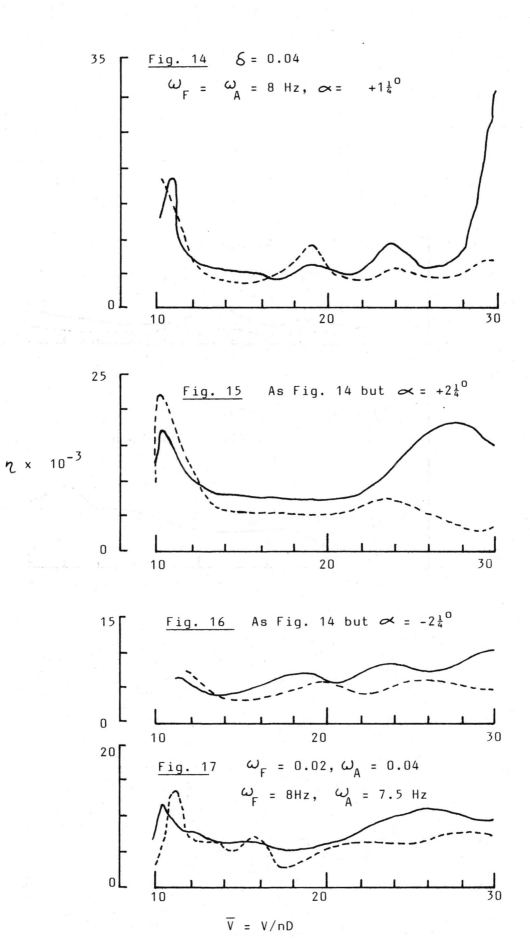

Fig. 14 $\delta = 0.04$
$\omega_F = \omega_A = 8$ Hz, $\alpha = +1\frac{1}{4}^{o}$

$\eta \times 10^{-3}$

Fig. 15 As Fig. 14 but $\alpha = +2\frac{1}{4}^{o}$

Fig. 16 As Fig. 14 but $\alpha = -2\frac{1}{4}^{o}$

Fig. 17 $\omega_F = 0.02$, $\omega_A = 0.04$
$\omega_F = 8$Hz, $\omega_A = 7.5$ Hz

$\overline{V} = V/nD$

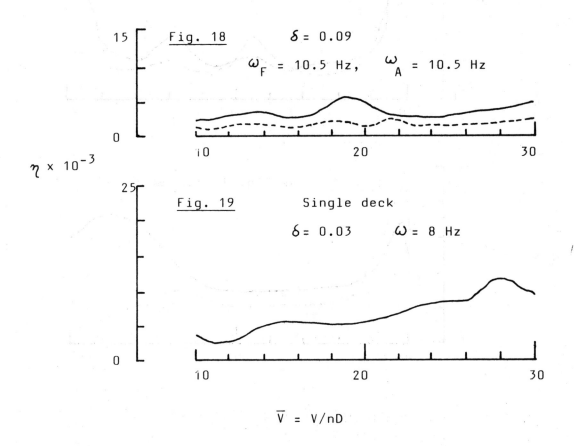

$\eta \times 10^{-3}$

Fig. 18 $\delta = 0.09$

$\omega_F = 10.5$ Hz, $\omega_A = 10.5$ Hz

Fig. 19   Single deck

$\delta = 0.03$ $\omega = 8$ Hz

$\overline{V} = V/nD$

FLOW INDUCED VIBRATIONS
IN FLUID ENGINEERING

Reading, England: September 14-16, 1982

RESPONSE OF A FREE-STANDING BRIDGE PIER
WITH BALANCED CANTILEVER SUPERSTRUCTURE;
MODEL TESTS IN WIND TUNNEL

Erik Hjorth-Hansen

The Norwegian Institute of Technology, Norway

Svein Halvorsen

Harboe og Leganger (Consulting Engineers), Norway

Summary

During construction the T-shaped units of a cantilevered bridge will be standing for
some time without support from (or continuity with) adjacent units which may not be finished
or connected. Due to the simple structural concept this period is likely to be a risky one
if high winds should occur, and temporary bracings and props may be needed.

A model shaped after a Norwegian bridge has been tested in a wind tunnel for bending
and torsional moments in the pier. These are shown in diagrams as function of airspeed for
three levels of turbulence and for three angles of azimuth. The response is shown in the form
of mean values and as dynamic peak values superimposed on the mean. In one test condition
a terrain model is included.

Organised and sponsored by
BHRA Fluid Engineering, Cranfield, Bedford MK43 0AJ, England

NOMENCLATURE

D = representative building dimension

f = natural frequency of building oscillation in its fundamental translational modes

Sn = mode generalized power spectral density function

V = wind velocity

ζ = modal damping

σ^2 = mean square responsive force

Subscripts

f = full scale values

m = model values

1. INTRODUCTION

The concrete bridges motivating this model test are built with a pre-stressed rectangular box girder (of variable depth), somewhat narrower than the overlying deck plate, and with reinforced piers of rectangular shape. These structural forms are known to generate forces due to the wake and to the turbulence in the incipient airflow. The tests were carried out to see whether periodic types of responses (from vortex shedding, galloping or flutter) are likely to take place when the basic shapes are combined the way they are in the structure at hand.

A single unit of Langangen Bridges (Axis 3 of the western viaduct of the twin bridges on road E-18 in Telemark County) built in 1977-79 is the full-scale case dealt with. The full-scale segment is 200 times larger than the model shown in Fig. 1, so the total length is 106 m. Except for omission of slope and curvature of the road-deck, the exterior was accurately reproduced in the model, but the highly complex development of stiffness and damping in a cracking, reinforced concrete structure greatly exceeded what was possible with a different model material. Neither did turbulence intensity or its spectral composition match that of the full-scale bridge. Consequently, the tests are to be seen as made more for diagnostic purposes than for detailed predictions. Reference to the full-scale structure is kept to a minimum.

Flow-induced vibration of a cantilever bridge model is dealt with by Davenport et al. in an experimental report (Ref. 1) on the bending moments in the completed state of a bridge. Meaas and Jensen (Ref. 2) have given results for the response of a full-scale, free-standing bridge unit. This pair of authors claim that the results match predictions from a numerical calculation of response due to turbulence, using a method of the Davenport type. Damping and torsional stiffness of a cyclicly loaded reinforced concrete pier has been studied experimentally by Jakobsen (Ref. 3).

2. MODEL AND TEST EQUIPMENT

The model shown in Figs. 1 and 2 was made of PVC plates (Trovidur produced by Dynamit Nobel) glued together (by Tangit produced by Henkel & Cie). For this PVC-material the density is 1370 kg/m^3 and Young's modulus is close to 3000 MPa, differing slightly for static and dynamic loading rates. As the PVC was of too low density for correct mass modelling of concrete, the plate thicknesses were increased to available dimensions. The modal masses still lacking then were represented by lumps (5.5 grammes each) attached in shielded positions at each end. In this way the requirement for identical average density within the contour of the structure was met with respect to modal quantities.

Foam blocks were attached at each end of the superstructure in order to represent a total of 90 tons of casting shields and moveable formwork of the full-scale structure. Of the three-layered footing shown in Fig. 1 only the upper part has to do with the full-scale structure.

In one test condition the model bridge was surrounded by a terrain model indicated on Fig. 3. The highest parts of the terrain were placed downwind of the pier.

Two pairs of strain gauges were attached 11 mm above the footing of the pier, and centrally on each face, in order to sense the bending strains and thereby the principal moments. A similar half bridge circuit was made up by gauges placed 24 mm from the centre of the shaft in order to sense the vertical bending moment near to the root of the superstructure. These two gauges were attached inside the box girder in a cross-section where the depth h (see Fig. 1b) is 38.5 mm. Wiring for the internal gauges was concealed inside the pier. Torsional strains (and moments) came from a full bridge circuit consisting of a pair of diagonally oriented crossgauges, placed 25 mm above the footing on the opposite sides of the pier being closest to one another. External gauges were wrapped with a muff of insulating tape in order to shield them and their leads from the airflow. The four circuits were connected independently to an eight channel strip-chart recorder (MFE M28) via 5 kHz carrier amplifiers (HB KWS 6A-5).

Three independent modes of vibration were detected. These are shown schematically in Fig. 4 along with the eigenfrequencies measured in still air.

Damping properties were determined through decay tests in still air. A complete log of these tests is shown in Fig. 5 along with the logarithmic decrements of modal damping, derived from the upper set of curves. Generally, the damping in each mode is close to independent of strain amplitude in the range covered. In a first trial plot (not shown here) the two highest modes showed lowering of damping with increase in amplitude, i.e. a performance typical for dry friction damping. As this could be an indication of imperfect bonding of the model, other explanations were sought for in order to disprove that there were a hidden fault in the model. There were however no symptoms like excessive cross-sensitivity between the channels nor change in eigenfrequency to confirm the suspicion. The charge had to be dropped at a later stage when it was made clear that the servomotors which drive the recorder's pens had been overloaded during the first cycles of (some of) the decay tests. This was proved in a separate test of the pen motors, where correction factors were determined as function of pen deflection when the driving frequency was equal to the natural frequency of the specific mode the channel was assigned to. In Fig. 5 the corrected results are shown.

Deemed on a background of experience with metallic models the damping levels in Fig. 5 are high. The limitation of resonant response in wind caused by high damping was however taken lightly because the damping levels were considered to be in reasonable ranges for similitude with a cracked full-scale structure. Relative to an uncracked prototype the model damping clearly is on the unsafe side.

The total length of the test floor is 5.5 m. This is divided into an upstream fetch of 4.2 m (from the end of the contraction to the model pier) and a downstream length of 1.3 m before the corner where the flow is deflected 90° in the closed-loop wind tunnel. The cross-section of the working section is 1.4 m wide and 1.1 m high. The model was fixed to a stiff steel footing cantilevered from a gravitational block weighing more than 650 kg and standing on the floor under the wind tunnel.

Three flow conditions were simulated, namely:

- Uniform, smooth flow; turbulence intensity (standard deviation of velocity divided by the mean value)

$$I \simeq 0.3\%$$

- Sheared profile (due to roughness elements (floor blocks) upstream and a grid at the entrance to the working section);

$$I \simeq 3\%$$

- Sheared profile from causes as mentioned above plus a terrain model representing the gorge and the hillsides;

$$I \simeq 6\%$$

The turbulence intensities above are averaged from three station values across the wind tunnel; within the influence zone of the model bridge and at the level of the superstructure. An edited log of the measurements made to describe the flow in the three cases is shown in Fig. 6. Turbulence was measured with a CTA hot wire equipment (DISA) with analog output through a RMS meter (Brüel & Kjær, Type 2425). This was used in its "slow" mode of operation which implies an averaging time of about 3 s. After further investigation it seems extremely unlikely that external capasitors were used to increase the averaging time beyond 3 s, although the original notes are lost.

The sheared profile shown in Fig. 6a is caused by a grid consisting of $10 \cdot 10$ mm horizontals and $10 \cdot 40$ mm verticals shown in Fig. 6b and the floor blocks shown in Fig. 3. These are 45 mm squares of height 32 mm mounted with the diagonal in the flow direction. In the same direction and normal to it the distance between rows of blocks is 120 mm. The blocks are glued onto 6 mm plywood plates of dimension $0.82 \cdot 1.22$ m, placed on the floor of the wind tunnel. Three such plates were used. Their leeward edges were respectively 0.8, 2.0 and 3.3 m upwind of the bridge pier.

Time series analysis of wind turbulence data was not carried out, but it is estimated that the integral length scale of turbulence was too low by a factor of about ten relative to the full-scale flow. The intensity of turbulence in the wind tunnel was also too low to fulfil the similarity requirement.

During the tests for structural response the mean value of airspeed was measured indirectly by a water manometer (DEBRO) which sensed the difference in static pressure over the contraction of the wind tunnel. Readings from the manometer were correlated with the reference velocity (Fig. 6a) for each of the three flow conditions.

3. TEST PROGRAMME AND DATA REDUCTION

Response recordings were made for a total of five test runs. These are identified by the flow condition (as mentioned in Section 2) and the wind incidence (azimuth) in the horizontal plane relative to the structure. The azimuth is measured from a plane through the axes of the pier and the superstructure. The case with azimut 90° was tested three times, i.e. in every flow condition. Azimuth of 45 and 0 were only used in combination with the sheared flow over a level terrain (except for the floor blocks). During the tests with azimuth 45 or 0 the instrumented cross-section of the box girder was on the upwind side of the pier.

The case with a terrain model was not followed up with variation in azimuth. For such variations to be useful, one would have needed a terrain model divided in several sectors in order to be able to rotate terrain and bridge as if they were a single unit, and at the same time respect the boundary condition set by the walls of the wind tunnel. At the time of testing it was decided to desist from a refined simulation of other wind incidences.

For each test condition the airspeed was reset about 20 times, and for each setting, the recorder was activated for a period ranging from 10 to 30 seconds to give continuous histories of strain. In the subsequent reduction of data, two quantities were selected for presentation from each recording period, i.e.

- the (temporal) mean pen deflection,
  and,
- the peak pen deflection measured relative to the mean,
  i.e. a dynamic quantity

The pen deflections were later converted to strains, taking into account that the output was the sum of contributions from two or four filaments. The nominal values were divided by 2.0 in case of the two bending channels of the pier, by 2.0 for the torsional channel in order to get a shear angle, and by 1.58 for the deck bending signal to correspond to a calculated strain on the underside of the box girder in the same cross-section.

A combined effect of a viscoelastic model material and incomplete equalization of temperature was that the strain base-lines before and after a full test run did not always coincide. In such cases it was assumed that the bias on the mean value had developed at a constant rate between the two known values. A large error is not likely to come from this choice, and a hypothetic error would appear as a deviation of the mean value strains from being proportional to velocity squared. The implied assumption here of drag coefficients being independent of Reynold's number is not controversial for a bluff object as the present bridge model.

4. RESULTS AND DISCUSSION

Overloading of pen motors mentioned in Section 2 is not likely to have biased the dynamic response as the recorder output was kept within a band of less than 60 per cent of the pen deflection range. It is also noted that the still air damping tests in Fig. 5 with safe margins cover the strain ranges excited by the airflow.

The results shown in Figs. 7 and 8 are identified by the recording channel numbers 1 through 4 used for the following quantities:

Ch. 1   Out-of-plane bending strain of the pier
         (11 mm above the footing)

Ch. 2   In-plane bending strain of the pier
         (11 mm above the footing)

Ch. 3   In-plane bending strain on the underside of the box girder
         (24 mm from the centre of the pier)

Ch. 4   Shear angle at the middle of the widest faces of the pier
         (25 mm above the footing)

Mean values are marked with discrete circles and an overbarred numeral indicating the channel number (e.g. $\bar{1}$). Dynamic peak values are designated with a channel number followed by an apostrophe(e.g. 1'). The peak values are measured relative to the mean. As the fluctuating strain was rather symmetric about the mean, no distinction between positive and negative peaks has been made. Peak values on different channels are commonly not reached at the same instant. In order to facilitate the comparison, diagram (a) in Fig. 7 has been repeated as Fig. 8a.

Model airspeed has been retained for the horizontal axes of Figs.7 and 8. Prototype velocities are about three times higher. Rescaling of the diagrams to full-scale velocity would have been misleading due to the defect in flow simulation.

The diagrams cover a ratio of ten in velocity and one hundred in strain. A quantity which is proportional to velocity squared should (in Figs. 7 and 8) appear as a straight line parallel to the diagonal from the lower left to the upper right corner. With close approximation this is seen to hold for the mean values of channels (1) and (2) which are the only cases where the mean is not trivially small. It is seen that the mean value on Ch. 1 is nearly unaffected by the type of flow, and it is only slightly smaller for the azimuth 45^0 than for 90^0.

By using the relationship between bending moments and strains determined in static tests before the wind tunnel tests began, the mean values of Figs.7 and 8 have been converted to overall drag coefficients. The two cases are Ch.1, azimuth 90^0 and Ch.2, azimuth 0^0, i.e. both are alongwind. In both cases the coefficient calculated is 1.6, provided that the reference area is taken as the projected area of the structure when seen along the main flow in each case.

The main trend of the dynamic peak values is that they are proportional to velocity squared as was the mean. However, the dynamic response is systematically somewhat higher than this within the range from 5 to (say) 10 m/s. This is taken as indication of cross-wind forces due to a preference periodicity in the wake, i.e. vortex shedding. Strouhal numbers known for similar structural forms make this assumption likely. However, this forcing mechanism is not particularly well tuned, and the response never approached a constant amplitude. This is seen in Fig.9 which is a reproduction of the strip-chart recordings corresponding to Fig.7a and 7b at about 7.5 m/s. The response hump attributed to shedding of vortices is no more pronounced in smooth flow (Fig.7c) than in the more turbulent cases. A rough comparison of Fig.7c with (a) and (b) shows that a factor 10 to 20 on turbulence intensity gave a factor 5 on the dynamic response.

The gust response factors ((mean+peak)/mean) for the out-of-plane mode calculated from data on Fig.7 are 1.1, 1.4 and 1.6 respectively for turbulence intensities of 0.3, 3 and 6%.

The dynamic, angular strain in Ch.(4) has been used to calculate an out-of-balance loading on the superstructure that would cause the same torsional moment in the pier. Attention is limited to the case with azimuth 90°, which seems to be the worst case for torsion. The calculation is based on response at 10 m/s. The static load model employed is one where a symmetric drag loading on the superstructure is perturbed by a factor (1+ε) on one side of the pier and by a factor (1-ε) on the other. For turbulence intensities of 0.3, 3 and 6% the corresponding values of ε were found to be 0.05, 0.22 and 0.38. (One of the factors that went into the calculation had a range of ±15% which is carried on to the figures above). Considering the very low levels of turbulence in the present cases, the redistribution factor ε is remarkably large, and it is likely to be even higher in full-scale turbulence.

In most cases the fluctuating, in-plane bending strain (Ch.2) has the highest peak. With azimuths 45 and 90 the dynamic part alone is of the same size, or larger, than the sum of the mean and the dynamic peak at azimuth 0°, i.e. when the main flow is in the plane of the structure.

The strain levels encountered during the tests are within reasonable limits and are not likely to cause fatigue failure in a short time for a welldesigned structure. However, the structure is of a flexible type where small strains may add up to considerable displacements.

5.   CONCLUSIONS

In relation to the full-scale bridge the test reported has shortcommings, namely, too little turbulence in the flow, and no allowance for the gradual change of damping and stiffness in a progressively cracking concrete structure. The damping however is not out of scale for a concrete structure having been strained to cracking before.

Within the limitations of the present model simulation the following conclusions are drawn:

- The dynamic response is predominantly caused by wind turbulence.

- Within a certain velocity range the dynamic response is amplified. This is thought to be a weak realization of shedding of vorties.

- The dynamic, in-plane response is higher than for the two other modes, except for azimuth 0° where it may be smaller.

- For the dynamic part of the torsional response to be predicted by a static loading system at turbulence intensity 6%, a symmetric loading on the two arms of the superstructure will have to be redistributed in such a manner that ~38% is taken off on one side and put on the other.

- The mean signals for bending of the pier are consistent with a drag coefficients of 1.6, referred to the projected area as seen in the flow direction (here: in, or normal to, the plane of the bridge unit).

6.   ACKNOWLEDGEMENTS

Mr. O. Haldorsen made the model and the other items necessary for the tests.

Mr. T. Meltzer (Laboratory for Building Materials and Structures) helped by setting up and checking the instruments.

Mr. P.-A. Krogstad (Division of Aero and Gas Dynamics) conducted the measurement of turbulence.

7. REFERENCES

1. Davenport, A.G., Isyumov, N., Fader, D.J. and Bowen, P.: "An Aeroelastic Study of the Northumberland Straits Bridge - Cantilevered Concrete Design". Research Report BLWT-4-69. The University of Western Ontario, Faculty of Engineering Science, Boundary Layer Wind Tunnel Laboratory, London, Canada, 1969.

2. Meaas, P. and Jensen, J.J.: "Wind response of cantilever bridges under construction". /In/ Holand, I. et al.(editors): "Safety of Structures under Dynamic Loading". Vol.II. Tapir Publishers, Trondheim, 1978.

3. Jakobsen, B.: "Cyclic Torsion Tests of Concrete Box Columns". Report No.80-2. Division of Structural Mechanics, The Norwegian Institute of Technology, Trondheim, 1980.

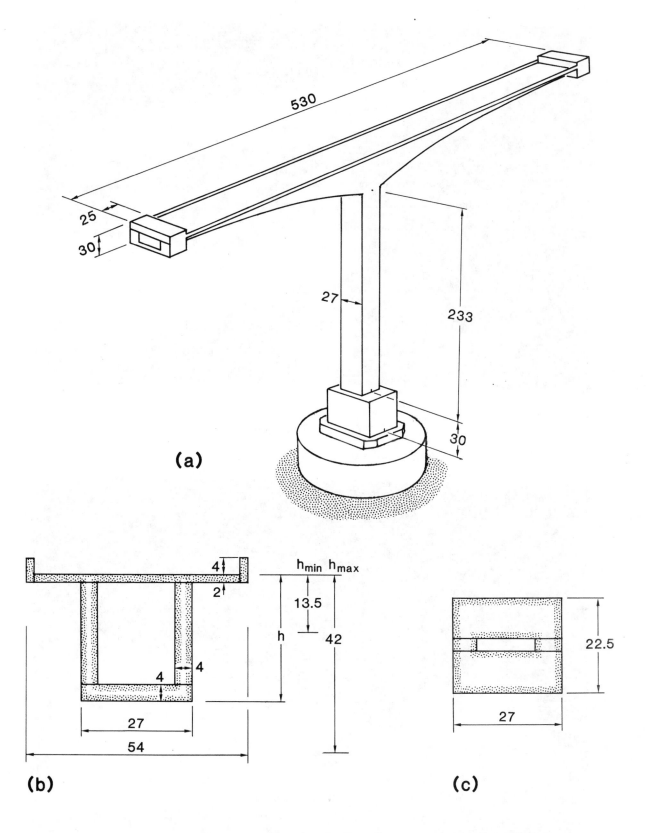

Fig. 1   The structure.   Model dimensions are shown in millimetres.

(a)   Perspective view

(b)   Cross - section of the superstructure

(c)   Cross - section of the pier

Fig. 2   The model

Fig. 3   Models of bridge and terrain in the wind tunnel

214

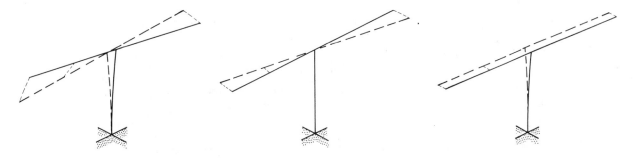

In-plane mode f_1 = 22.9 Hz   Torsional mode f_2 = 26.6 Hz   Out-of-plane mode f_3 = 34.2 Hz

Fig. 4   Modes and eigenfrequencies of the model structure

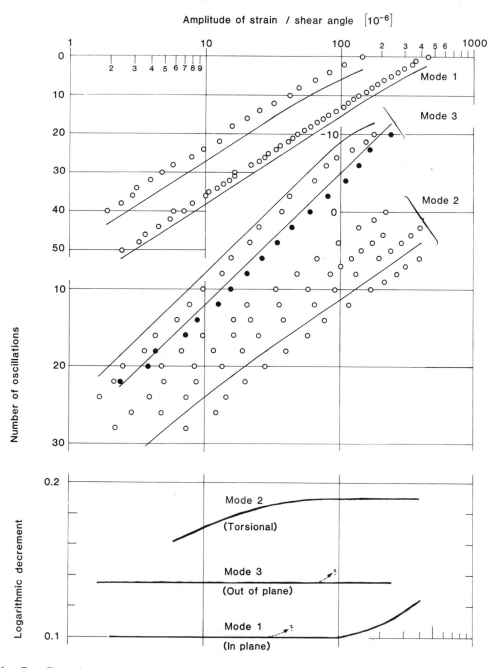

Fig. 5   Damping test results.   (Points show readings.   Upper curves are smoothed
representations.   The three lower curves are for logarithmic decrement vs. amplitude.
Note the broken scales.)

215

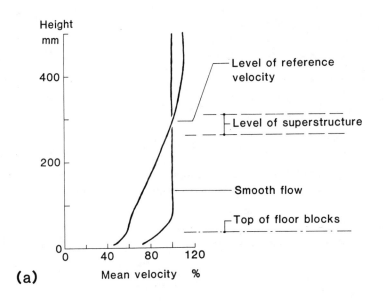

(a)

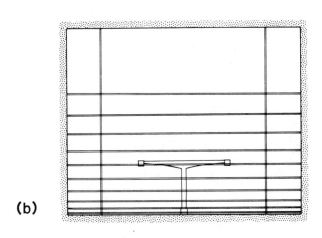

(b)

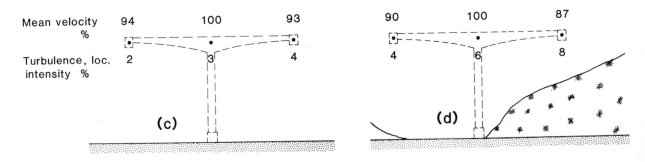

(c)   **(d)**

Fig. 6   Flow characteristics.   (a) Mean velocity profiles over the floor of an empty tunnel (smooth flow) and over the rougher fetch.   (b) Grid at the entrance to the working section.   (c) & (d) Mean velocity and turbulence before and after introduction of the terrain model shown in Fig. 3.

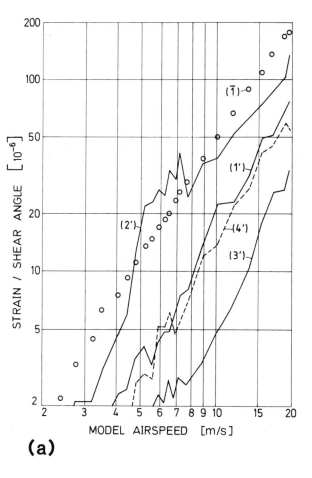

(a)

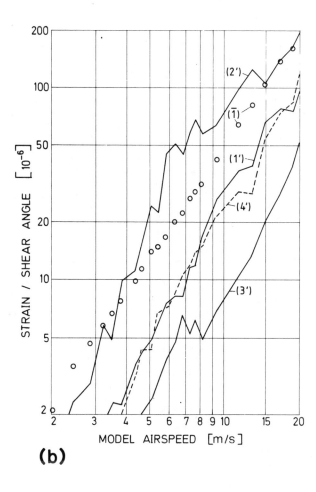

(b)

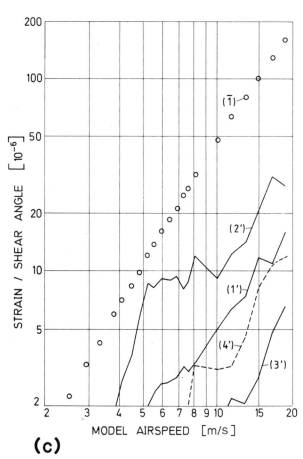

(c)

Fig. 7 Response to airflow normal to the plane of the structure (azimut 90°), with

(a) shear flow over level terrain,
(b) terrain according to Fig. 3,
(c) uniform, smooth flow.

Legend:

($\bar{n}$) Mean value of Ch. n
(n') Dynamic, peak value of Ch. n

Ch. 1 is for out-of-plane bending of pier
Ch. 2 is for in-plane bending of pier
Ch. 3 is for bending of superstructure
Ch. 4 is for torsion of pier

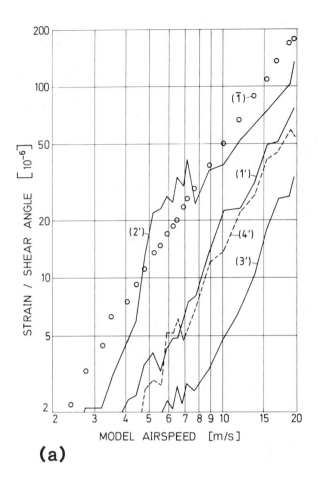

(a)

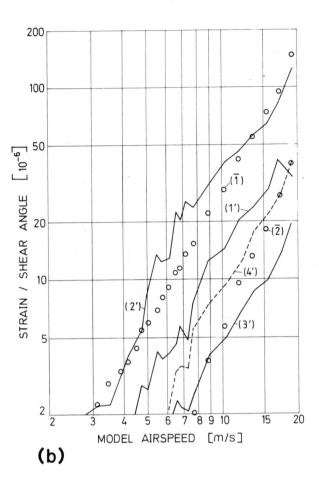

(b)

Fig. 8 Response to airflow at variable incidence in the horizontal plane of the structure

(a) Azimuth 90° (= flow normal to the plane of the structure)

(b) Azimuth 45°

(c) Azimuth 0°

Legend as for Fig. 7

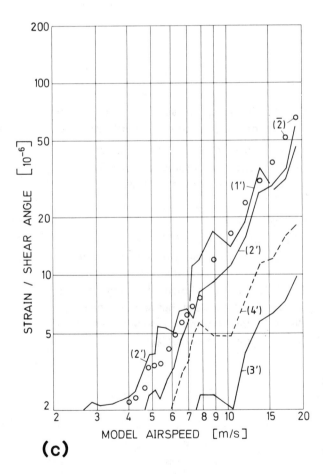

(c)

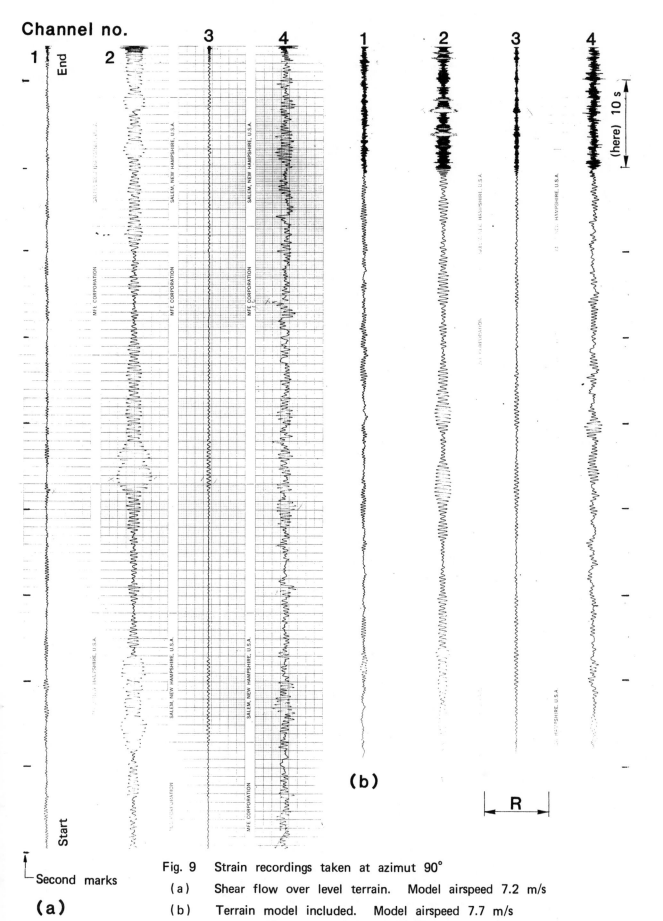

Channel no.

1   End   2   3   4   1   2   3   4

(here) 10 s

(b)

R

Start

Second marks

(a)

Fig. 9   Strain recordings taken at azimut 90°

(a)   Shear flow over level terrain.   Model airspeed 7.2 m/s

(b)   Terrain model included.   Model airspeed 7.7 m/s

The pen deflection range (R shown above) correspond to the following strains mentioned from left to right: 154, 154, 97, 31,   154, 307, 97, 77 $\left[\mu m/m\right]$ or $\left[\mu rad/rad\right]$

219

CRITICAL FLOW VELOCITIES AND STABILITY ZONES
FOR WAKE-INDUCED OSCILLATIONS

A. R. E. Oliveira and W. M. Mansour

Federal University of Rio de Janeiro, Brazil

Summary

A nonlinear analysis is adopted to investigate subspan oscillations. Stability zones and critical wind velocities are identified for the non-resonant case. Conditions for auto-oscillations are established.

Organised and sponsored by
BHRA Fluid Engineering, Cranfield, Bedford MK43 0AJ, England

1. INTRODUCTION

Subspan oscillations were studied in the past by several researchers using different approaches. Linearized analyses are attempted in (ref. 1 to 5) . An energy-balance approach is described in (ref.6) to determine the amplitude of the limit-cycle oscillations. Wind-tunnel tests are reported in (ref.7). Aerodynamic models are presented in (ref.8) . Stability analysis of the motion is given in (ref. 9 and 10). A finite-element approach is applied in (ref. 11).

Recently Oliveira and Mansour (ref. 12 to 14) used nonlinear analysis to study the wake-induced oscillations. Ref. 12 reports a mathematical model for the coefficients of lift and drag based on the published experimental data of Cooper (ref.7). References 13 and 14 discuss a strategy to evaluate the radii of limit cycles using the asymptotic approximation of Krylof and Bogolyobof. This paper reports on some other aspects regarding the critical wind velocities and the associated stability zones using nonlinear analysis.

2. THE MATHEMATICAL MODEL

In the analysis that follows, a bar underneath a letter indicates a vector quantity and a bar above the letter indicates a complex-conjugate.

The windward cylinder is considered fixed in inertial space and is taken as the origin of the system of axes . The x-axis is aligned with the flow velocity $\underline{V}$. Fig. 1 shows the forces acting on the leeward cylinder.

A force-balance for the moving cylinder gives:

$$\left. \begin{array}{l} m\,\ddot{x} + k_x\,x = f_x \\ m\,\ddot{y} + k_y\,y = f_y \end{array} \right\} \tag{1.a}$$

where:

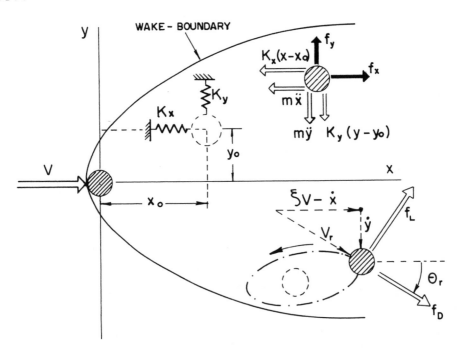

Fig. 1 : Forces on leeward cylinder

m = mass of leeward cylinder

k_x, k_y = horizontal and vertical stiffness of supports.

f_x, f_y = aerodynamic forces along the x and y-directions.

x_o, y_o = coordinates of the moving cylinder in the undisturbed position.

x ,y = displacements of the downstream cylinder measured from x_o, y_o.

Dots on the letters refer to differentiation with respect to time t. Referring to Fig. (1) ,one can write:

$$f_x = f_L \sin\theta_r + f_D \cos\theta_r$$
$$f_y = f_L \cos\theta_r - f_D \sin\theta_r$$
(1.b)

where f_L and f_D are the lift and drag forces on the moving cylinder and θ_r is the angle of approach of the relative velocity $\underline{V}_r$.

The local flow velocity in the vicinity of the moving cylinder is given by $\zeta\underline{V}$ where ζ is a constant which takes account of the wake's aerodynamics. The lift and drag forces are given by:

$$f_L = \frac{1}{2} \rho \ell d V_r^2 \, C_L^* = \frac{\rho \ell d V_r^2}{2\zeta^2} \, C_L$$
$$f_D = \frac{1}{2} \rho \ell d V_r^2 \, C_D^* = \frac{\rho \ell d V_r^2}{2\zeta^2} \, C_D$$
(1.c)

where:

ℓ, d = length and diameter of moving cylinder.

ρ = air density

C_L^* , C_D^* = lift and drag coefficients based on local velocity.

C_L , C_D = lift and drag coefficients based on free flow velocity $\underline{V}$.

Since $\zeta V - \dot{x}$ is $>> \dot{y}$,one can write:

$$V_r \simeq \zeta V - \dot{x}$$
(1.d)

Substituting (1.c) in (1.b),making use of (1.d) and dropping second order terms ($\dot{x}^2$ and $\dot{x}\dot{y}$) ,one obtains :

$$f_x = \frac{\rho \ell d V}{2\zeta} \{ C_L \dot{y} + C_D(\zeta V - 2\dot{x}) \}$$
$$f_y = \frac{\rho \ell d V}{2\zeta} \{ C_L(\zeta V - 2\dot{x}) - C_D \dot{y} \}$$
(1.e)

Combining (1.a) and (1.e) ,one can write the model in the following dimensionless form :

$$X'' + X = \mu\{ C_L Y' + C_D(\alpha - 2X') \}$$
$$Y'' + K^2 Y = \mu\{ C_L(\alpha - 2X') - C_D Y' \}$$
(1)

where:

$X = x/d$	; $Y = y/d$	; $X_o = x_o/d$	; $Y_o = y_o/d$
$\omega_x^2 = k_x/m$	; $\omega_y^2 = k_y/m$	; $K = \omega_y/\omega_x$	
$\tau = \omega_x t$	; $\mu = \rho \ell d V / 2m\zeta\omega_x$		; $\alpha = \zeta V/d\omega_x$

(2)

Primes here refer to differentiation with respect to the dimensionless time τ .

3. LIFT AND DRAG COEFFICIENTS

The experimental results published by Cooper (ref.6) are widely accepted by researchers in this area. Typical plots are shown in Fig.(2) for C_L and C_D .

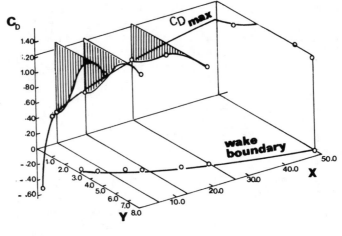

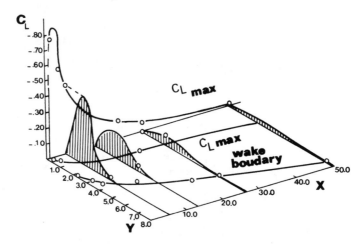

Fig. 2 : Lift and Drag coefficients[6]

Several attempts were made by previous investigators (ref.2,6 and 8) to construct models to fit C_L and C_D. The published models are too difficult for direct implementation in equation (1). The authors of this paper developed an adequate model[12] in the following form:

$$C_L = \frac{1}{X+X_o} \{A_1(Y+Y_o)^3 - A_2(Y+Y_o) \}$$

$$C_D = B_1(Y+Y_o)^4 + B_2(Y+Y_o)^2 + B_3\sqrt{X+X_o} + B_4$$

(3)

Curve fitting showed that :
$A_1 = 0.07$; $A_2 = 1.07$
$B_1 = -0.004$; $B_2 = 0.10$; $B_3 = 0.04$; $B_4 = 0.47$

yield the best representation for C_L and C_D in the range:

$$10 < X_o < 30 \quad \text{and} \quad Y_o < 4$$

As a first approximation ,the C_L and C_D coefficients can be expressed in the form :

$$C_L = a_o + a_1 X + a_2 Y$$

$$C_D = b_o + b_1 X + b_2 Y$$

(4)

where the a and b-coefficients are obtained by expanding (3) in Taylor's series about X_o, Y_o to obtain:

$$
\begin{array}{ll}
a_o = (A_1 Y_o^3 - A_2 Y_o)/X_o & ; \quad b_o = B_1 Y_o^4 + B_2 Y_o^2 + B_3 \sqrt{X_o} + B_4 \\[2mm]
a_1 = -(A_1 Y_o^3 - A_2 Y_o)/X_o^2 & ; \quad b_1 = B_3/2\sqrt{X_o} \\[2mm]
a_2 = (3A_1 Y_o^2 - A_2)/X_o & ; \quad b_2 = 4B_1 Y_o^3 + 2B_2 Y_o
\end{array}
\tag{5}
$$

4. LINEAR TRANSFORMATION OF THE GOVERNING MODEL

If one substitutes (4) in (1), the resulting equations will contain constant terms. To free the model from these terms, one can adopt the following linear transformation:

$$
X = u_1 + X_* \qquad ; \qquad Y = u_2 + Y_*
\tag{6}
$$

where X_* and Y_* are constants yet to be determined. Substituting (6) in (1) and (4) and equating the free terms to zero, one obtains :

$$
\left.
\begin{array}{l}
X_* = \{ a_o b_2 - b_o \left[a_2 - \dfrac{K^2}{\mu\alpha} \right] \} / D \\[4mm]
Y_* = \{ b_o a_1 - a_o \left[b_1 - \dfrac{1}{\mu\alpha} \right] \} / D \\[4mm]
D = \left[b_1 - \dfrac{1}{\mu\alpha} \right] \left[a_2 - \dfrac{K^2}{\mu\alpha} \right] - a_1 b_2 \qquad ; \quad D \neq 0
\end{array}
\right\}
\tag{7}
$$

The governing model takes the following form:

$$
\left.
\begin{array}{l}
u_1'' - \gamma_1 u_2 + \omega_1^2 u_1 = \mu g_1(u_1, u_1', u_2, u_2') \\[2mm]
u_2'' - \gamma_2 u_1 + \omega_2^2 u_2 = -\mu g_2(u_1, u_1', u_2, u_2')
\end{array}
\right\}
\tag{8}
$$

The mathematical model given by (8) is based on a linearization of the expressions of C_L and C_D as given by (4). The model itself is nonlinear and will be referred to as the first-approximation model. The coefficients and functions appearing in (8) are given by:

$$
\begin{array}{ll}
\omega_1^2 = 1 - \mu\alpha b_1 & ; \quad \omega_2^2 = K^2 - \mu\alpha a_2 \\[2mm]
\gamma_1 = \mu\alpha b_2 & ; \quad \gamma_2 = \mu\alpha a_1 \\[2mm]
\hat{v}_1 = X_*/\mu\alpha & ; \quad \hat{v}_2 = K^2 Y_*/\mu\alpha \\[2mm]
g_1(u_1, u_1', u_2, u_2') = (a_1 u_1 + a_2 u_2)u_2' - 2(b_1 u_1 + b_2 u_2)u_1' + \\
\qquad\qquad\qquad\qquad\quad \hat{v}_2 u_2' - 2\hat{v}_1 u_1' \\[2mm]
g_2(u_1, u_1', u_2, u_2') = 2(a_1 u_1 + a_2 u_2)u_1' + (b_1 u_1 + b_2 u_2)u_2' + \\
\qquad\qquad\qquad\qquad\quad \hat{v}_1 u_2' + 2\hat{v}_2 u_1'
\end{array}
\tag{9}
$$

For the majority of practical applications, the coefficient μ lies in the range :

$$0.015 > \mu > 0.0015$$

Hence μ can be considered a small parameter. On the other hand $\mu\alpha$ is not.

5. MULTIPLE TIME SCALE APPROACH FOR THE FIRST APPROXIMATION MODEL

A solution for (8) is assumed in the form:

$$
\left.
\begin{array}{l}
u_1 = \varepsilon u_{11}(T_o, T_1) + \varepsilon^2 u_{12}(T_o, T_1) \\[2mm]
u_2 = \varepsilon u_{21}(T_o, T_1) + \varepsilon^2 u_{22}(T_o, T_1)
\end{array}
\right\}
\tag{10}
$$

where :
$$T_o = \tau \qquad ; \qquad T_1 = \varepsilon\tau$$

$u_{ij}(T_o, T_1)$ are unknown functions and ε is a small but finite parameter.

The operators $(d/d\tau)$ and $(d^2/d\tau^2)$ are written in the following form:

$$\frac{d}{d\tau} = D_o + \varepsilon D_1$$

$$\frac{d^2}{d\tau^2} = D_o^2 + \varepsilon\{2D_oD_1\} + \varepsilon^2 D_1^2$$

where

$$D_o = \frac{\partial}{\partial T_o} \qquad ; \qquad D_1 = \frac{\partial}{\partial T_1}$$

Considering $\hat{\nu}_1 = \varepsilon\nu_1 \quad ; \quad \hat{\nu}_2 = \varepsilon\nu_2$, substituting (10) in (8) and equating the coefficients of ε and ε^2 to zero , one obtains :

$$\left. \begin{array}{l} D_o^2 u_{11} - \gamma_1 u_{21} + \omega_1^2 u_{11} = 0 \\[2mm] D_o^2 u_{21} - \gamma_2 u_{11} + \omega_2^2 u_{21} = 0 \end{array} \right\} \tag{11}$$

and

$$\left. \begin{array}{l} D_o^2 u_{12} - \gamma_1 u_{22} + \omega_1^2 u_{12} = \mu\{(a_1 u_{11} + a_2 u_{21})(D_o u_{21}) - 2(b_1 u_{11} + b_2 u_{21})(D_o u_{11}) + \\[2mm] \qquad\qquad \nu_2(D_o u_{21}) - 2\nu_1(D_o u_{11})\} - 2D_o D_1 u_{11} \\[4mm] D_o^2 u_{22} - \gamma_2 u_{12} + \omega_2^2 u_{22} = -\mu\{2(a_1 u_{11} + a_2 u_{21})(D_o u_{11}) + (b_1 u_{11} + b_2 u_{21})(D_o u_{21}) + \\[2mm] \qquad\qquad \nu_1(D_o u_{21}) + 2\nu_2(D_o u_{11})\} - 2D_o D_1 u_{21} \end{array} \right\}$$
$$\tag{12}$$

The solution for (11) takes the form:

$$\left. \begin{array}{l} u_{11} = \{A_1^*(T_1)e^{j\Omega_1 T_o} + A_2^*(T_1)e^{j\Omega_2 T_o}\} + cc \\[4mm] u_{21} = \{\Lambda_1 A_1^*(T_1)e^{j\Omega_1 T_o} + \Lambda_2 A_2^*(T_1)e^{j\Omega_2 T_o}\} + cc \end{array} \right\} \tag{13}$$

where A_n^* and Λ_n ,(n=1,2), are generally complex quantities, and cc denotes "complex conjugate". Substituting (13) in (11) and simplifying, one obtains the following characteristic equation:

$$\Omega^4 - (\omega_1^2 + \omega_2^2)\Omega^2 + (\omega_1^2 \omega_2^2 - \gamma_1\gamma_2) = 0 \tag{14.a}$$

which yields:

$$\Omega_{1,2}^2 = \frac{1}{2}\{(\omega_1^2 + \omega_2^2) \pm \sqrt{(\omega_1^2 - \omega_2^2)^2 + 4\gamma_1\gamma_2}\} \tag{14.b}$$

$$\Lambda_n = \frac{\omega_1^2 - \Omega_n^2}{\gamma_1} = \frac{\gamma_2}{\omega_2^2 - \Omega_n^2} \quad ; \quad (\gamma_1 \neq 0, \gamma_2 \neq 0) \tag{14.c}$$

Substituting (13) in (12) one obtains:

$$D_o^2 u_{12} - \gamma_1 u_{22} + \omega_1^2 u_{12} =$$

$$j<[-2D_1 A_1^* - \mu(2\nu_1 - \nu_2\Lambda_1)A_1^*]\Omega_1 e^{j\Omega_1 T_o} + [-2D_1 A_2^* - \mu(2\nu_1 - \nu_2\Lambda_2)A_2^*]\Omega_2 e^{j\Omega_2 T_o} +$$

$$[\Lambda_1(a_1 + a_2\Lambda_1) - 2(b_1 + b_2\Lambda_1)]\mu A_1^{*2}\Omega_1 e^{j2\Omega_1 T_o} +$$

$$[\Lambda_2(a_1+a_2\Lambda_2)-2(b_1+b_2\Lambda_2)]\mu A_2^{*2}\Omega_2 e^{j2\Omega_2 T_o} + [-\Omega_1\{\overline{\Lambda}_1(a_1+a_2\Lambda_2)-2(b_1+b_2\Lambda_2)\}+$$

$$\Omega_2\{\Lambda_2(a_1+a_2\overline{\Lambda}_1)-2(b_1+b_2\overline{\Lambda}_1)\}]\mu\overline{A}_1^{*} A_2^{*} e^{j(\Omega_2-\Omega_1)T_o} > +$$

other non-contribuiting terms.
and $\qquad\qquad\qquad\qquad\qquad\qquad\qquad\qquad\qquad\qquad$ (15.a)

$$D_o^2 u_{22}-\gamma_2 u_{12}+\omega_2^2 u_{22} =$$

$$j<[-2\Lambda_1 D_1 A_1^{*} -\mu(2\nu_2+\nu_1\Lambda_1)A_1^{*}]\Omega_1 e^{j\Omega_1 T_o} + [-2\Lambda_2 D_1 A_2^{*}-\mu(2\nu_2+\nu_1\Lambda_2)A_2^{*}]\Omega_2 e^{j\Omega_2 T_o} +$$

$$[-2(a_1+a_2\Lambda_1)-\Lambda_1(b_1+b_2\Lambda_1)]\mu A_1^{*2}\Omega_1 e^{j2\Omega_1 T_o} +$$

$$[-2(a_1+a_2\Lambda_2)-\Lambda_2(b_1+b_2\Lambda_2)]\mu A_2^{*2}\Omega_2 e^{j2\Omega_2 T_o} + [\Omega_1\{2(a_1+a_2\Lambda_2)+\overline{\Lambda}_1(b_1+b_2\Lambda_2)\}-$$

$$\Omega_2\{2(a_1+a_2\overline{\Lambda}_1)+\Lambda_2(b_1+b_2\overline{\Lambda}_1)\}]\mu\overline{A}_1^{*} A_2^{*} e^{j(\Omega_2-\Omega_1)T_o} > +$$

other non-contributing terms .

$$\qquad\qquad\qquad\qquad\qquad\qquad\qquad\qquad\qquad\qquad (15.b)$$

6.   STABILITY BOUNDARIES FOR THE NONRESONANT CASE

For the nonresonant case, i.e. Ω_2 is not near $2\Omega_1$ or near $\frac{1}{2}\Omega_1$, one can seek a particular solution for equations (15) in the form:

$$\left. \begin{array}{l} u_{12} = P_{11} e^{j\Omega_1 T_o} + P_{12} e^{j\Omega_2 T_o} \\ \\ u_{22} = P_{21} e^{j\Omega_1 T_o} + P_{22} e^{j\Omega_2 T_o} \end{array} \right\} \qquad (16)$$

Substituting this solution in (15.a) and (15.b) and equating the coefficients of $e^{j\Omega_1 T_o}$ and $e^{j\Omega_2 T_o}$ on both sides ,one obtains :

$$\left. \begin{array}{l} (\omega_1^2-\Omega_n^2) P_{1,n} + (-\gamma_1) P_{2,n} = j R_{1,n}^{*} \\ \\ (-\gamma_2) P_{1,n} +(\omega_2^2-\Omega_n^2) P_{2,n} = j R_{2,n}^{*} \end{array} \right\} \qquad (17)$$

where

$$\left. \begin{array}{l} R_{1,n}^{*} = \{-2D_1 A_n^{*} -\mu(2\nu_1-\nu_2\Lambda_n) A_n^{*}\}\Omega_n \\ \\ R_{2,n}^{*} = \{-2\Lambda_n D_1 A_n^{*} -\mu(2\nu_2+\nu_1\Lambda_n)A_n^{*}\}\Omega_n \end{array} \right\} \qquad (18)$$

The determinant of the coefficients of the L.H.S. of (17) is essentially identical to (14.a) . Consequently one can write:

$$\begin{vmatrix} \omega_1^2 - \Omega_n^2 & j R_{1,n}^{*} \\ \\ -\gamma_2 & j R_{2,n}^{*} \end{vmatrix} = 0$$

which yields :

$$R_{1,n}^{*} = -(\frac{\gamma_1}{\gamma_2}) \Lambda_n R_{2,n}^{*} \qquad (19)$$

Substituting (18) in (19) ,one obtains:

$$\tau_n D_1 A_n^{*} + A_n^{*} = 0 \qquad (20.a)$$

where

$$\tau_n = \left(\frac{2}{\mu}\right)\left\{\frac{\left(\frac{\gamma_1}{\gamma_2}\right)\Lambda_n^2 + 1}{\left(\frac{1}{\gamma_2}\right)\Lambda_n(2\nu_2+\nu_1\Lambda_n)+(2\nu_1-\nu_2\Lambda_n)}\right\} \quad ; \qquad (20.b)$$

Denominator $\neq 0$

If τ_n is > 0 , the leeward cylinder will tend to its stable equilibrium position if disturbed. It will be unstable if τ_n is < 0 . The boundaries separating the stable and unstable regions are obtained by equating the numerator in (2o.b) to zero. This implies that $(\Omega_1 = \Omega_2)$. Equation (14.b) yields the following condition in this case:

$$(\omega_1^2 - \omega_2^2)^2 + 4\gamma_1\gamma_2 = 0$$

which can be written in the following form:

$$X_o = \left\{\frac{-\beta_1 \pm \sqrt{\beta_1^2 + 4\beta_o\beta_2}}{2\beta_o}\right\}^2$$

where

$$V_o = \sqrt{\mu\alpha} = (\sqrt{\rho\ell/2k_x})V \quad ; \quad \beta_o = (K^2-1)/V_o^2 \quad ; \quad \beta_1 = B_3/2 \qquad (21)$$

$$\beta_2 = (3A_1 Y_o^2 - A_2) \pm 2Y_o\sqrt{(A_1 Y_o^2 - A_2)(4B_1 Y_o^2 + 2B_2)}$$

An examination of (21) shows that for every level of Y_o in the $X_o : Y_o$ plane ,there is a possibility of obtaining up to four values for X_o which determine the boundaries of the stability zone. The boundaries are functions of the specified dimensionless wind velocity V_o. Typical plots are shown in Fig. (3). The following data are used :

$m = 3.13$ Kg $\quad ; \quad \ell = 0.91$ m $\quad , \quad d = 0.04$ m $\quad ; \quad \omega_x = 6.91$ rd/sec.
$\omega_y = 7.23$ rd/sec.; $\rho = 1.225$ Kg/m^3

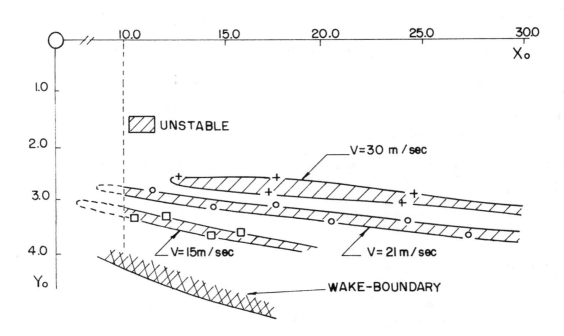

Fig. 3 : Stability boundaries for
a specified wind velocity

228

7. AUTO-OSCILLATIONS FOR THE NONRESONANT CASE

Sustained oscillations, i.e $A_n^* $ = constant , can occur if $(1/\tau_n)=0$. The denominator in (20.b) is equated to zero to obtain the following conditions :

$$(\frac{\gamma_1}{\gamma_2}) \Lambda_n (2\nu_2 + \nu_1 \Lambda_n) + (2\nu_1 - \nu_2 \Lambda_n) = 0 \tag{22}$$

There are four cases to be examined as shown after :

7.1 case 1 : $\nu_1 \neq 0$ and $\nu_2 \neq 0$

Equation (22) can be satisfied by any of the following three conditions :

$$(\frac{\gamma_1}{\gamma_2}) \Lambda_1 \Lambda_2 = 2 \qquad (a)$$
$$\Lambda_1 = \Lambda_2 \qquad (b) \tag{23}$$
$$2 \gamma_1 = \gamma_2 \qquad (c)$$

The first relation in (23) can be dismissed since it implies that $\gamma_1 \gamma_2 = 0$ which contradicts condition (14.c). The second relation gives the boundaries (21) . These are the loci of all points in the $X_o : Y_o$ plane which are associated with sustained oscillations for a critical dimensionless wind velocity $V_c = (\sqrt{\rho \ell / 2 k_x}) V$ given by:

$$V_c = \sqrt{\frac{K^2 - 1}{a_2 - b_1 \pm 2\sqrt{-a_1 b_2}}} \tag{24}$$

The last relation in (23) yields :

$$Y_o = 0 \quad \text{and}$$
$$X_o = \pm \frac{1}{2} \sqrt{\frac{A_2 - A_1 Y_o^2}{B_2 + 2 B_1 Y_o^2}} \tag{25}$$

The curves in (25) represent the loci of all points with sustained oscillations which are not dependent on a specific critical velocity to induce them.

7.2 case 2 : $\nu_1 = 0$ and $\nu_2 \neq 0$

Equations (22) yield the condition:

$$2\gamma_1 = \gamma_2$$

which gives rise to sustained oscillations along the curves given by (25) . But in this case , each point on the curve is associated with a critical wind velocity V_c given by:

$$V_c = K \sqrt{\frac{b_o}{b_o a_2 - a_o b_2}} \tag{26}$$

7.3 case 3 : $\nu_1 \neq 0$ and $\nu_2 = 0$

Equations (22) yield the condition:

$$\Lambda_1 = \Lambda_2$$

which gives rise to sustained oscillations for points on the boundary curves (21) where the critical velocity V_c is given by:

$$V_c = \sqrt{\frac{a_o}{a_o b_1 - b_o a_1}} \tag{27}$$

7.4 case 4 : $\nu_1=0$ and $\nu_2=0$

Combining relations (26) and (27) ,one obtains the following curve in the $X_o:Y_o$ plane:

$$b_o K^2(a_o b_1 - b_o a_1) - a_o(b_o a_2 - a_o b_2) = 0 \qquad (28)$$

Fig. (4) shows points in the $X_o:Y_o$ plane with the same critical wind velocity.

8.   CONCLUSIONS

The studies conducted on the first-approximation nonlinear model revealed the following zones in the wake of the fixed cylinder :

(a) For every given wind velocity ,there exists instability zones where the amplitudes will only be bounded by the mechanical constraints of the real system . The latter were not accounted for in the model.

(b) For every point in the field ,there exists a critical wind-velocity which can induce sustained oscillations.

(c) A higher order approximation for C_L and C_D is required to evaluate the radii of the limit cycles.

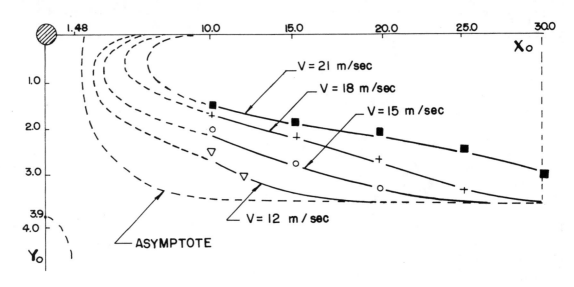

Fig. 4 : Iso-critical wind velocity curves

(d) The unstable zones were found to grow in size as the wind velocity increases . They tend to appear near the boundaries of the wake. These findings coincide with those reported by Cooper[6].

(e) The iso-critical velocity curves are bounded by the asymptotic curves given by equation (25) . The iso-curves tend to move towards the asymptotes as the wind velocity decreases.

(f) Sustained oscillations can occur above or below the x_o-axis because the mathematical model used here neglected cross-coupling. This is in agreement with the conclusions of Tsui[10].

9. REFERENCES

1. Smith I.P.;"The Aeroelastic Stability of Twin Bundled Conductors";
   CERL Laboratory Note No. RD/L/N 170-66,April 1967
2. Simpson A. and Lawson T.V.;"Oscillations of Twin Power Transmission
   Lines";Royal Aeronautical Society paper No.25,vol.2,April 1968.
3. Zaffanella L.E.;"Investigation of Bundle Conductor Oscillations";
   IEEE Conference paper C69 23-PWR,1969
4. Simpson A.;"Stability of Subconductors of Smooth Circular Cross
   Section";Proceedings IEE,Vol.117,No.4,April 1970,pp 741-750
5. Ikegami R.,Salus W.L. and Watanabe R.K.;"Structural Analysis ,High
   Voltage Power Transmission Systems";Boeing Co. Seattle,BPA Contract
   No. 14-03-1362 N,November 1971.
6. Diana G. and Gasparetto M. ;"Energy Method for Computing the Amplitude
   of Vibration of Conductor Bundles due to Wake Effect";L'Energia
   Elettrica,No.8,1972 pp 532-539
7. Cooper K.R.;"Wind Tunnel and Theoretical Investigations into the
   Aerodynamic Stability of Smooth and Stranded Twin Bundled Power
   Conductors";NRC (CANADA) Tec. Report LA-115 and LA-117 ,1973
8. Simpson A. and Price S.J.;"On the use of Damped and Undamped Quasi-
   Static Aerodynamic Models in the Study of Wake Induced Flutter";IEEE
   Conference Paper C 74-378-6 ,1974

9 Price S.J.;"Wake-Induced Flutter of Power Transmission Conductors";
   Journal of Sound and Vibration 38 (1),pp 125-147 ,1975
10.Tsui Y.T.;"Two-Dimensional Stability Analysis of a Circular Conductor
   in the Wake of another";IEEE paper A75 576-579 ,1975
11.Curami A.,Diana G.,Riva R.,DiGiacomo G. and Nicolini P.;"Wake-
   Induced Oscillations in Bundle Systems;Finite Element Method"; IEEE
   paper A77 218-221 ,1977
12.Oliveira A.R.E. and Mansour W.M. ;"Vibrações de Sub-Vão em Linhas
   de Transmissao", Proceedings of the second Brasilian Energy Congress
   CBE April 1981,Rio de Janeiro pp 1193-1204
13.Oliveira A.R.E. and Mansour W.M.;" Calculo das Amplitudes dos Ciclos-
   Limite";Proceedings of sixth Brasilian Congress COBEM,pp1-9,Dec.1981
14.Oliveira A.R.E. and Mansour W.M.;"Nonlinear Analysis and Simulations
   of Auto-Oscillations of Twin Bundle Conductors of Transmission Lines";
   Proceedings of the 10'th IMAC's Congress,Montreal,Canada,August 1982

**FLOW INDUCED VIBRATIONS
IN FLUID ENGINEERING**

Reading, England: September 14-16, 1982

THE INSTABILITY FLOW VELOCITY OF
TUBE ARRAYS IN CROSSFLOW

Shoei-sheng Chen

Argonne National Laboratory, U.S.A.

Summary

Fluid flowing across a tube array can cause dynamic instability. Once large-amplitude oscillations occur, severe damage may result in a short time. Such instability must be avoided in design. This paper presents two methods for calculating the instability flow velocity: an analytical method, which requires fluid-force coefficients, and empirical correlations, which are based on published experimental data. The analytical method is more difficult to apply in practice because of the difficulty to compute fluid-force coefficients. However, empirical correlations can be applied with ease once the flow-velocity distribution and damping value are determined.

Organised and sponsored by
BHRA Fluid Engineering, Cranfield, Bedford MK43 0AJ, England

NOMENCLATURE

c_i	Tube damping coefficient in the x direction
D	Tube diameter
e_i	Tube damping coefficient in the y direction
$E_i I_i$	Flexural rigidity of tube i
f	f_v or f_c
f_c	Natural frequency of coupled mode in still fluid
f_f	Natural frequency of coupled mode in flow
f_i	Fluid force per unit length acting on tube i in the x direction
f_u	Natural frequency of uncoupled mode in still fluid
f_v	Natural frequency in vacuum
g_i	Fluid force per unit length acting on tube i in the y direction
ℓ	Tube length
m	m_v or m_c
m_c	Effective mass per unit length in still fluid
m_f	Effective mass per unit length in flow ($= m_c$)
m_i	Mass per unit length of tube i
m_u	Effective mass per unit length of uncoupled motion
m_v	Tube mass per unit length ($= m_i$)
n	Number of tubes in an array
P	Tube pitch (see Fig. 1)
R	Tube radius
t	Time
T	Tube pitch in the direction 90° from that associated with the tube pitch P
U	Gap flow velocity
U_a	Approach flow velocity
U_m	Mean gap flow velocity
u_i	Displacement of tube i in the x direction
U_f	Reduced flow velocity ($= U_m/f_f D$)
U_r	Reduced flow velocity ($= U_m/f D$)
v_i	Displacement of tube i in the y direction
x,y,z	Cartesian coordinates
α_v, β_v	Instability functions based on in-vacuum parameters
α_c, β_c	Instability functions based on in-fluid parameters
$\alpha_{ij}, \beta_{ij}, \sigma_{ij}, \tau_{ij}$	Fluid added mass coefficients
$\alpha'_{ij}, \beta'_{ij}, \sigma'_{ij}, \tau'_{ij}$	Flow-velocity-dependent damping coefficients
$\bar{\alpha}_{ij}, \bar{\beta}_{ij}, \bar{\sigma}_{ij}, \bar{\tau}_{ij}$	Fluid viscous damping coefficients
$\bar{\alpha}'_{ij}, \bar{\beta}'_{ij}, \bar{\sigma}'_{ij}, \bar{\tau}'_{ij}$	Fluid-damping coefficients
$\alpha''_{ij}, \beta''_{ij}, \sigma''_{ij}, \tau''_{ij}$	Fluidelastic stiffness coefficients
λ	Eigenvalues of Eq. 17

234

μ	Eigenvalues of added mass matrix
δ_m	Mass-damping parameter $(= 2\pi\zeta m/\rho D^2)$
ζ	ζ_v or ζ_c
ζ_c	Modal damping ratio of coupled mode in still fluid
ζ_f	Modal damping ratio of coupled mode in flow
ζ_u	Modal damping ratio of uncoupled mode in still fluid
ζ_v	Modal damping ratio in vacuum
ρ	Fluid density
$\phi(z)$	Gap-flow velocity-distribution function
$\psi_m(z)$	Orthonormal modal function of the mth mode in vacuum
ω	$\sqrt{-1}\,\lambda$
ω_{im}	Natural frequency of the mth mode of tube i in vacuum

1. INTRODUCTION

Fluid flowing across a tube array can induce tube vibration and instability. Small tube vibration due to flow excitations always exists but may not cause detrimental effect. In contrast, dynamic instability may result in large-amplitude oscillations, which can cause extensive damage in a short time.

The characteristics of tube vibration and instability vary widely in different tube arrays. In general, at low flow velocity, tube vibration is induced by turbulence and, in some cases, by vortex shedding. Once the flow velocity is increased to a certain value, tube-oscillation amplitude increases rapidly with flow; this flow velocity is called the instability flow velocity or critical flow velocity.

The critical flow velocity of a tube array depends on many system parameters. Various formulas have been proposed for calculating the critical flow velocity, and stability maps have been employed to evaluate various equipment design. There is a significant scattering in the experimental data obtained by different investigators, and some phenomena cannot be explained using the original stability criterion developed by Connors (Ref. 1). Now, it is recognized that this is attributed to the fact that there are different instability phenomena.

Most recently, Chen (Refs. 2,3) used a mathematical model to show that there are two basic instability mechanisms: velocity mechanism and displacement mechanism. Experiments have also been performed to verify the existence of different instability mechanisms and the transition from one to the other (Ref. 4). Based on the model, the discrepancies among different stability criteria and experimental data obtained by different investigators can be resolved. At present, the instability phenomena of tube arrays in crossflow are much better understood than before.

The objective of this paper is to summarize the available data in a systematic manner. First, the general stability theory is reviewed to illustrate the significance of various system parameters. Then, stability maps are developed based on the published experimental data. Finally, future research needs are pointed out.

2. A MATHEMATICAL MODEL

2.1 Flow Velocity.

The detailed flow-velocity distribution in a tube array is difficult to measure or calculate. In this paper, the average flow velocity will be applied for all cases. Flow velocity depends on tube arrangement and tube pitch. The most frequently encountered tube arrays are shown in Fig. 1. Average flow velocities defined by different investigators are not consistent. In this study, regardless of tube arrangement, the average gap flow velocity U is defined as follows:

$$U = \frac{P/D}{(P/D - 1)} U_a, \tag{1}$$

where U_a is the approach flow velocity, P is tube pitch, and D is tube diameter.

2.2 Equations of Motion.

Consider an array of n tubes vibrating in a flow as shown in Fig. 2. The axes of the tubes are parallel to the z axis, and each tube has the same radius R. The fluid is flowing with a gap velocity $U(z)$. The displacement components of tube i in the x and y directions are u_i and v_i, respectively. The motion-dependent, fluid-force components acting on tube i in the x and y directions are f_i and g_i, respectively; f_i and g_i are (Ref. 2)

$$f_i = -\rho\pi R^2 \sum_{j=1}^{n} (\alpha_{ij} \frac{\partial^2 u_j}{\partial t^2} + \sigma_{ij} \frac{\partial^2 v_j}{\partial t^2})$$

$$- \sum_{j=1}^{n} [(\rho R U \alpha'_{ij} + \bar{\alpha}_{ij}) \frac{\partial u_j}{\partial t} + (\rho R U \sigma'_{ij} + \bar{\sigma}_{ij}) \frac{\partial v_j}{\partial t}]$$

$$+ \rho U^2 \sum_{j=1}^{n} (\alpha''_{ij} u_j + \sigma''_{ij} v_j) \tag{2}$$

and

$$g_i = -\rho\pi R^2 \sum_{j=1}^{n} (\tau_{ij} \frac{\partial^2 u_j}{\partial t^2} + \beta_{ij} \frac{\partial^2 v_j}{\partial t^2})$$

$$- \sum_{j=1}^{n} [(\rho R U \tau'_{ij} + \bar{\tau}_{ij}) \frac{\partial u_j}{\partial t} + (\rho R U \beta'_{ij} + \bar{\beta}_{ij}) \frac{\partial v_j}{\partial t}]$$

$$+ \rho U^2 \sum_{j=1}^{n} (\tau''_{ij} u_j + \beta''_{ij} v_j). \tag{3}$$

The force components proportional to tube accelerations $\partial^2 u_j/\partial t^2$ or $\partial^2 v_j/\partial t^2$ are called fluid-inertial forces, those proportional to tube velocity $\partial u_j/\partial t$ or $\partial v_j/\partial t$ are called fluid-damping forces, and those proportional to tube displacements u_j or v_j are called fluidelastic forces. When the flow velocity is not equal to zero, fluid viscous damping coefficients and flow-velocity-dependent damping coefficients can also be combined as follows:

$$\rho R U \bar{\alpha}'_{ij} = \rho R U \alpha'_{ij} + \bar{\alpha}_{ij},$$

$$\rho R U \bar{\sigma}'_{ij} = \rho R U \sigma'_{ij} + \bar{\sigma}_{ij},$$

$$\rho R U \bar{\tau}'_{ij} = \rho R U \tau'_{ij} + \bar{\tau}_{ij}, \tag{4}$$

$$\rho R U \bar{\beta}'_{ij} = \rho R U \beta'_{ij} + \bar{\beta}_{ij}.$$

The equations of motion for tube i in the x and y directions are

$$E_i I_i \frac{\partial^4 u_i}{\partial z^4} + c_i \frac{\partial u_i}{\partial t} + m_i \frac{\partial^2 u_i}{\partial t^2} = f_i$$

and $\tag{5}$

$$E_i I_i \frac{\partial^4 v_i}{\partial z^4} + e_i \frac{\partial v_i}{\partial t} + m_i \frac{\partial^2 v_i}{\partial t^2} = g_i.$$

In an array of n tubes, there are 2n coupled equations of motion.

236

2.0 Analysis.

In most practical situations, all tubes are of the same length and have the same type of boundary conditions. In this case the modal functions for tubes vibrating in the x and y directions will be the same; thus, let

$$u_i(z,t) = \sum_{m=1}^{\infty} a_{im}(t)\psi_m(z)$$

and

$$v_i(z,t) = \sum_{m=1}^{\infty} b_{im}(t)\psi_m(z),$$

(6)

where $\psi_m(z)$ is the mth orthonormal function of the tubes in vacuum, i.e.,

$$\frac{1}{\ell} \int_0^{\ell} \psi_m \psi_n dz = \delta_{mn},$$

(7)

where ℓ is the length of the tubes. Assume that the flow velocity is given by

$$U(z) = U_m \phi(z),$$

where

$$\frac{1}{\ell} \int_0^{\ell} \phi(z)dz = 1.$$

(8)

Using Eqs. 2-8 yields

$$m_i \omega_{im}^2 a_{im} + c_i \dot{a}_{im} + m_i \ddot{a}_{im} + \rho \pi R^2 \sum_{j=1}^{n} (\alpha_{ij} \ddot{a}_{jm} + \sigma_{ij} \ddot{b}_{jm})$$

$$+ \sum_{j=1}^{n} [(\bar{\alpha}_{ij} + \rho R U_m \alpha'_{ijm})\dot{a}_{jm} + (\bar{\sigma}_{ij} + \rho R U_m \sigma'_{ijm})\dot{b}_{jm}]$$

$$- \rho U_m^2 \sum_{j=1}^{n} (\alpha''_{ijm} a_{jm} + \sigma''_{ijm} b_{jm}) = 0$$

(9)

and

$$m_i \omega_{im}^2 b_{im} + e_i \dot{b}_{im} + m_i \ddot{b}_{im} + \rho \pi R^2 \sum_{j=1}^{n} (\tau_{ij} \ddot{a}_{jm} + \beta_{ij} \ddot{b}_{jm})$$

$$+ \sum_{j=1}^{n} [(\bar{\tau}_{ij} + \rho R U_m \tau'_{ijm})\dot{a}_{jm} + (\bar{\beta}_{ij} + \rho R U_m \beta'_{ijm})\dot{b}_{jm}]$$

$$- \rho U_m^2 \sum_{j=1}^{n} (\tau''_{ijm} a_{jm} + \beta''_{ijm} b_{jm}) = 0,$$

(10)

where ω_{im} is the mth natural frequency of tube i in vacuum, and

$$\alpha'_{ijm} = \frac{1}{\ell} \int_0^{\ell} \alpha'_{ij} \phi \psi_m^2 dz, \qquad \alpha''_{ijm} = \frac{1}{\ell} \int_0^{\ell} \alpha''_{ij} \phi^2 \psi_m^2 dz,$$

$$\sigma'_{ijm} = \frac{1}{\ell} \int_0^{\ell} \sigma'_{ij} \phi \psi_m^2 dz, \qquad \sigma''_{ijm} = \frac{1}{\ell} \int_0^{\ell} \sigma''_{ij} \phi^2 \psi_m^2 dz,$$

(11)

$$\tau'_{ijm} = \frac{1}{\ell} \int_0^\ell \tau'_{ij} \phi \psi_m^2 \, dz, \qquad \tau''_{ijm} = \frac{1}{\ell} \int_0^\ell \tau''_{ij} \phi^2 \psi_m^2 \, dz,$$

(11)
(Contd.)

$$\beta'_{ijm} = \frac{1}{\ell} \int_0^\ell \beta'_{ij} \phi \psi_m^2 \, dz, \qquad \beta''_{ijm} = \frac{1}{\ell} \int_0^\ell \beta''_{ij} \phi^2 \psi_m^2 \, dz.$$

Note that Eqs. 9 and 10 can be applied to all values of m. For each m, there are 2n coupled equations. However, there is no coupling among the equations for different m. This is true for a tube array having the same length and the same type of boundary conditions, no matter whether the tubes are single span or multiple spans.

2.4 Effective Mass, Natural Frequency, and Modal Damping Ratio.

Equations 9 and 10 can be used to obtain effective mass, natural frequencies, mode shapes and modal damping in different conditions.

2.4.1 In-Vacuum Parameters.

For tubes vibrating in vacuum, set all parameters related to fluid equal to zero in Eq. 9 or 10; then

$$m_v = m_i,$$

$$\zeta_v = \frac{c_i}{2m_i \omega_{im}} \quad \text{or} \quad \frac{e_i}{2m_i \omega_{im}},$$

(12)

$$f_v = \frac{\omega_{im}}{2\pi}.$$

In general, c_i and e_i are equal and m_v, ζ_v, and f_v are the same for all tubes.

2.4.2 Uncoupled Vibration in Still Fluid.

Setting $U_m = 0$ in Eqs. 9 and 10 and considering a single tube vibration in a particular direction only, one can obtain the parameters of this type of motion. Since the fluid coupling is neglected, it is called uncoupled vibration. For example, consider tube 1 in the x direction only. Equation 9 becomes

$$\ddot{a}_{im} + 2\zeta_u \omega_u \dot{a}_{1m} + \omega_u^2 a_{1m} = 0,$$

where

$$\zeta_u = \frac{c_1 + \bar{\alpha}_{11}}{2m_u \omega_u},$$

$$2\pi f_u = \omega_u = \frac{\omega_{im}}{(m_u/m_v)^{1/2}},$$

(13)

$$m_u = m_1 + \rho \pi R^2 \alpha_{11}.$$

These parameters depend on tube location and direction of oscillation.

2.4.3 Coupled Vibration in Still Fluid.

Set $U_m = 0$ in Eqs. 9 and 10 and solve for the coupled equations; natural frequencies, f_c, modal damping ratio, ζ_c, and effective mass, m_c, can be obtained. For a tube array consisting of identical tubes, a closed-form solution is obtained for m_c and f_c (Ref. 5):

$$m_c = m_i + \rho \pi R^2 \mu$$

and

$$2\pi f_c = \omega_c = \frac{\omega_{im}}{(m_c/m_v)^{1/2}} ,$$

(14)

where μ is the eigenvalue of the added mass matrix. In an array of n tubes, there are 2n coupled natural frequencies corresponding to a single frequency in vacuum.

2.4.4 In-flow Vibration.

Equations 9 and 10 may be written in the matrix form

$$[M]\{\ddot{W}\} + [C]\{\dot{W}\} + [K]\{W\} = \{0\},$$

(15)

where [M], [C], and [K] are the mass, damping, and stiffness matrices, respectively, and {W} is the displacement vector. Let

$$\{W\} = \{\overline{W}\}\exp(\lambda t).$$

(16)

Substituting Eq. 16 into Eq. 15 yields

$$[\lambda^2 M + \lambda C + K]\{\overline{W}\} = 0.$$

(17)

The eigenvalue λ and the corresponding eigenvector can be calculated from Eq. 17. Oscillation frequency is designated by ω. The natural frequency and modal damping ratio are given by

$$f_f = \frac{Re(\omega)}{2\pi}$$

and

$$\zeta_f = \frac{Im(\omega)}{Re(\omega)} .$$

(18)

The effective mass m_f is the same as that in still fluid m_c.

Different system parameters in different conditions are summarized in Table 1. Some of these parameters will be used in the stability criteria.

2.5 Instability Mechanisms and Stability Criteria.

The role of various fluid-force coefficients can be demonstrated using the mathematical model.

1. Fluid inertial forces, proportional to fluid added mass coefficients α_{ij}, σ_{ij}, τ_{ij}, and β_{ij}, affect system natural frequencies and introduce coupled oscillations among different tubes.

2. Fluid-damping forces, proportional to $\overline{\alpha}'_{ij}$, $\overline{\sigma}'_{ij}$, $\overline{\tau}'_{ij}$ and $\overline{\beta}'_{ij}$, contribute to system damping or may cause instability; in general, the effect on natural frequencies is not very significant.

3. Fluidelastic forces, proportional to α''_{ij}, σ''_{ij}, τ''_{ij}, and β''_{ij}, may affect both system damping and natural frequencies and may cause instability.

Stability of a tube array can be determined by the characteristics of ω: (1) when ω is real, the system performs undamped oscillations; (2) when ω is complex, having a positive imaginary part, the system is stable and performs damped oscillations; (3) when ω is complex, having a negative imaginary part, the system becomes unstable dynamically; and (4) when ω is imaginary, the system loses stability by divergence. The computation procedures are straightforward based on Eq. 17.

Dynamic instability may be caused by velocity mechanism or displacement mechanism.

1. Velocity Mechanism: The dominant fluid force is proportional to the velocity of the tubes. Depending on the reduced flow velocity, fluid-damping force may

act as an energy-dissipation mechanism or an excitation mechanism for tube oscillations. When it acts as an excitation mechanism, the system damping is reduced. Once the modal damping of a mode becomes negative, the tubes lose stability. This type of instability is called fluid-damping-controlled instability. The instability criterion is (Ref. 2) given by

$$\frac{U_m}{f_v D} = \alpha_v \frac{2\pi\zeta_v m_v}{\rho D^2} \tag{19}$$

or

$$\frac{U_m}{f_c D} = \alpha_c \frac{2\pi\zeta_c m_c}{\rho D^2} . \tag{20}$$

where α_v is a function of fluid damping coefficients $\bar{\alpha}'_{ij}$, $\bar{\sigma}'_{ij}$, $\bar{\tau}'_{ij}$, and $\bar{\beta}'_{ij}$, and α_c is a function of flow-velocity-dependent damping coefficients α'_{ij}, σ'_{ij}, τ'_{ij}, and β'_{ij}.

2. Displacement Mechanisms: The dominant fluid force is proportional to the displacements of the tubes. The fluidelastic force may affect natural frequencies as well as modal damping. As the flow velocity increases, the fluidelastic force may reduce the modal damping. When the modal damping of a mode becomes negative, the tubes become unstable; this type of instability is called fluidelastic-stiffness-controlled instability. The stability criterion is given by (Ref. 2)

$$\frac{U_m}{f_v D} = \beta_v \left(\frac{2\pi\zeta_v m_v}{\rho D^2}\right)^{0.5} \tag{21}$$

or

$$\frac{U_m}{f_c D} = \beta_c \left(\frac{2\pi\zeta_c m_c}{\rho D^2}\right)^{0.5} , \tag{22}$$

where β_v and β_c are functions of fluidelastic stiffness coefficients α''_{ij}, σ''_{ij}, τ''_{ij}, and β''_{ij}.

In general cases, the two mechanisms are superimposed on each other. Then the stability criterion can be written

$$\frac{U_m}{f_v D} = F \left(\frac{2\pi\zeta_v m_v}{\rho D^2} , \frac{m_v}{\rho D^2} , \frac{P}{D} , \text{turbulence characteristics}\right) \tag{23}$$

or

$$\frac{U_m}{f_c D} = G \left(\frac{2\pi\zeta_c m_c}{\rho D^2} , \frac{m_c}{\rho D^2} , \frac{P}{D} , \text{turbulence characteristics}\right). \tag{24}$$

The main differences of the two basic instability mechanisms are summarized in Table 2.

It should be emphasized that both fluid-damping coefficients and fluidelastic-stiffness coefficients are functions of the reduced flow velocity U_f (= $U_m/f_f D$). Therefore, the parameters α_v and α_c in Eqs. 19 and 20, and β_v and β_c in Eqs. 21 and 22, are functions of the reduced flow velocity U_f.

3. EMPIRICAL STABILITY CRITERIA

3.1 Various Published Stability Criteria.

Various stability criteria have been proposed based on experimental data. Most of these criteria can be grouped into two classes:

1. The critical flow velocity U_m/fD is a function of the mass damping parameter,

240

$$\frac{U_m}{fD} = \alpha_1 \left(\frac{2\pi m \zeta}{\rho D^2}\right)^{\alpha_2}. \tag{25}$$

2. The critical flow velocity is a function of mass ratio $(m/\rho D^2)$ and damping $(2\pi\zeta)$,

$$\frac{U_m}{fD} = \beta_1 \left(\frac{m}{\rho D}\right)^{\beta_2} (2\pi\zeta)^{\beta_3}. \tag{26}$$

Those models have been adopted by various investigators; the studies are summarized in Tables 3 and 4 (see Refs. 1, 6-17).

Most recently, Price and Paidoussis (Ref. 18) proposed another stability criterion:

$$\frac{U_m}{fD} = 16.9 \left(\frac{\sqrt{T^2 + P^2}}{D} - 1\right)^{1.7} \left\{1 + \left(1 + 0.365 \frac{2\pi m \zeta}{\rho D^2}\right)^{0.5}\right\}. \tag{27}$$

This criterion is applicable for tube arrays losing stability at relatively large U_r.

Experimental data for the critical flow velocity obtained by various experimentalists are not in agreement, and various stability criteria do not correlate well. This is attributed to the following reasons:

1. Different parameters are used by different investigators; some use in-vacuum parameters, and some use in-fluid parameters or in-flow parameters. Even with the same stability criterion, the results will be different using two different sets of parameters, as illustrated in Table 1.

2. Instability may be caused by different instability mechanisms. In the past, the fluidelastic-stiffness-controlled instability mechanism has been used exclusively. It is not expected that the stability criterion for fluidelastic-stiffness-controlled instability can be used to correlate data for fluid-damping-controlled instability.

3. The gap flow velocities defined by different investigators are not consistent with one another, and the critical flow velocities are not determined with the same method.

4. Critical flow velocities of tube arrays depend on tube arrangement, spacing, and other parameters.

It is apparent that to develop a universal stability criterion applicable to all cases will be difficult, if not impossible.

3.2 Dimensionless Parameters.

In the empirical correlations for critical flow velocities, different system parameters measured in vacuum or in still fluid are used. Most investigators use the modal damping value ζ, mass per unit length m, and natural frequency f measured in still fluid, but some use those measured in vacuum (actually in air), or in flowing fluid. From the analysis of the mathematical model, it is clear that one can use those parameters determined either in vacuum or in still fluid, provided that appropriate fluid-damping coefficients are used. However, the stability criteria based on the in-vacuum or in-fluid values are not identical. For practical applications, it is more convenient to use the in-vacuum parameters, since they are well defined. In-fluid parameters are more difficult to determine. In particular, for heavy fluid, inertia and viscous coupling become important; coupled-mode frequencies f_c, effective mass m_c, and modal damping ζ_c are more difficult to measure.

The stability criteria, Eqs. 19, 21, and 23, are expressed in terms of in-vacuum parameters, and Eqs. 20, 22, and 24 are expressed in terms of in-fluid parameters. The effects of various parameters are briefly discussed as follows:

241

1. $2\pi\zeta_v m_v/\rho D^2$ or $2\pi\zeta_c m_c/\rho D^2$: This is the most important parameter. The critical flow velocity increases with this parameter.

2. $m_v/\rho D^2$ or $m_c/\rho D^2$: This parameter determines the role of added mass.

3. P/D: Fluid-force coefficients depend on tube arrangement; therefore, the critical flow velocity will depend on P/D.

4. Turbulence Characteristics: Fluid-force coefficients depend on incoming turbulence characteristics (intensity and scale). Again, the critical flow velocity will depend on the turbulence characteristics.

3.3 Empirical Stability Diagrams.

Available experimental data for the reduced flow velocity U_r ($= U_m/fD$) will be plotted as a function of the mass-damping parameter δ_m ($= 2\pi\zeta m/\rho D^2$) for different tube arrangements. It was shown in Sec. 2.5 that either the in-vacuum parameters m_v, f_v and ζ_v, or the in-fluid parameters, m_c, f_c and ζ_c, can be used in the stability criteria. In most experiments, the in-vacuum parameters are actually measured in air. The effect of air on those parameters is small. Therefore, m_v, f_v, and ζ_v will be based on those measured in air. In liquid-flow tests, the values of in-fluid parameters m_u, f_u, and ζ_u are generally measured and, in most cases, in-vacuum parameters m_v, f_v, and ζ_v, and in-fluid parameters m_c, f_c, and ζ_c, are not measured. Under such circumstances, the in-fluid parameters of uncoupled modes m_u, f_u, and ζ_u are used.

In the stability diagrams, the data for air flow will be denoted by open symbols, liquid flow by solid symbols, and two-phase flow by semi-solid symbols. The stability diagrams summarize the published data for different tube arrangements obtained by different investigators.

3.3.1 Tube Row. (Refs. 1, 4, 8, 11, 20-26)

The critical flow velocity for a tube row depends on the pitch-to-diameter ratio P/D. Several studies have been made to investigate the effect of P/D on the critical flow velocity. Blevins (Ref. 19) shows that

$$\alpha_1 = \frac{2(2\pi)^{0.5}}{[(\frac{D}{P})^2 \{2(\frac{D}{P})^3 - (\frac{D}{P})^2\}]^{0.25}} \cdot \tag{28}$$

At D/P = 0.5, α_1 becomes infinite; this is not consistent with the experimental results. Equation 28 is applicable for P/D < 2 only. In a systematic investigation, Ishigai et al. (Ref. 20) show that, for a tube row,

$$\alpha_1 = 8\left(\frac{P}{D} - 0.375\right) \quad . \tag{29}$$

Therefore, the critical flow velocity is proportional to (P/D - 0.375). A new reduced critical flow velocity incorporating the effect of tube spacing is defined as follows:

$$\bar{U}_r = \frac{\dfrac{U_m}{fD}}{\dfrac{P}{D} - 0.375} \cdot \tag{30}$$

Then, $\bar{U}_r$ is independent of tube spacing P/D.

Figure 3 shows the critical flow velocity of $\bar{U}_r$ as a function of the mass-damping parameter $2\pi\zeta m/\rho D^2$. The in-vacuum parameters are used whenever these data are available. Experimental data obtained in air correlate reasonably well for tube rows with different pitch-to-diameter ratios ranging from 1.19 to 2.68. In liquid flow, with the exception of Chen and Jendrzejczyk (Ref. 4), who use in-vacuum parameters, other investigators use different parameters: Connors (Ref. 11) and Halle and Lawrence (Ref. 23) use f_u, m_u, and ζ_u, while Heilker and Vincent (Ref. 26) use f_f, m_u and ζ_f.

3.3.2 Square Array (90°). (Refs. 8, 9, 11, 12, 14, 17, 22, 25-28)

Square tube arrays with different spacing are tested in air by Soper (Ref. 28), and in water by Chen and Jendrzejczyk (Ref. 14). The results of these tests show that the critical flow velocity is not very sensitive to the variation of tube spacing. Therefore, for square arrays, the critical flow velocity U_r will be plotted as a function of δ_m regardless of the spacing. The results are given in Fig. 4.

Although different tests are performed for different flow conditions and tube spacing, the data correlate fairly well. In particular, for in-water tests, different investigators use different parameters: Tanaka and Takahara (Ref. 17) use in-vacuum parameters, Heilker and Vincent (Ref. 26) use in-flow parameters, and the others use in-fluid parameters.

3.3.3 Rotated Square Array (45°). (Refs. 9, 12, 22, 26, 28)

Based on Soper's data (Ref. 28), the critical flow velocity is approximately proportional to (P/D - 0.5). Note that in Eq. 30, $U_r = \bar{U}_r$ for P/D = 1.375. In order to make $\bar{U}_r = U_r$ at P/D = 1.375, for this case, $\bar{U}_r$ is defined as follows:

$$\bar{U}_r = \frac{\dfrac{U_m}{fD}}{1.143\left(\dfrac{P}{D} - 0.5\right)} . \tag{31}$$

The results are given in Fig. 5.

3.3.4 Triangular Array (30°). (Refs. 8, 9, 12, 14, 22, 26-30)

Following the same procedure as that for rotated square arrays, $\bar{U}_r$ is defined as follows:

$$\bar{U}_r = \frac{\dfrac{U_m}{fD}}{2.105\left(\dfrac{P}{D} - 0.9\right)} . \tag{32}$$

At P/D = 1.375, $\bar{U}_r = U_r$. Figure 6 summarizes the results. There is more scattering of the data at low values of δ_m. This is attributed to different parameters used by different investigators and may be caused by different spacings. Note that Eq. 32 is based on Soper's data obtained in a wind tunnel. The variation of the critical flow velocity with tube spacing in water may be different from that in air. For example, the variation of $\bar{U}_r$ with tube spacing in the data by Zukauskas and Katinas (Ref. 27) is different from those by Soper (Ref. 28).

3.3.5 Rotated Triangular Arrays (60°). (Refs. 9, 12, 13, 22, 26, 28, 29, 30)

Soper's data show that the critical flow velocity varies insignificantly with tube spacing. All available experimental data are plotted in Fig. 7 regardless of the tube spacing. They agree reasonably well.

4. DESIGN EVALUATION METHODS TO AVOID INSTABILITY

Critical flow velocities can be predicted by using the analytical method or by using empirical stability diagrams.

4.1 Analytical Method.

The analysis is described in Sec. 2. The stability of a tube array can be determined from Eq. 17. The following steps can be taken:

4.1.1 Fluid-Force Coefficients.

To carry out the analysis, various fluid-force coefficients must be known. Fluid-added mass coefficients α_{ij}, β_{ij}, σ_{ij}, and τ_{ij} can be calculated based on the potential flow solution; a computer program to calculate those coefficients is available (Ref. 31). At present, no analytical method is available to calculate the fluid-damping coefficients $\bar{\alpha}'_{ij}$, $\bar{\beta}'_{ij}$, $\bar{\sigma}'_{ij}$, and $\bar{\tau}'_{ij}$, and fluidelastic stiffness coefficients α''_{ij}, β''_{ij}, σ''_{ij}, and τ''_{ij}. There are only a few sets of experimental data.

4.1.2 Tube Properties.

Tube mass m_i and tube natural frequencies in vacuum ω_{im} can be measured or calculated relatively easily (Refs. 32, 33). Tube damping coefficients c_i or e_i are calculated as follows:

$$c_i = 2m_i \omega_{im} \zeta_i. \tag{33}$$

In most cases, ζ_i is measured or estimated. If the tubes in an array are not identical, it is conservative to assume all tubes have the same properties as the one with lower ζ_i and ω_{im}.

4.1.3 Number of Tubes.

For an array of n tubes, there are 2n equations. In practical situations, it is not necessary to include all tubes; a finite number of tubes will be sufficient in computing the critical flow velocity. It is suggested the tube array used in computation should include the following:

> Tube row, five tubes
> Square and rotated square arrays, nine tubes
> Triangular and rotated triangular arrays, seven tubes.

In some cases, a small number of tubes can be used to estimate roughly the critical flow velocity. For example, using one tube only will enable one to calculate the critical flow velocity for fluid-damping-controlled type instability.

4.1.4 Parametric Study.

Once the analysis is carried out, it is straightforward to conduct a parametric study. This information may be useful to alleviate instability.

4.2 Empirical Stability Diagrams.

The stability diagrams given in Figs. 3-7 can be used in design evaluation.

1. <u>Calculation of the Mass-Damping Parameter</u> δ_m $(= 2\pi m_v \zeta_v / \rho D^2)$

Fluid density ρ, tube diameter D, and tube mass per unit length m_v, are relatively easy to determine. The modal damping ratio ζ_v can be estimated or measured; it is recommended to use the in-vacuum value.

2. <u>Determination of the Lower Bound of the Critical Flow Velocity</u>

The lower bounds for different tube arrays are given in Figs. 3-7 by solid lines. These are summarized in Table 5. The critical flow velocity calculated from Table 5 can then be compared with the actual flow velocity.

5. CONCLUDING REMARKS

The instability of a tube array subjected to crossflow may be attributed to fluid-damping force, fluidelastic force, or a combination of both. Since a different fluid force will be dominant in different parameter ranges, a single stability criterion cannot be applied for all cases. In general, at low values of mass-damping parameter, instability is often of the fluid-damping-controlled type; and at high values of the mass-damping parameter, instability is frequently of the fluidelastic-stiffness-controlled type. In most cases, both fluid-damping and fluidelastic forces contribute to the instability of a tube array.

In the past, the instability was considered to be attributed to the fluidelastic force only and the instability criterion was that developed by Connors (Ref. 1). Different investigators have used Connors' stability criterion to correlate all experimental data and found the correlation did not collapse the data well. Now it is recognized that some of the instability is attributed to the fluid-damping force, and naturally, we don't expect that Connors' criterion can be applied to these cases.

Based on different instability mechanisms and published experimental data, the stability criteria for five different tube arrangements are summarized in this paper. I realize that these stability criteria in some cases may not represent true critical flow velocities for tube arrays under different conditions. Nevertheless, these criteria have accounted for the available information at present and will be improved as soon as more theoretical and experimental results become available.

Ideally, the critical flow velocity should be calculated based on the procedures discussed in Sec. 2. Because of the difficulty to calculate the fluid-damping coefficients and fluidelastic-stiffness coefficients, it is not possible to perform such computations in general. Only a few cases, in which these fluid coefficients have been measured, permit us to predict the critical flow velocity in a rigorous manner. In design assessment, it is almost imperative to rely on empirical correlations. The approach to use these correlations is straightforward. For example, consider a square array with a pitch-to-diameter ratio of 1.5. Assume that the tube natural frequency in air is 50 Hz, tube diameter is 2.54 cm, and the mass-damping parameter is 0.2. The critical flow velocity calculated from Table 5 is as follows:

$$U_m = fD(2.1\delta_m^{0.15})$$

$$= 50 \times 2.54(2.1 \cdot 0.2^{0.15})$$

$$= 2.09 \text{ m/s}. \tag{34}$$

To improve the stability criteria given in Table 5, the key step is to predict the fluid-force coefficients. The fluid-inertia coefficients, α_{ij}, β_{ij}, σ_{ij}, and τ_{ij}, can be calculated based on the potential flow theory (Ref. 5). In most practical applications, the results from the potential flow theory will be acceptable. However, the potential flow solutions for fluid-damping coefficients and fluidelastic-stiffness coefficients are unacceptable. Therefore, the main task is to develop an analytical method to compute these coefficients. This is one of the problems that is certain to be pursued in the field of computational fluid dynamics. These fluid-force coefficients can also be measured using the technique demonstrated by Tanaka and Takahara (Ref. 17). This is a very tedious process. Furthermore, fluid-force coefficients are a function of geometry. It will require a large number of experiments before one can quantify the fluid force for all practical tube arrangements.

In addition to the prediction techniques, understanding of the basic fluid dynamics for flow across a vibrating tube array remains a difficult task. Detailed flow measurements and theoretical study of the flow field have to be carried out before one can identify the basic flow effect and the effect of tube motion on flow field. The interaction process of tube array and crossflow is certain to receive more attention in the future.

In the past decade, the original work by Connors (Ref. 1) has given great impetus to the numerous, innovative studies of tube arrays subjected to crossflow. The mechanism described by Connors has been used to interpret different phenomena, and Connors' stability criterion has been used extensively and misused occasionally. Furthermore, erroneous interpretations of Connors' criterion by some investigators have been published in different journals. These illustrate the lack of understanding in this subject. Based on the model described in Sec. 2, the inconsistency among experimental data obtained by different investigators as well as different phenomena reported in literature can now be resolved reasonably well. Although it is still not possible to predict the critical flow velocity analytically, because of the difficulty to calculate the fluid-force coefficients, there is a sound basis for further development to quantify the instability flow velocity.

ACKNOWLEDGMENTS

This work was performed under the sponsorship of the Office of Reactor Research and Technology, U. S. Department of Energy.

I am indebted to Dr. M. W. Wambsganss for his review and suggestions of the original manuscript and Dr. B. M. H. Soper of AERE Harwell, England, for providing his data for the critical flow velocity as a function of the mass-damping parameter.

REFERENCES

1. Connors, H. J.: "Fluidelastic vibration of tube arrays excited by cross flow." Flow-induced vibration of heat exchangers, ASME, Dec. 1970, pp. 42-56.

2. Chen, S. S.: "Instability mechanisms and stability criteria of a group of circular cylinders subjected to cross flow; part I: theory." ASME Paper No. 81-DET-21, to appear in J. Mech. Design, ASME.

3. Chen, S. S.: "Instability mechanisms and stability criteria of a group of circular cylinders subjected to cross flow, part II: numerical results and discussions." ASME Paper No. 81-DET-22, to appear in J. Mech. Design, ASME.

4. Chen, S. S., and Jendrzejczyk, J. A.: "Experiment and analysis of instability of tube rows subject to liquid crossflow." to be published.

5. Chen, S. S.: "Vibration of nuclear fuel bundles." Nuclear Engineering and Design 35, 1975, pp. 399-422.

6. Blevins, R. D.: "Fluid elastic whirling of a tube row." Trans. ASME, J. Pressure Vessel Technology 96, 1974, pp. 263-267.

7. Chen, Y. N.: "The orbital movement and the damping of the fluidelastic vibration of tube banks due to vortex formation, part 2, criterion for the fluidelastic orbital vibration of tube arrays." Trans. ASME 96, Series B, 1974, pp. 1065-1071.

8. Gross, H.: "Investigations in aeroelastic vibration mechanisms and their application in design of tubular heat exchangers." Technical University of Hanover, Ph.D. dissertation, 1975.

9. Gorman, D. J.: "Experimental development of design criteria to limit liquid cross-flow-induced vibration in nuclear reactor heat exchange equipment." Nuclear Science and Engineering, 61, 1976, pp. 324-336.

10. Savkar, S. D.: "A brief review of flow induced vibration of tube arrays in cross-flow." Trans. ASME, Journal of Fluids Engineering, 99, 1977, pp. 517-519.

11. Connors, H. J.: "Fluidelastic vibration of heat exchanger tube arrays." Trans. ASME, Journal of Mechanical Design 100, 1978, pp. 347-353.

12. Pettigrew, M. J., Sylvestre, Y., and Campagna, A. O.: "Vibration analysis of heat exchanger and steam generator designs." Nuclear Engineering and Design 48, 1978, pp. 97-115.

13. Weaver, D. S., and Grover, L. K.: "Cross-flow induced vibrations in a tube bank - turbulent buffeting and fluid elastic instability." Journal of Sound and Vibration, 59, 2, 1978, pp. 277-294.

14. Chen, S. S., and Jendrzejczyk, J. A.: "Experiments on fluid elastic instability in tube banks subjected to liquid cross flow." Journal of Sound and Vibration, 78, 3, 1981, pp. 355-381.

15. Paidoussis, M. P.: "Flow-induced vibrations in nuclear reactors and heat exchangers, practical experiences and state of knowledge." Sym. on Practical Experiences with Flow-Induced Vibration (Karlsruhe, Germany: Sept. 3-6, 1979).

16. Paidoussis, M. P.: "Fluidelastic vibration of cylinder arrays in axial and cross flow, state of the art." In: Flow-Induced Vibration Design Guidelines, ASME, 1980, pp. 11-46.

17. Tanaka, H., and Takahara, S.: "Fluid elastic vibration of tube array in cross flow." Journal of Sound and Vibration, 77, 1981, pp. 19-37.

18. Price, S. J., and Paidoussis, M. P.: "Fluidelastic instability of an infinite double row of circular cylinders subjected to a uniform crossflow." ASME Paper No. 81-DET-24. To appear in J. Mech. Design, ASME.

19. Blevins, R. D.: "Fluid elastic whirling of tube rows and tube arrays." Trans. ASME, Journal of Fluids Engineering, 99, 1977, pp. 457-461.

20. Ishigai, S., Nishikawa, E., and Yagi, E.: "Structure of gas flow and vibration in tube bank with tube axes normal to flow." Int. Symp. on Marine Engineering, (Tokyo: 1973) pp. 1-5-23 to 1-5-33.

21. Tanaka, H.: "A study on fluid elastic vibration of a circular cylinder array (one-row cylinder array)." Trans. of the Japan Society of Mechanical Engineers 46, 408 (Section B), 1980, pp. 1398-1407.

22. Hartlen, R. T.: "Wind tunnel determination of fluidelastic vibration thresholds for typical heat exchanger tube patterns." 74-309-K, Ontario Hydro, Toronto, Canada, Aug. 1974.

23. Halle, H., and Lawrence, W. P.: "Crossflow-induced vibration of a row of circular cylinders in water." Presented at the ASME-IEEE Joint Power Generation Conference (Long Beach, CA: 1977) ASME Paper No. 77-JPGC-NE-4.

24. Southworth, P. J., and Zdravkovich, M. M.: "Cross-flow-induced vibrations of finite tube banks in in-line arrangements." Journal of Mechanical Engineering Science, 17, 4, 1975, pp. 190-198.

25. Blevins, R. D., Gibert, R. J., and Viliard, B.: "Experiments on vibration of heat exchanger tube arrays in cross flow." 6th SMiRT, 1981, Paper No. B6/9.

26. Heilker, W. J., and Vincent, R. Q.: "Vibration in nuclear heat exchangers due to liquid and two-phase flow." Presented at the Century 2 Nuclear Engineering Conference (San Francisco, CA: Aug. 19-21, 1980) ASME Paper No. 80-C2/NE-4.

27. Zukauskas, A., and Katinas, V.: "Flow-induced vibration in heat-exchanger tube banks," Symp. on Practical Experiences with Flow-Induced Vibration (Karlsruhe, Germany: Sept. 3-6, 1979).

28. Soper, B. M.: "The effect of tube layout on the fluidelastic instability of tube bundles in cross flow." Flow-Induced Heat Exchanger Tube Vibration - 1980, ASME, Ed. J. M. Chenoweth and J. R. Stenner, 1980, pp. 1-9.

29. Connors, H. J.: "Fluidelastic vibration of tube arrays excited by nonuniform cross flow." Flow-Induced Vibration of Power Plant Components, ASME, 1980, pp. 93-107.

30. Yeung, H., and Weaver, D. S.: "The effect of approach flow direction on the flow induced vibrations of a triangular tube array." ASME Paper No. 81-DET-25, to appear in Journal of Mechanical Design.

31. Chen, S. S., and Chung, H.: "Design guide for calculating hydrodynamic mass, part I: circular cylindrical structures." ANL-CT-76-45, June 1976.

32. Chen, S. S., and Wambsganss, M. W.: "Design guide for calculating natural frequencies of straight and curved beams on multiple supports," ANL-CT-74-06, June 1974.

33. Blevins, R. D.: "Formulas for natural frequency and mode shape." Van Nostrand Reinhold Co., 1979.

34. Chen, S. S., and Jendrzejczyk, J. A.: "Flow velocity dependence of damping in tube arrays subjected to liquid crossflow." Trans. ASME, Journal of Pressure Vessel Technology 103, 1981, pp. 130-135.

Table 1. Effective mass, natural frequency, and modal
damping ratio in different conditions

Parameters	In Vacuum	In Still Fluid		In Flow
		Uncoupled Vibration	Coupled Vibration	
Effective Mass (m)	m_v	m_u	m_c	m_f
Natural Frequency (f)	f_v	f_u	f_c	f_f
Modal Damping Ratio (ζ)	ζ_v	ζ_u	ζ_c	ζ_f

Table 2. Comparison of two different instability mechanisms

Mechanisms	Fluid-Damping-Controlled Instability (Velocity Mechanism)	Fluidelastic-Stiffness-Controlled Instability (Displacement Mechanism)
Instability Criteria	$\dfrac{U}{f_v D} = \alpha_v(U_f)\left(\dfrac{2\pi m_v \zeta_v}{\rho D^2}\right)$	$\dfrac{U}{f_v D} = \beta_v(U_f)\left(\dfrac{2\pi m_v \zeta_v}{\rho D^2}\right)^{0.5}$
Dominant Fluid Force	Flow-Velocity-Dependent Damping Force	Fluidelastic Force
Theory	Unsteady Fluid Dynamic Forces	Quasi-Steady Fluid Dynamic Forces
Fluid Coupling	Not necessary	Necessary
Phase Relationship of Tube Oscillations	0, 180°	0, ±90°, 180°
Effect of Detuning	Less significant	More significant

Table 3. Values of α_1 and α_2

Investigators	α_1	α_2	Remarks
Connors (Ref. 1) (1970)	9.9	0.5	Tube row with P/D = 1.42
Blevins (Ref. 6) (1974)	$\dfrac{2(2\pi)^{0.5}}{(C_x C_y)^{0.25}}$	0.5	C_x and C_y are fluidelastic force coefficients
Y. N. Chen (Ref. 7) (1974)	$\beta Re^{-0.25}$	0.5	Re = Reynolds number β = constant
Gross (Ref. 8) (1975)	4/k	1.0	For square array, and k determined from fluid force
Gorman (Ref. 9) (1976)	3.3	0.5	Suggested design guideline
Savkar (Ref. 10) (1977)	$4.95(P/D)^2$	0.5	For triangular arrays
Connors (Ref. 11) (1978)	$0.37 + 1.76 \dfrac{P}{D}$	0.5	For square array $1.41 \leqslant \dfrac{P}{D} \leqslant 2.12$
Pettigrew et al. (Ref. 12) (1978)	3.3	0.5	Suggested design guideline
Weaver and Grover (Ref. 13) (1978)	7.1	0.21	Rotated triangular array P/D = 1.375
Chen and Jendrzejczyk (Ref. 14) (1981)	2.49 to 6.03	0.2 to 1.08	For various rectangular arrays and mixed array in water flow

Table 4. Values of β_1, β_2, and β_3

Investigators	β_1	β_2	β_3	Remarks
Paidoussis (Ref. 15) (1979)		0.5	0.25	Using published data
Paidoussis (Ref. 16) (1980)	a. $2.3\left(\frac{P}{D} - 1\right)$ b. $5.8\left(\frac{P}{D} - 1\right)$	0.4	0.4	a. Including all data b. Excluding some data
Tanaka and Takahara (Ref. 17) (1981)		a. 0.5 b. 0.333	a. 0.5 b. 0.2	a. Low-density fluid b. High-density fluid

Table 5. Lower bound on critical flow velocity

Array	Parameter Range for δ_m	$\dfrac{U_m}{fD}$
Tube Row	$0.05 < \delta_m < 0.3$	$1.35(P/D - 0.375)\delta_m^{0.06}$
	$0.3 < \delta_m < 4.0$	$2.30(P/D - 0.375)\delta_m^{0.5}$
	$4.0 < \delta_m < 300$	$6.00(P/D - 0.375)\delta_m^{0.5}$
Square (90°)	$0.03 < \delta_m < 0.7$	$2.10\ \delta_m^{0.15}$
	$0.7 < \delta_m < 300$	$2.35\ \delta_m^{0.48}$
Rotated Square (45°)	$0.1 < \delta_m < 300$	$3.54(P/D - 0.5)\delta_m^{0.5}$
Triangular (30°)	$0.1 < \delta_m < 2$	$3.58(P/D - 0.9)\delta_m^{0.1}$
	$2 < \delta_m < 300$	$6.53(P/D - 0.9)\delta_m^{0.5}$
Rotated Triangular (60°)	$0.01 < \delta_m < 1$	$2.8\ \delta_m^{0.17}$
	$1 < \delta_m < 300$	$2.8\ \delta_m^{0.5}$

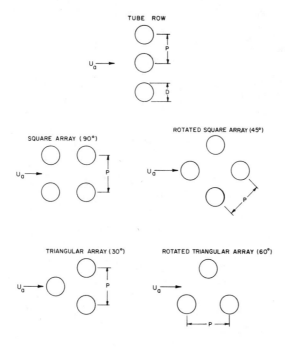

Fig. 1. Tube arrangement

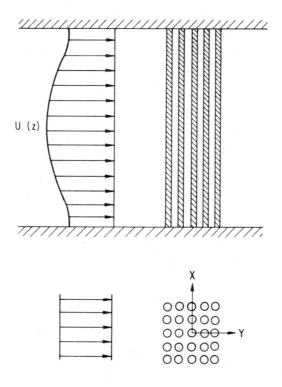

Fig. 2. An array of tubes in crossflow

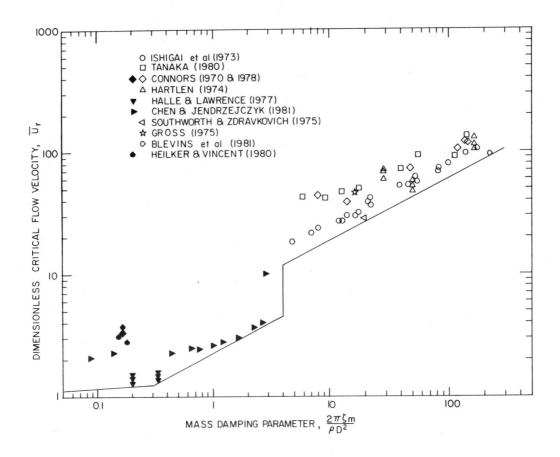

Fig. 3. Stability diagram for tube rows

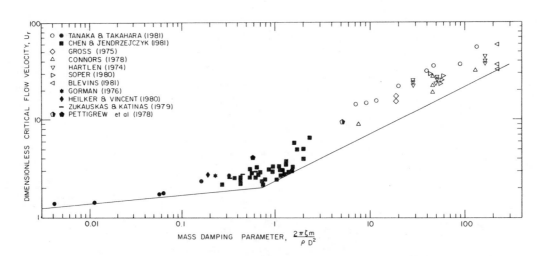

Fig. 4. Stability diagram for square arrays

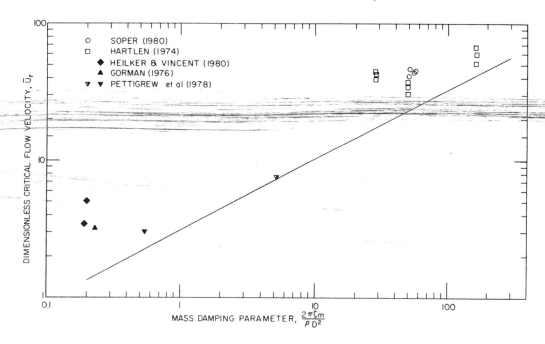

Fig. 5. Stability diagram for rotated square arrays

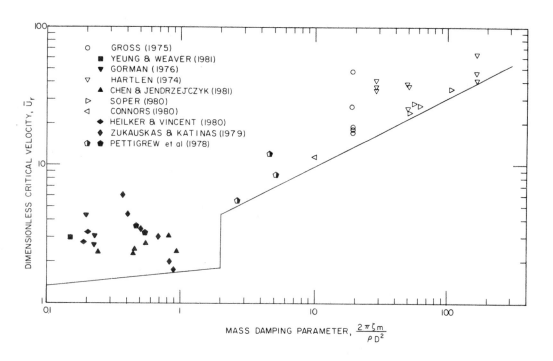

Fig. 6. Stability diagram for triangular arrays

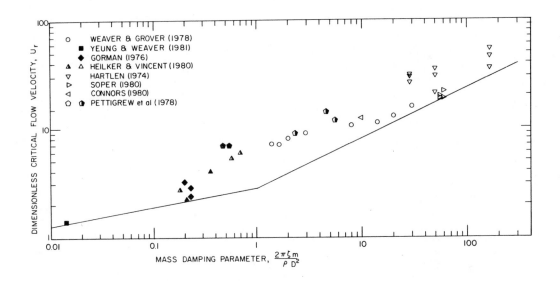

Fig. 7. Stability diagram for rotated triangular arrays

FLOW INDUCED VIBRATIONS
IN FLUID ENGINEERING

Reading, England: September 14-16, 1982

COMPARATIVE ANALYSIS OF CROSS-FLOW INDUCED
VIBRATIONS OF TUBE BUNDLES

F.N. Remy and D. Bai

Electricité de France, France

Summary

Flow induced vibrations can be found in a lot of heat exchangers such as steam generators, water heaters, condensers.

In this text, the results of several studies, carried out at Electricité de France Research and Development Laboratories, concerning the forecasting of vibrational motion of tube bundles, are presented. These tests are covering a large range of the reduced damping $m\delta/\rho d^2$. They are done on very different tube bundles submitted to one phase flow or two phase flow; the density of the flow is either high or low.

The comparison of all these results lead to propose a modelling which allows to represent the tubes response in terms of reduced parameters.

So a simplified criterium can be set up, which allows to avoid large vibrations in tube bundles.

Organised and sponsored by
BHRA Fluid Engineering, Cranfield, Bedford MK43 0AJ, England

NOMENCLATURE :

A : Acceleration of the tube

C_F : Force coefficient

D : Diameter of the tube

δ : Logaritmic decrement of the damping

f : Frequency of the tube

F : Force acting on the tube

k : Connor Constant

m : Lineic mass of the tube

M : Apparent generalised mass $M = m \dfrac{\int^{\text{total length}} \varphi^2 \, dl}{\int_{\text{excited length}} \varphi^2 \, dl}$

ρ : Mass volumic of the fluid

U : Gap velocity between tubes

x : Amplitude of the tube

φ : Modal deformation of the tube

1. INTRODUCTION

Flow Induced Vibrations in tube bundles can have destructive effects on heat exchangers of nuclear plants. For that reason, a lot of research studies are being done on that phenomena, the aim of all there studies being to be able to predict and so to avoid large vibrations of tube bundles.

Geometric characteristics of heat exchangers can be very different in nature, as flow characteristics : it's therefore difficult to predict the behaviour of all the different types of heat exchangers from only one or two types of tests made on models.

It's the reason for which we undertook a series of tests including a large variety of fluids - air,water,air and water mixture, low pressure steam - and a large variety of tube bundles - one span or multispan-, on scale models or full scale models. The data obtained from all these tests have been reduced to make a synthetic approach allowing a comparison of all the tests and so a confident prediction of vibration amplitude.

Five series of tests have been carried out, each of them concerning one type of tube bundles.

2. TEST BUNDLES

Bundles number 1 to 3 are mainly connected to P.W.R. steam generators, but cannot correctly take into account the problems linked to the U bend part an to the two phase nature of the flow, the air + water mixture being too different in nature from the steam + water mixture.

Bundles number 4 and 5 are connected to large condensers. The tube arrangements are either square either triangular, but the pitch to diameter ratio is in all cases about 1.4. The main geometric data on these bundles are shown on Table 1.

Bundle n°1 (Fig.1). Composed of 49 tubes : 33 of them are rigid tubes.
. 16 are elastically mounted
. Submitted to water flow.O.D. = 22 mm.
. Tube length : 200 mm.
. Pitch to diameter ratio : 1.44.
. Squarred arrangement. Frequency in water flow : 34 Hz (Ref.1).

Bundle n°2 The same as bundle n°1, but the arrangement is a triangular one.
Pitch to diameter ratio : 1.44 (Ref.1).

Bundle n°3 (Fig.2) Composed of 100 tubes. O.D.: 10 mm.
. Tube length : 1 m.
. Squarred arrangement.
. Pitch to diameter ratio : 1,44.
. Submitted to two phase flow (air + water mixture).
. Tube frequency : 56 Hz in air flow, 40 Hz in water flow.(Ref.2)
. Flow on a window in the center part of the bundle.

Bundle n°4 (Fig.3) Overall tube length : 10,9 m.
. 12 support plates and 2 tube plates O.D. : 19 mm.
. Tube material : Titanium.
. Triangular arrangement.
. Pitch to diameter ratio : 1.37.
. Submitted to low pressure steam flow.

Bundle n°5 Overall tube length : 14,2 m., with 15 support plates.
. Titanium tubes O.D. 19 mm.
. Triangular arrangement.

. Pitch to diameter ratio : 1.42.
. Submitted to low pressure steam flow (Ref.3).

3. TUBE DAMPING

Criteria usually used to predict vibration risks are not very reliable. This is mainly due to the use of a standard value for the damping δ of the tubes. But it appears that the damping value is very variable, specially as far as multispanned tubes are concerned : in fact, the tube hole clearances produce non linearities in the behaviour of the tubes. Futhermore, flow around tubes induces a damping varying with the flow velocity, so with the tube amplitude.

The measurements of the damping in the lift direction of a tube wibrating in the middle of bundles 1 and 2 are shown on Fig.4 The fluid excitation has a turbulence shaped spectrum. The damping values are higher than the values obtained in quiet fluid.

4. COMPARATIVE ANALYSIS

4.1. Results

Results obtained on the five test bundles are presented first. In all cases, it's the velocity in the gap between tubes which is taken into account to evaluate the dynamic head. For bundle n°3, the slip between the two phases is taken into account ; it is measured by an optical probe. Bundles n°1 and 2 are instrumented by means of strain gages, bundles 3 to 5 by biaxial accelerometers.

The movement of the tubes (acceleration or amplitude) versus the gap velocity or the dynamic head of the flow is shown on Fig.5. Great differences in the response of the tubes can be observed, even in the same bundle, where the amplitude level (in drag or lift direction) can be different from tube to tube.

Furthermore, according to the tube bundle, the evolution of tube amplitude can be different, even if flow characteristics or geometric conditions seem to be quite the same : this is true for example for bundles n° 1 and 3.

Consequently, for a better analysis of the tube response, it seems interesting to use non dimensional parameters.

4.2 Non dimensional parameters.

If use assume that the tube movement in one direction is the response of an oscillator, we can describe the excitation by the fluid as a force F, such as $F = C_F.D.\rho U^2$. C_F beeing the excitation force coefficient, including also the influence of the correlation length of the spanwise fluid excitation. At the resonance, the amplitude of the tube can be described as follows :

$$\frac{C_F}{4\pi\delta} = \frac{f^2 MX}{D\rho U^2} = \frac{M A}{4\pi^2 D\rho U^2}$$

The last parameter represents the ratio of the vibration amplitude (or acceleration) by the flow dynamic head ; it is equal to the ratio of the excitation force coefficient by the virtual damping of the tube (Ref.4).

If there is no coupling between tubes and if the excitation is due only to turbulence, the coefficient C_F is quite a constant over the range of Reynolds numbers considered. So far, the C_F/δ parameter will tend to decrease as the flow velocity increases because of the increase of the tube damping with the movement amplitude.

The second non dimensional parameter is the ratio of the dynamic head by the generalised stiffness of the tubes. Using the criterium proposed by CONNORS (Ref.5) $U_c/fD = K (m\delta/\rho D^2)^{\frac{1}{2}}$ this parameter is equal, when the flow velocity reaches the critical value, to the product of the square value of K by the logarithmic decrement of the vibration.

It's then possible to analyse and compare all the data.

4.3. Synthesis of the results

The results elaborated as defined earlier are shown on Fig.6 Several remarks can be done :

. The curves related to bundles submitted to one phase flow have all the same shape : after decreasing or lightly increasing values, there is an abrupt increasing. The first area corresponds to an excitation of the tubes by the flow turbulence, which can be amplified by an organization of the tubes by the the gap between tubes, which lead to a narrowband excitation spectrum.

For values of $\rho U^2/f^2 M$ greater than 1, for the bundles number 2 and 3 submitted to one phase flow, the abrupt increase of the values is due to a decrease of the apparent damping because of energy transfer by a coupling phenomenon with neighbouring tubes.

For the bundles number 4 and 5, the amplitude levels are one order of magnitude lower. This is perhaps due to the special structure of the flow or more probably to an important damping value of the multispanned tubes.

For these bundles 4 and 5, impacting of adjacent tubes is obtained whereas the amplitude computed from R.M.S. values of the acceleration is equal to several one tenth of millimeters. The time movement of the tubes is in fact composed of very low amplitude motion followed by bursts : tubes impact during these bursts.

It's the reason why we have developed our criterium which allows to avoid a non confident damping measurement.

. For the bundle number 3, submitted to two phase flow, the mixture character of the flow avoids a coupling between tubes. The excitation level is higher, but it has a wideband spectrum. For dynamic head values higher than those corresponding to $K\sqrt{\delta} = 1$, the difference between the smallest movements and the largest ones is greater.

4. CONCLUSION

The comparative analysis of several tube bundles submitted to flows of different nature allows to define simple criteria for evaluating tube amplitude :

- The critical value of the dynamic head for which large vibrations can be observed corresponds to a $K^2\delta$ value of about 0.5. For $K^2\delta$ values between 0.5 and 1.5, a specific study has to be done to evaluate exactly the risks. For $K^2\delta$ values larger than 1.5, the existence of large vibrations can be considered as certain.

- For lower flow velocities, the amplitude level can be computed by adopting a $C_F/4\pi\delta$ value of about 10^{-2}.

- For a bundle submitted to two phase flow, no fluidelastic instability is observed if the $K^2\delta$ value is lower than 5 ; but in that case the excitation value is greater, so the value of the $C_F/4\pi\delta$ parameter to take into account is about 5.10^{-2}.

5. REFERENCES

(1) REMY F.N. Vibration Study of a Tube Bank in Liquid Cross Flow.
SMIRT 5 Berlin 1979 Paper n° B5/7.

(2) BAI D. Tube Vibrations in Large Condenser. Symposium IUTAM 1979 Session A p.132.

(3) BAI D. Flow Induced Vibrations of multi-span tube bundles of large condensers :
experimental studies on full scale models in steam cross-flow.
Keswick 1982. Paper 68.

(4) REMY F.N. Flow Induced Vibration of tube bundles in two-phase cross-flow
Keswick 1982. Paper 68-1.

(5) CONNORS H.J. ASME Winter Annual Meeting 1970.

MAIN DATA ON TESTED BUNDLES

		Bundle 1	Bundle 2	Bundle 3	Bundle 4	Bundle 5
O.D.	mm	22.22	22.22	10	19	19
Tube length	m	0.2	0.2	1	10.8	14.2
number of support plates		0	0	0	12	15
arrangement						
pitch/diameter ratio		1.44	1.44	1.44	1.37	1.42
Flow nature		water	water	air + water	steam	steam
Frequency	Hz	34	34	49.7	38.5	34.2
Virtual mass	Kg/m	1.32	1.32	0.31 air 0.55 water	0.385	0.385
reduced damping $m \delta / \rho \, d^2$ (computed with $\delta = 0.063$		0.42	0.42	0.188		

Table 1

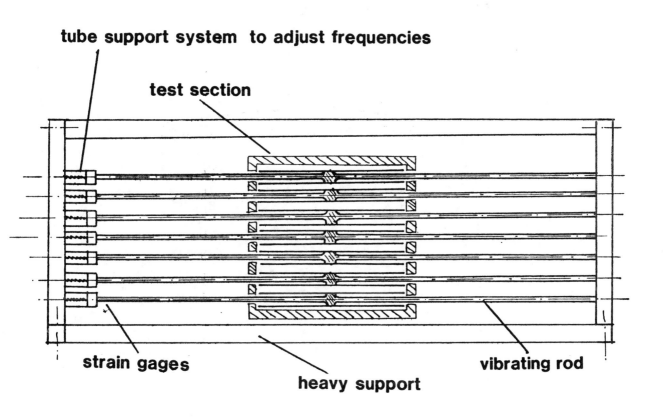

tube support system to adjust frequencies

test section

strain gages

heavy support

vibrating rod

Figure 1- Schematic drawing of bundles 1&2

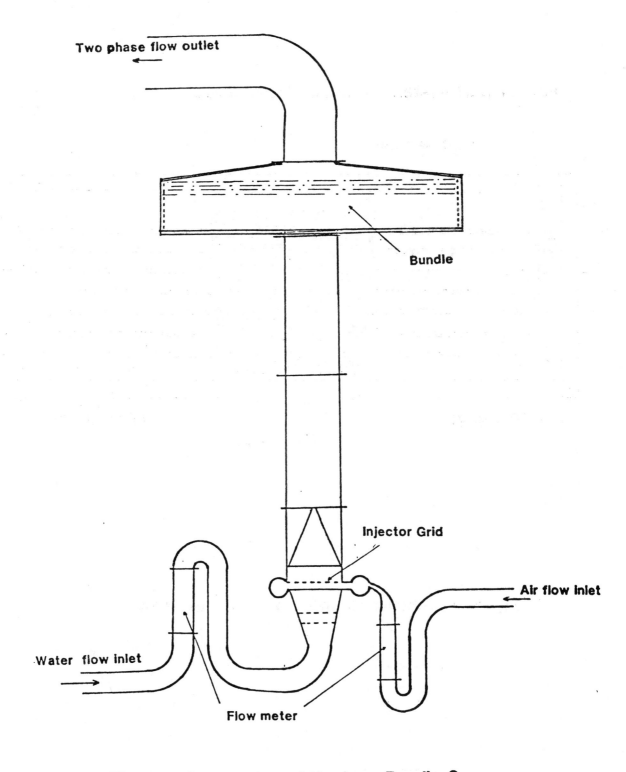

Figure 2-Description of the loop Bundle 3

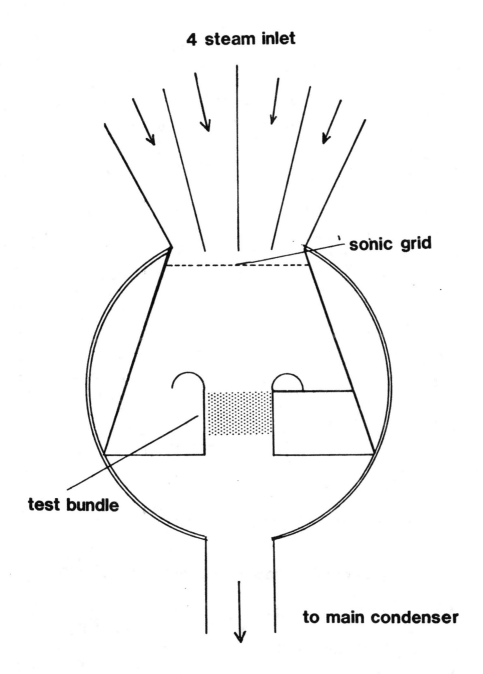

4 steam inlet

sonic grid

test bundle

to main condenser

Figure 3- Schematic diagram of the test rig for Bundles 4&5

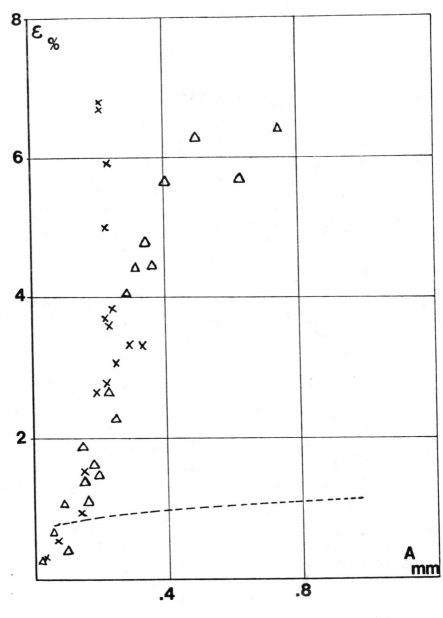

---   non moving flow.excitation by a shaker

x   drag direction ⎫
                ⎬ excitation by the fluid
△   lift direction ⎭

Figure 4- Evolution of the damping with the amplitude.Bundle n° 2

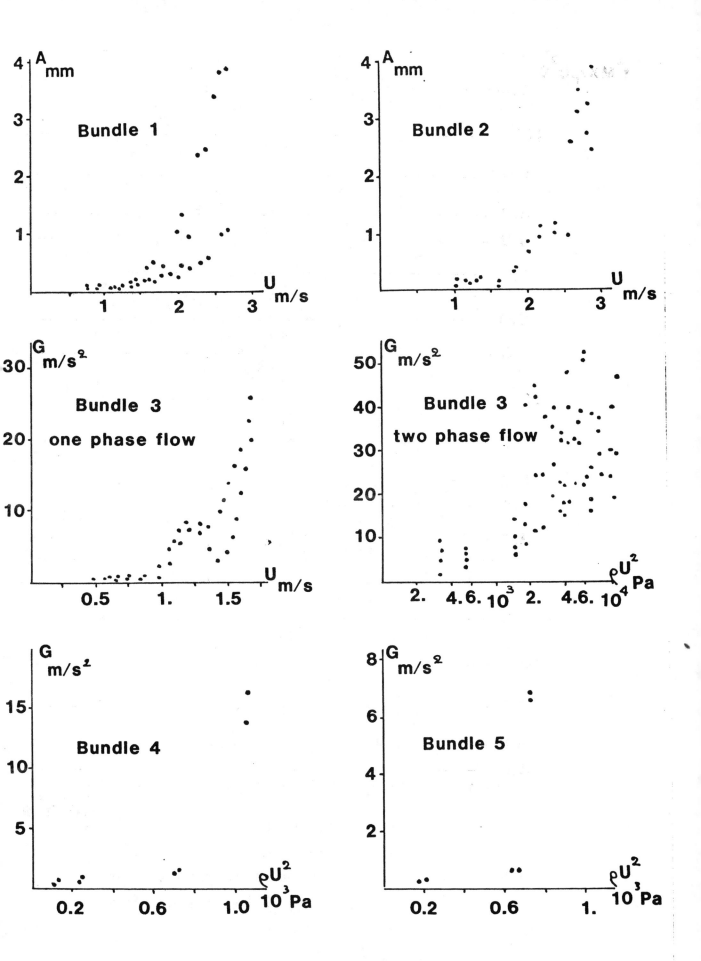

Figure 5- Evolution of the amplitude with flow characteristics

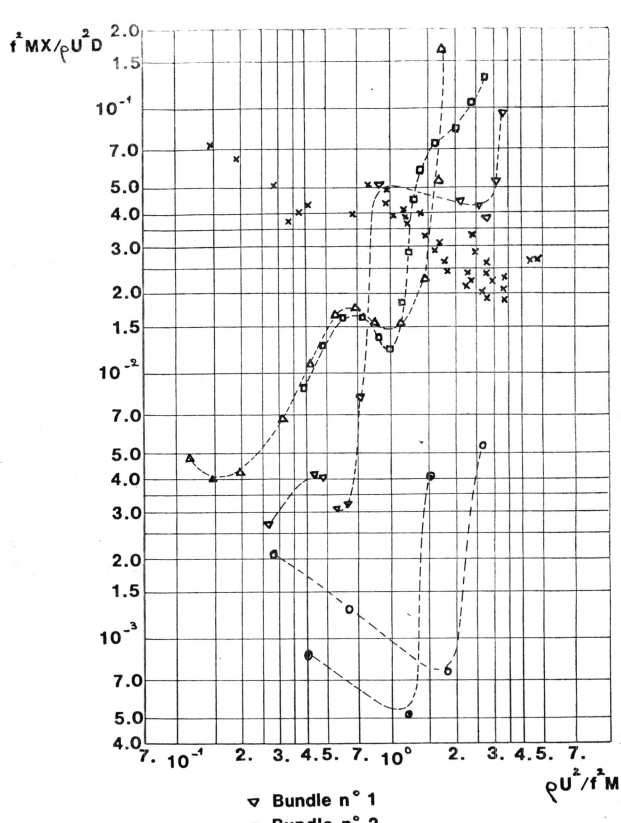

$f^2MX/\rho U^2D$

$\rho U^2/f^2M$

▽ **Bundle n° 1**

▫ **Bundle n° 2**

△ **Bundle n° 3 one phase flow**

✗ **Bundle n° 3 two phase flow**

○ **Bundle n° 4**

◉ **Bundle n° 5**

Figure 6- Comparative Analysis

**FLOW INDUCED VIBRATIONS
IN FLUID ENGINEERING**

Reading, England: September 14-16, 1982

A NONLINEAR APPROACH FOR THE DETERMINATION OF
LIMIT-CYCLE AMPLITUDES OF SUBSPAN OSCILLATIONS

A. R. E. Oliveira and W. M. Mansour

Federal University of Rio de Janeiro, Brazil

Summary

Lift and drag coefficients of the leeward conductor are represented by second order polynomials in X and Y. A harmonic balance approach is used to obtain the four relations governing the amplitudes of motion. Limit cycle amplitudes are obtained by evaluating the coordinates of the local minima of a surface defined by a quadratic function which is constructed using the nonlinear equations. Typical results are presented.

Organised and sponsored by
BHRA Fluid Engineering, Cranfield, Bedford MK43 0AJ, England

1. INTRODUCTION

Several investigators tried to evaluate the amplitudes of the limit cycles for wake-induced oscillations of transmission lines using numerical, experimental, finite-element and energy-balance methods. Reference (1) gives a comprehensive survey of the publications in this area. This paper presents a nonlinear analysis approach. It is based on the work reported by the same authors in (Ref.2). In that reference it was shown that linearized models for C_L and C_D are not sufficient to determine the amplitudes of limit-cycles.

2. NOMENCLATURE

a_{ij} = constants as defined in (1) and (4)

A_j = constants for C_L ; $A_1 = 0.07$; $A_2 = 1.07$

b_{ij} = constants as defined in (1) and (4)

B_j = constants for C_D ; $B_1 = -0.004$; $B_2 = 0.10$; $B_3 = 0.04$;$B_4 = 0.47$

C_L, C_D = lift and drag coefficients of the leeward cylinder.

d = diameter of the moving conductor (m)

k = ω_y/ω_x ; K_x, k_y = stiffness of suspension of moving cylinder

ℓ = length of moving cylinder (m)

m = mass of moving cylinder (kg)

t = time (sec.)

V = free wind velocity (m/sec.)

x, y = deviations of leeward cylinder measured from (x_o, y_o) (m.).

x_o, y_o = coordinates of leeward cylinder in the undisturbed position (m.)

X, Y = (x/d) , (y/d)

X_o, Y_o = (x_o/d) ,(y_o/d) ; X_*, Y_* = defined by (2)

α = $\zeta V/d\omega_x$

γ_1, γ_2 = $(\mu\alpha b_{12}^*)$, $(\mu\alpha a_{21}^*)$

ζ = factor accounting for the aerodynamics in the vicinity of the moving cylinder.

μ = $\rho\ell dV/2\zeta m\omega_x$

ν_1, ν_2 = $(X_*/\mu\alpha)$;$(k^2 Y_*/\mu\alpha)$

ρ = air density (kg/m^3)

τ = dimensionless time = $\omega_x t$

ω_x^2, ω_y^2 = (k_x/m) , (k_y/m)

ω_1^2, ω_2^2 = $(1-\mu\alpha b_{21}^*)$;$(k^2 - \mu\alpha a_{12}^*)$

3. A HIGHER APPROXIMATION MODEL

The governing equations for the leeward cylinder were shown to be given by[2]:

$$X'' + X = \mu\{C_L Y' + C_D(\alpha - 2X')\}$$
$$Y'' + k^2 Y = -\mu\{C_D Y' - C_L(\alpha - 2X')\}$$

where C_L and C_D are approximated by [2]:

$$C_L = \{A_1(Y+Y_o)^3 - A_2(Y+Y_o)\}/(X+X_o)$$

$$C_D = B_1(Y+Y_o)^4 + B_2(Y+Y_o)^2 + B_3\sqrt{(X+X_o)} + B_4$$

Expanding the expressions of C_L and C_D in Taylor's series about X_o and Y_o and retaining the linear and quadratic terms, one obtains:

$$C_L = \sum_{j=1}^{3} \sum_{i=1}^{4-j} a_{ij} X^{i-1} Y^{j-1}$$

$$C_D = \sum_{j=1}^{3} \sum_{i=1}^{4-j} b_{ij} X^{i-1} Y^{j-1}$$

where the a and b-coefficients are given by:

$$
\left.
\begin{aligned}
&a_{11} = (A_1 Y_o^3 - A_2 Y_o)/X_o \quad ; \quad a_{21} = -a_{11}/X_o \quad ; \quad a_{12} = (3A_1 Y_o^2 - A_2)/X_o \quad ; \\
&a_{31} = -a_{21}/X_o \qquad\qquad ; \quad a_{22} = -a_{12}/X_o \quad ; \quad a_{13} = 3A_1 Y_o/X_o \quad ; \\
&b_{11} = B_1 Y_o^4 + B_2 Y_o^2 + B_3 \sqrt{X_o} + B_4 \qquad ; \quad b_{21} = B_3/2\sqrt{X_o} \quad\qquad ; \\
&b_{12} = 4B_1 Y_o^3 + 2B_2 Y_o \qquad ; \quad b_{31} = -b_{21}/4X_o ; \quad b_{22} = 0 \qquad ; \\
&b_{13} = 6B_1 Y_o^2 + B_2
\end{aligned}
\right\} \tag{1}
$$

To free the model from the constant terms, one adopts the following linear transformation:

$$X = u_1 + X_* \qquad \text{and} \qquad Y = u_2 + Y_*$$

where X_* and Y_* are determined by solving the following two equations iteratively :

$$
\left.
\begin{aligned}
X_* &= \mu\alpha \sum_{j=1}^{3} \sum_{i=1}^{4-j} b_{ij} X_*^{i-1} Y_*^{j-1} \\
Y_* &= \left(\frac{\mu\alpha}{k^2}\right) \sum_{j=1}^{3} \sum_{i=1}^{4-j} a_{ij} X_*^{i-1} Y_*^{j-1}
\end{aligned}
\right\} \tag{2}
$$

The higher-approximation model takes the following form:

$$
\left.
\begin{aligned}
u_1'' &- \gamma_1 u_2 + \omega_1^2 u_1 \\
&= \mu\{ u_2' [\, a_{21}^* u_1 + a_{12}^* u_2 + a_{31} u_1^2 + a_{22} u_1 u_2 + a_{13} u_2^2 \,] - \\
&\quad 2u_1' [\, b_{21}^* u_1 + b_{12}^* u_2 \,] + (\alpha - 2u_1')[\, b_{31} u_1^2 + \\
&\quad b_{22} u_1 u_2 + b_{13} u_2^2 \,] + \nu_2 u_2' - 2\nu_1 u_1' \}
\end{aligned}
\right\} \tag{3.a}
$$

$$
\left.
\begin{aligned}
u_2'' &- \gamma_2 u_1 + \omega_2^2 u_2 \\
&= -\mu\{ u_2' [\, b_{21}^* u_1 + b_{12}^* u_2 + b_{31} u_1^2 + b_{22} u_1 u_2 + b_{13} u_2^2 \,] + \\
&\quad 2u_1' [\, a_{21}^* u_1 + a_{12}^* u_2 \,] - (\alpha - 2u_1')[\, a_{31} u_1^2 + \\
&\quad a_{22} u_1 u_2 + a_{13} u_2^2 \,] + 2\nu_2 u_1' + \nu_1 u_2' \}
\end{aligned}
\right\} \tag{3.b}
$$

where

$$
\left.
\begin{aligned}
a_{12}^* &= a_{12} + a_{22} X_* + 2a_{13} Y_* \quad ; \quad a_{21}^* = a_{21} + 2a_{31} X_* + a_{22} Y_* \\
b_{12}^* &= b_{12} + b_{22} X_* + 2b_{13} Y_* \quad ; \quad b_{21}^* = b_{21} + 2b_{31} X_* + b_{22} Y_*
\end{aligned}
\right\} \tag{4}
$$

4. A HARMONIC BALANCE APPROACH

It was shown[2] that μ is a small parameter. For $\mu \approx 0$, the governing equations read:

$$u_1'' - \gamma_1 u_2 + \omega_1^2 u_1 = 0$$
$$u_2'' - \gamma_2 u_1 + \omega_2^2 u_2 = 0$$

The natural undamped frequencies Ω_1 and Ω_2 of the linearized system are the roots of the characteristic equation given by:

$$\Omega^4 - (\omega_1^2 + \omega_2^2)\Omega^2 + (\omega_1^2 \omega_2^2 - \gamma_1 \gamma_2) = 0$$

We assume a solution in the form:

$$
\left.
\begin{aligned}
u_1 &= \sum_{n=1}^{2} (P_n \sin\Omega_n\tau + Q_n \cos\Omega_n\tau) \\[2ex]
u_2 &= \sum_{n=1}^{2} (R_n \sin\Omega_n\tau + S_n \cos\Omega_n\tau)
\end{aligned}
\right\}
\tag{5}
$$

where (P_n, Q_n, R_n, S_n) are unknown constants.

Substituting (5) and its derivatives in (3.a) and (3.b), and equating the coefficients of the sine and cosine terms, we obtain the following sets of nonlinear algebraic relations:

$$
\begin{bmatrix}
r_{1n} & r_{2n} & r_{3n} & r_{4n} \\
\tilde{r}_{2n} & \tilde{r}_{1n} & \tilde{r}_{4n} & \tilde{r}_{3n} \\
s_{1n} & s_{2n} & s_{3n} & s_{4n} \\
\tilde{s}_{2n} & \tilde{s}_{1n} & \tilde{s}_{4n} & \tilde{s}_{3n}
\end{bmatrix}
\begin{Bmatrix} P_n^2 \\ Q_n^2 \\ R_n^2 \\ S_n^2 \end{Bmatrix}
+
\begin{bmatrix}
P_n R_n & 0 & 0 & 0 \\
0 & Q_n S_n & 0 & 0 \\
0 & 0 & P_n R_n & 0 \\
0 & 0 & 0 & Q_n S_n
\end{bmatrix}
\begin{Bmatrix} r_{5n} \\ \tilde{r}_{5n} \\ s_{5n} \\ \tilde{s}_{5n} \end{Bmatrix}
$$

$$
+
\begin{bmatrix}
\rho_{1n} & \rho_{2n} & \rho_{3n} & \rho_{4n} \\
\rho_{2n} & -\rho_{1n} & \rho_{4n} & -\rho_{3n} \\
\sigma_{1n} & \sigma_{2n} & \sigma_{3n} & \sigma_{4n} \\
\sigma_{2n} & -\sigma_{1n} & \sigma_{4n} & -\sigma_{3n}
\end{bmatrix}
\begin{Bmatrix} P_n \\ Q_n \\ R_n \\ S_n \end{Bmatrix}
=
\begin{Bmatrix} 0 \\ 0 \\ 0 \\ 0 \end{Bmatrix}
$$

$$n = 1,2 \tag{6}$$

where

$$r_{1n} = 2b_{31}Q_n - 3a_{31}S_n \qquad ; \quad r_{2n} = 2b_{31}Q_n - a_{31}S_n$$

$$r_{3n} = (a_{22} + 6b_{13})Q_n - a_{13}S_n \qquad ; \quad r_{4n} = (2b_{13} - a_{22})Q_n - a_{13}S_n$$

$$r_{5n} = 2a_{31}Q_n - 2(a_{22} + 2b_{13})S_n \qquad ;$$

$$s_{1n} = 2a_{31}Q_n - (3b_{31} + 2a_{22})S_n \qquad ; \quad s_{2n} = 2a_{31}Q_n + (2a_{22} - b_{31})S_n$$

$$s_{3n} = 6a_{13}Q_n - b_{13}S_n \qquad ; \quad s_{4n} = 2a_{13}Q_n - b_{13}S_n$$

$$s_{5n} = 2(b_{31} + 2a_{22})Q_n - 4a_{13}S_n \qquad ;$$

$$\rho_{1n} = 4(\Omega_n^2 - \omega_1^2)/\mu\Omega_n \quad ; \quad \rho_{2n} = 8\nu_1 \quad ; \quad \rho_{3n} = 4\gamma_1/\mu\Omega_n \quad ; \quad \rho_{4n} = -4\nu_2$$

$$\sigma_{1n} = -4\gamma_2/\mu\Omega_n \quad ; \quad \sigma_{2n} = 8\nu_2 \quad ; \quad \sigma_{3n} = 4(\omega_2^2 - \Omega_n^2)/\mu\Omega_n \quad ; \quad \sigma_{4n} = 4\nu_1$$

$$n = 1, 2 \qquad\qquad (7)$$

A tilt $\tilde{}$ above a letter indicates that its value is obtained by exchanging (Q for P) and (S for R) in the expressions given by (7). For example $\tilde{r}_{1n}$ is obtained from r_{1n} as follows :

$$\tilde{r}_{1n} = 2b_{31}P_n - 3a_{31}R_n$$

and $\tilde{s}_{4n}$ is given by :

$$\tilde{s}_{4n} = 2 a_{13}P_n - b_{13}R_n \quad ; \quad \ldots\text{etc.}$$

Relations (6) represent two separate sets of equations; each consists of four nonlinear algebraic relations in four unknowns: P_n, Q_n, R_n and S_n . It is obvious that $(P_n = Q_n = R_n = S_n = 0)$ is a solution which satisfies the above relations . It pertains to the undisturbed position of the moving conductor. Any non-zero quadruple that satisfies (6) represents a limit cycle.

5. AMPLITUDES OF LIMIT CYCLES

Relations (6) can be written in the form:

$$f_i(P_n, Q_n, R_n, S_n) = 0 \qquad ; \quad i = 1, 2, 3, 4$$
$$n = 1, 2$$

The amplitudes of the limit cycles are essentially the coordinates of local minima of the surface given by :

$$\Psi = \sum_{i=1}^{4} f_i^2 \qquad\qquad (8)$$

A number of standard optimization routines are available to minimize quadratics similar to Ψ . The majority of them were tried and gave oscillatory results around the local minima.

A "direct search" program was developed for this case. A flow-diagram depicting the strategy of search is shown in Fig.(1). One element of the initial quadruple is perturbed at a time. For each new quadruple the function Ψ is evaluated and compared with the previous value Ψ_* . The old value of the element is replaced by the perturbed value only if Ψ is less than Ψ_* .

The perturbation is imparted both ways (the positive and the negative) using increments Δ_k (k=1,2,3,4) . The Δ's are chosen equal to 5% of the absolute value of the respective elements, with a lower bound equal to 0.01d.

A 4x1 flag vector {L} is zeroed at the beginning of each cycle of iteration and is tested at the end of the cycle. The appropriate entry of {L} is changed to 1 if a minimization is achieved ; i.e. $\Psi < \Psi_*$. Iteration is terminated if {L} is found {0} at the end of any cycle.

Several trials were attempted to establish a strategy for choosing the initial quadruple. It was found that:

$$P_n = 0.9 X_o \quad ; \quad Q_n = -2.0 Y_o \quad ; \quad R_n = -2.0 Y_o \quad ; \quad S_n = 0.9 X_o$$

are adequate for most of the cases tried with this approach.

The strategy of Fig. (1) was implemented in a general-purpose computer routine . A typical case was analized for different wind velocities and for different points in the wake. Convergence was found slower than the standard routines but the numerical oscillations vanished. Typical results are shown in a later section.

6. STABILITY CONSIDERATIONS

To assess the stability of the limit cycles obtained in the previous section one imparts small disturbances to the solution which satisfies relations (3.a) and (3.b).

Let $(\delta_1, \delta_2, \delta_3, \delta_4)$ denote perturbations imparted to (P,Q,R,S) respectively. The δ's are considered functions of τ. To simplify the notations in the analysis that follows we denote $(\sin \Omega_n \tau)$ by s and $(\cos \Omega_n \tau)$ by c. One can show that the variables (u_1) and (u_2) and their derivatives will assume the following changes after applying the disturbance:

$$
\left.
\begin{aligned}
u_k &\rightarrow u_k + (\delta_{2k-1})s + (\delta_{2k})c \\
u_k' &\rightarrow u_k' + (\delta_{2k-1}' - \delta_{2k}\Omega_n)s + (\delta_{2k}' + \delta_{2k-1}\Omega_n)c \\
u_k'' &\rightarrow u_k'' + (\delta_{2k-1}'' - 2\delta_{2k}'\Omega_n - \delta_{2k-1}\Omega_n^2)s + (\delta_{2k}'' + 2\delta_{2k-1}'\Omega_n - \delta_{2k}\Omega_n^2)c
\end{aligned}
\right\}
$$

$$
\begin{aligned}
k &= 1,2 \\
n &= 1,2
\end{aligned}
\tag{9}
$$

Substituting (9) in equations (3.a) and (3.b), dropping higher order terms in the δ's and equating the sine and cosine terms to zero, one obtains the following matrix relation:

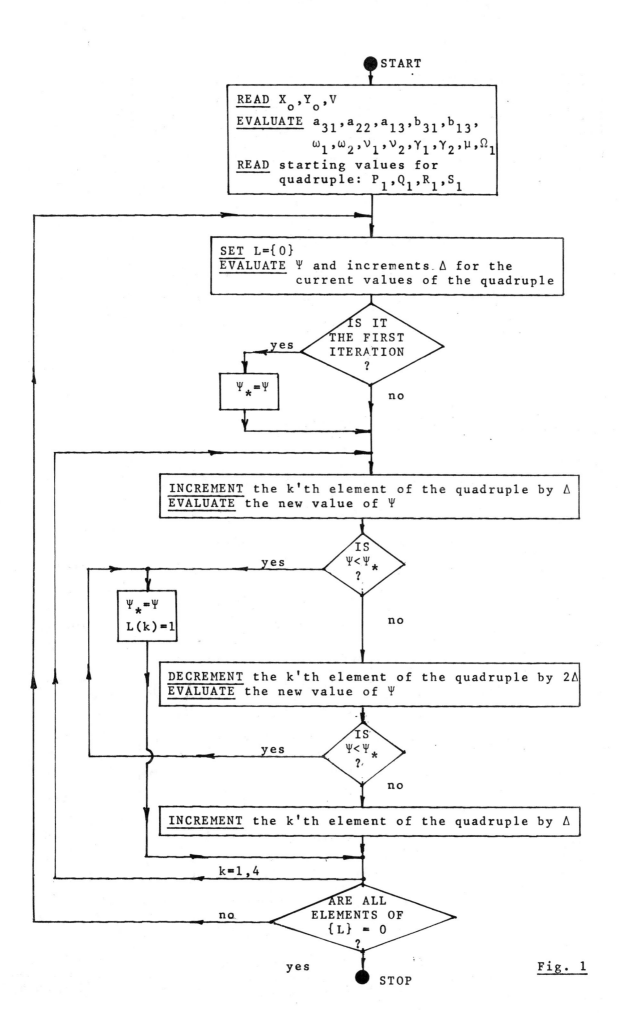

Fig. 1

273

$$\begin{Bmatrix} \delta_1'' \\ \delta_2'' \\ \delta_3'' \\ \delta_4'' \end{Bmatrix} + \begin{bmatrix} g_{11} & g_{12}+\Gamma & g_{13} & g_{14} \\ \tilde{g}_{12} & \tilde{g}_{11} & \tilde{g}_{14} & \tilde{g}_{13} \\ -2g_{13} & -2g_{14} & \frac{1}{2}g_{11} & \frac{1}{2}g_{12}+\Gamma \\ -2\tilde{g}_{14} & -2\tilde{g}_{13} & \frac{1}{2}\tilde{g}_{12}-\Gamma & \frac{1}{2}\tilde{g}_{11} \end{bmatrix} \begin{Bmatrix} \delta_1' \\ \delta_2' \\ \delta_3' \\ \delta_4' \end{Bmatrix} +$$

$$\begin{bmatrix} h_{11}+\lambda_1 & h_{12} & h_{13}+\lambda_2 & h_{14} \\ -\tilde{h}_{12} & -\tilde{h}_{11}+\lambda_1 & -\tilde{h}_{14} & -\tilde{h}_{13}+\lambda_2 \\ h_{31}+\lambda_3 & h_{32} & h_{33}+\lambda_4 & h_{34} \\ -\tilde{h}_{32} & -\tilde{h}_{31}+\lambda_3 & -\tilde{h}_{34} & -\tilde{h}_{33}+\lambda_4 \end{bmatrix} \begin{Bmatrix} \delta_1 \\ \delta_2 \\ \delta_3 \\ \delta_4 \end{Bmatrix} = \{0\}$$

(10)

where

$$g_{11} = (\mu/2)\{3b_{31}P^2+b_{31}Q^2+3b_{13}R^2+b_{13}S^2+4\nu_1\}$$

$$g_{12} = (\mu)\{b_{31}PQ+b_{13}RS\}$$

$$g_{13} = (-\mu/4)\{3a_{31}P^2+a_{31}Q^2+3a_{13}R^2+a_{13}S^2+3a_{22}PR+a_{22}QS+4\nu_2\}$$

$$g_{14} = (-\mu/4)\{2a_{31}PQ+a_{22}QR+2a_{13}RS+a_{22}SP\}$$

$$h_{11} = (\mu\Omega/2)\{-2b_{31}PQ-a_{31}QR+(2b_{13}+a_{22})RS+3a_{31}SP\}$$

$$h_{12} = (\mu\Omega/4)\{-2b_{31}P^2-6b_{31}Q^2-(6b_{13}+a_{22})R^2-(2b_{13}-a_{22})S^2-2a_{31}(PR-QS)-8\nu_1\}$$

$$h_{13} = (\mu\Omega/2)\{-a_{31}PQ-(6b_{13}+a_{22})QR+a_{13}RS+(2b_{13}+a_{22})SP\}$$

$$h_{14} = (\mu\Omega/4)\{3a_{31}P^2+a_{31}Q^2+a_{13}R^2+3a_{13}S^2+2(2b_{13}+a_{22})PR-2(2b_{13}-a_{22})QS+4\nu_2\}$$

$$h_{31} = (\mu\Omega/2)\{-2a_{31}PQ+(b_{31}-2a_{22})QR+2a_{13}RS+(2a_{22}-3b_{31})SP\}$$

$$h_{32} = (\mu\Omega/2)\{-a_{31}P^2-3a_{31}Q^2-3a_{13}R^2-a_{13}S^2+(b_{31}-2a_{22})PR-(b_{31}+2a_{22})QS-4\nu_2\}$$

$$h_{33} = (\mu\Omega/2)\{(b_{31}-2a_{22})PQ-6a_{13}QR-b_{13}RS+2a_{13}SP\}$$

$$h_{34} = (\mu\Omega/4)\{(3b_{31}-2a_{22})P^2-(b_{31}+2a_{22})Q^2-b_{13}R^2-3b_{13}S^2+4a_{13}PR-4a_{13}QS-4\nu_1\}$$

$$\Gamma = -2\Omega \quad ; \quad \lambda_1 = \omega_1^2-\Omega^2 \quad ; \quad \lambda_2 = -\gamma_1 \quad ; \quad \lambda_3 = -\gamma_2 \quad ; \quad \lambda_4 = \omega_2^2-\Omega^2$$

(11)

The tilt ~ above a given quantity indicates that it can be obtained from the original expression given by (11) by interchanging the variables as shown after:

$$(P \underset{\leftarrow}{\overset{\rightarrow}{}} Q) \quad \text{and} \quad (R \underset{\leftarrow}{\overset{\rightarrow}{}} S)$$

Equation (10) can be written in a compact form as follows:

$$\{\delta''\}+[G]\{\delta'\}+[H]\{\delta\} = \{0\}$$

For a stable limit cycle, the linear set of equations given by (10) should also be stable. The characteristic equation can be obtained and the roots examined using one of many available approaches including the Routh-Hurwitz criterion.

7. TYPICAL RESULTS

To test the nonlinear approach outlined in the previous sections a system was chosen with the following parameters:

$$m = 3 \text{ Kg.} \quad ; \quad \ell = 1 \text{ m.} \quad ; \quad d = 0.04 \text{ m.}$$
$$\omega_x = 10 \text{ rad./sec}; \quad \omega_y = 12 \text{ rad/sec} \quad ; \quad \rho = 1.225 \text{ Kg./m}^3$$

Different wind velocities were tried and the wake was investigated for different points in the range :

$$10 < X_o < 30 \qquad ; \quad 0.5 < Y_o < 4.0$$

Typical results are shown in Fig. 2 for a wind velocity (V=20 m/sec) . Fig.(2.a) demonstrates the variation of the quadruple (P,Q,R,S) in a direction perpendicular to the wind-velocity-vector at a given fixed value for X_o . Fig. (2.b) shows the variation along the wind-vector for a fixed value of Y_o .

Other computer-runs for this case showed that Fig.(2.a) does not change significantly for other values of X_o in the range specified above. Peaks for the quadruple occur around $Y_o = 2.0$ for other values of X_o . The amplitudes die out as one approaches the boundary of the wake on one side and the X_o-axis on the other

Fig. (3.a) gives a plot for the inclination (Θ-degrees) of the major axis of the elliptic path of the moving conductor. It is measured from the X-axis. The plot is in reasonable agreement with the published experimental data [1].

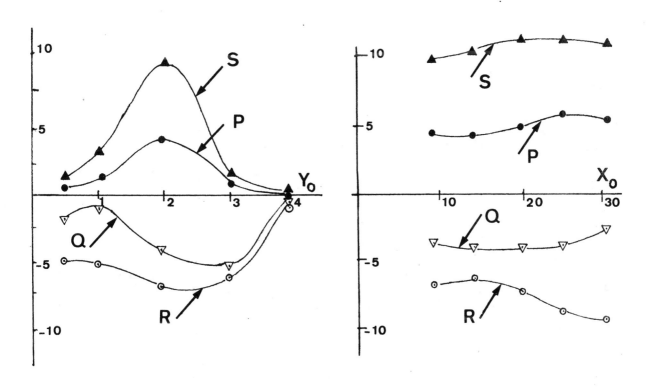

(a) $\underline{X_o = 10}$ (b) $\underline{Y_o = 2.0}$

Fig.2 : Limit-cycles of the moving conductor
(V = 20 m/sec.)

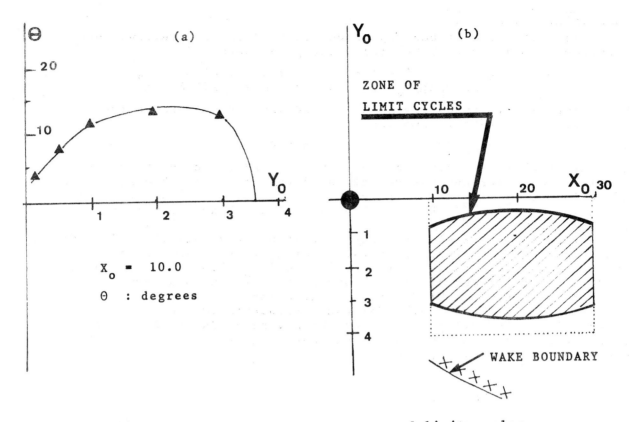

Fig. 3 : Geometry and existence-zones of limit-cycles

Fig.(3.b) shows the zones ,in the wake,where limit cycles are likely to occur in this case.

8. CONCLUSIONS

A nonlinear analysis is presented for the determination of the amplitudes of the limit cycles of wake-induced oscillations of transmission lines. A harmonic-balance is used to yield a set of nonlinear algebraic relations. A"direct-search" routine was used to obtain the local minima. The assessment of the stability of the limit cycles is : achieved by investigating the roots of the characteristic equation of the linear differential equations describing the perturbed motion.

It is obvious that the convergence to a local minimum depends very much on the initial quadruple. It is possible to obtain other local minima which will satisfy the set of algebraic relations governing the amplitudes of the motion. Of course not all of them are stable or even feasible. The surface represented by Ψ has a high degree of complexity and suggests the possible simultaneous existence of a number of close stable limit cycles. This perhaps explains the hard-self-excited nature of the phenomenon and the fact that the leeward conductor seldom follows a definite path once set in motion.

The authors tried a Kryloff and Bogoliuboff approach to the same problem[3]. The method gave a prediction of a single-configuration for the limit cycle. Nevertheless the findings of both approaches are qualitatively in agreement with the published experimental data.

9. REFERENCES

1. Gilbert and Jackson,"Transmission Line Reference Book", EPRI,1979
2. Oliveira A.R.E and Mansour W.M. ,"Critical Flow Velocities and Stability Zones for Wake-Induced Oscillations", Proceedings Int. Conference on Flow induced Vibrations ,BHRA,September 1982.
3. Oliveira A.R.E and Mansour W.M., "Nonlinear Analysis and Simulation of Twin Bundle Conductors of Transmission Lines", Proceedings of IMAC's Conference ,Montreal-Canada,August 1982.

FLOW INDUCED VIBRATIONS
IN FLUID ENGINEERING

Reading, England: September 14-16, 1982

SELF-INDUCED PULSATING LIQUID FLOW
IN A HYDRAULIC SYSTEM

A.P. Szumowski

Warsaw Technical University, Poland

G.E.A. Meier

Max-Planck-Institut für Strömungsforschung, F.R. Germany

Summary

Pulsating liquid flow of large amplitudes can appear in hydraulic lines as a result of the interaction between flow and valves. In this paper such a flow in a system containing a tank, a tube and a valve, which has an elastically supported plate, is investigated. The functions of the flow properties versus some non-dimensional parameters have been found. These parameters depend upon supply pressure, inertia of the liquid and of the valve, as well as the elasticity of the plate support and the dimensions of the system. The occurrence of cavitation zones in the liquid flow has been taken into account. Theoretical results have been compared with experimental ones.

Organised and sponsored by
BHRA Fluid Engineering, Cranfield, Bedford MK43 0AJ, England

NOMENCLATURE

A	bore area of the tube
A_p	area of the valve plate
a	velocity of sound in liquid
a_r	velocity of sound in liquid filling an elastic tube
a_m	velocity of sound in mixture of liquid and saturated vapour
D	tube diameter
D_p	diameter of the valve plate
E	Young's modulus
K	bulk modulus of liquid
$k, \bar{k}$	spring constant, $4k/(\pi D \rho a^2)$
$L, \bar{L}$	length of the tube, L/D
$M, \bar{M}$	mass of the valve plate, $M/(A D \rho)$
$P, \bar{P}$	liquid force, $\bar{P}/(\rho a^2 A)$
$P_s, \bar{P}_s$	liquid force in the case of quasisteady liquid flow in the tube, $P_s/(\rho a^2 A)$
$P_d, \bar{P}_d$	liquid force in the case when a simple compression wave occurs, $P_d/(\rho a^2 A)$
$p, \bar{p}$	pressure, $p/(\rho a^2)$
$p_0, \bar{p}_0$	supply pressure, $p_0/(\rho a^2)$
$p_w, \bar{p}_w$	mean pressure of the impulse (mean pressure on the valve plate in the period of time when the valve is closed), $p_w/\rho a^2$
$p_v, \bar{p}_v$	pressure of saturated vapour, $p_v/(\rho a^2)$
$p_\emptyset$	pressure in the minimum cross-section of the valve slit
$S, \bar{S}$	spring force, $S/(\rho a^2 A)$
$t, \bar{t}$	time, ta/D
$u, \bar{u}$	velocity of the valve plate, u/a
$v, \bar{v}$	flow velocity, v/a
$v_\emptyset$	flow velocity in the minimum cross-section of the valve slit
x	coordinate along the tube
$y, \bar{y}$	coordinate of the valve plate position, y/D

$y_*, \bar{y}_*$ $\quad$ $[1 - (D_p/D)^2]/(4D_p/D^2)$, $\quad y_*/D$

$\Delta l, \Delta \bar{l}$ $\quad$ nominal travel of the valve plate (spring deflection during valve closure), $\quad \Delta l/D$

β $\quad$ discharge coefficient

$\emptyset$ $\quad$ relative area of the minimum cross-section of the valve slit

$\emptyset_*$ $\quad$ $1 - (D_p/D)^2$

ρ $\quad$ density of liquid

δ $\quad$ thickness of the tube wall

Suffixes

A, B $\quad$ cross-sections as shown in Fig. 1c.

s $\quad$ slit

All mentioned above pressures except p_v are determined in relation to the ambient pressure p_{at}.

1. INTRODUCTION

Waterhammer in hydraulic lines can emerge in a self-induced manner as a result of interaction between the liquid and the fixtures, especially valves (Ref. 1-4). When this process repeats itsself periodically a pulsating liquid flow of high amplitude appears.This flow usually shows cavitation.

In this paper a pulsating liquid flow in a system shown in Fig. 1c, containing a buffer tank, a tube and a valve, which has an elastically supported plate, is investigated. The mechanism of that flow is similar to that in hydraulic ram (Ref. 12) and it is as follows. The system is supplied with the liquid under pressure. When the pressure is increased the flow velocity in the pipe grows and exerts a stronger pressure force on the plate of the valve towards the spring. The force overcomes the resistance of the spring and the valve closes gradually, stopping the flow. At the moment the flow stops, appears a relatively strong increase of the pressure in the vicinity of the valve. This pressure, which considerably exceeds the supplied pressure (pressure in the buffer tank) induces a pressure wave propagating upstream as shown in Fig. 1. When this wave reaches the tank, a reflected wave appears and reduces the pressure in the tube to the supplied one. Behind the front of the reflected wave the flow is directed towards the tank. The moment the reflected wave reaches the valve, the pressure on the plate decreases and the spring opens the valve. The liquid outflows through the valve and the flow velocity in the tube increases again as a consequence of the further wave process. Again, it exerts stronger pressure on the plate, the valve closes and the flow stops. In this way the cycle is repeated. The pressure and flow velocity at any cross-section of the tube vary in time and are non-uniform along the pipe as a result of the flow pulsations.

The aim of this paper is to give a mathematical model of the above described wave process, and to find the conditions for its appearance. Beside the theoretical also experimental investigations have been performed. In the theory the one-dimensional flow of an elastic inviscid and heat non-conducting liquid has been assumed. A special significance in the mathematical model of the investigated process has been attached to the boundary conditions at the end of the tube connected with the valve. These conditions express the interaction between the flow of the liquid and the motion of the valve. Therefore, they are represented by equations of the liquid flux through the valve as well as by the equation of motion of the valve plate simultaneously.

Most hydraulic transients are accompanied by cavitation. P. Schweitzer and V. Szebehely (Ref. 5) proved, that cavitation is always associated with microscopic bubbles of entrained gas, which generally appear in a turbulent liquid flow. When the pressure decreases, the dissolved gas is released into the microscopic bubbles with the result that they grow and become visible. This process is known as gaseous cavitation. However, the evoluation and solution processes of the gas (air) in the liquid (water) are relatively slow. For example, the so-called "half-life" of evolution defined as the time required to complete half of the air evolution in destilled water amounts to about 4 s (Ref. 5). Therefore, the gaseous cavitation can influence hydraulic transiens in long tubes i.e., when the time for return propagation of disturbance is of the order of some tenths seconds or seconds. It was the case in the most experiments (Ref. 6-11). For short tubes considered in this paper the time for return propagation of disturbance is of the order of some milliseconds. Thus, in this case the gaseous cavitation can be neglected.

When the pressure decreases to the saturated vapour pressure evaporation into air nuclei emerges: vapourous cavitation. This process is a relatively fast one. One can assume, that the vapourous cavitation emerges immediately when the pressure reaches the value equal to the saturated vapour pressure.

2.   GOVERNING EQUATIONS

With the assumptions mentioned above the liquid flow outside of the cavitation zones is described by the following equations:

continuity equation

$$\frac{\partial \rho}{\partial t} + \frac{\partial (\rho v)}{\partial x} = - \frac{\rho}{A}\left(\frac{\partial A}{\partial t} + u\frac{\partial A}{\partial x}\right) \tag{1}$$

and momentum equation

$$\frac{\partial v}{\partial t} + v\frac{\partial v}{\partial x} + \frac{1}{\rho}\frac{\partial p}{\partial x} = 0, \tag{2}$$

The right-hand side of Eq. (1) expresses the influence of a deformability of the tube on the flow properties.

Equations (2) and (3) have the following characteristics

$$dx/dt = v \pm a_r, \tag{3}$$

where $\quad a_r = a/\left(1 + \frac{D}{\delta}\frac{K}{E}\right)^{\frac{1}{2}}.$

The following compatibility equations correspond to them

$$dp \overset{+}{_-} \rho \, a_r \, dv = 0. \tag{4}$$

In the case when $v \ll a_r$, both ρ and a_r depend weakly on the flow properties. Thus, one can neglect v in Eqs. (3) and assume Eqs. (3) and (4) to be linear.

3. BOUNDARY AND INITIAL CONDITIONS

The boundary conditions at the ends of the tube result from the presence of the buffer tank at one side and of the valve at the other. For small dimensions of these arrangements with respect to the length of the tube the quasisteady boundary conditions can be applied.

Assuming the constant pressure in the buffer tank to be equal to the supplied pressure, we have at the tube end connected to the tank $p_o = p_A + \rho \, v_A^2/2$ for the case when liquid flows from the tank into the tube and $p_A = p_o$ for the opposite case.

The valve opens and closes periodically. When it is closed $v_B = 0$. When it is opened the boundary conditions are determined by the equations for the flow through the valve, and by **equation** of motion of the valve plate. The flow through the valve is complicated. It is non-uniform and includes dead-water regions. That flow is usually directed from the tube to the surroundings. However, the return flow also appears in the time portion which corresponds to the initial phase of the valve opening.

In the present analysis the flow through the valve is assumed to be one-dimensional. In this connexion an equivalent convergent nozzle as shown in Fig. 2, with variable outlet cross-section equal to the minimum cross-section area of the valve slit has been assumed. The nozzle moves in the tube, has the same mass as the plate and is supported by the same spring. The flow through the nozzle is described by the following equations in the coordinate system related to the nozzle:

continuity equation

$$v_B + u = \beta \, \phi \, (v_\phi + u) \tag{5}$$

and energy equation

$$p_B + \rho \, (v_B + u)^2/2 = \rho \, (v_\phi + u)^2 /2 , \tag{6}$$

where ϕ is the relative area of outlet cross-section of the nozzle.

The value of ϕ depends upon the position of the valve plate in the following manner:

$$\phi = \begin{cases} 1 - (D_p/D)^2 & \text{for } y \geqslant y_*, \\ 4 \, y \, D_p/D^2 & \text{for } y \leqslant y_*, \end{cases} \tag{7}$$

in which y_* is the plate position, where the given above values of ϕ are equal. Equation (6) corresponds to the flow from the tube to the surroundings. For the return flow we have (see Fig. 2c)

$$p_B + \rho \, (v_\phi + u)^2/2 = p_1 + \rho \, (v_1 + u)^2/2 . \tag{8}$$

It is assumed in Eq. (8) that pressure in the minimum cross-section of the nozzle is equal to p_B. Moreover, we have $p_1 = -\rho v_1^2/2$ and $v_1 = v_B$.

The velocity of the valve plate and its momentary position is determined by the equation of motion of the plate

$$M \frac{d^2 y}{dt^2} - k(\Delta l - y) = -P, \qquad (9)$$

where P designates the interaction force between the fluid and the plate. That force results from the momentum equation for the flow through the valve.

$$P = A[p_B + \rho(v_B + u)|v_B + u|] - (A - A_p)[p_s + \rho(v_s + u)|v_s + u|] \quad (10)$$

where p_s and v_s are pressure and flow velocity in the slit between the valve plate and the tube. For $y \geqslant y_*$, $p_s = 0$ in the case of the flow to the surroundings or $p_s = p_B$ in the opposite case. Moreover, $v_s = v_{\emptyset}$ in both cases.

For $y \leqslant y_*$, v_s is calculated from the continuity equation (see Fig. 2a)

$$v_B + u = \beta \emptyset_* (v_s + u) \qquad (11)$$

and p_s from the energy equation

$$p_B + \rho(v_B + u)^2/2 = p_s + \rho(v_s + u)^2/2 \qquad (12)$$

in the case of the flow to the surroundings or $p_s = p_B$ in the opposite case.

Equations (5-12) enable the determination of the function $f(v_B, p_B, \emptyset)$, which compose the boundary conditions for the opened valve.

In the initial conditions steady flow for the fully opened valve has been assumed. These conditions are as follows

$$y_i = \Delta l,$$

$$v(x, 0) = v_i = \beta \emptyset \sqrt{2p_0/\rho}, \qquad (13)$$

$$p(x, 0) = p_i = p_0 - \rho v_i^2/2.$$

These conditions express a certain non-equilibrium state, in which the liquid force on the valve plate is not balanced.

4. NUMERICAL SOLUTION

The governing equations are solved by the method of characteristics, in which the occurrence of vapourous cavitation is taking into account. The exemplary net of characteristics is shown in Fig. 3, in which two flow regions are distinguished: the region of the liquid flow (region L) and the region of flow of the liquid-saturated vapour mixture (cavitation zone - region M). In the investigated flow

$a_m \approx v \ll a_r$ (Ref. 12, 13). Therefore, one can assume the characteristics in the region L to be straight and parallel to each other as well as the characteristics in the region M to be parallel to the t-axis.

The cavitation zone in the investigated flow usually emerges at the valve and propagates along the tube. The boundaries of that zone displace with different velocities, which vary in time due to the disturbances moving in the liquid. Therefore the length of the cavitation zone as well as its location change.

On the other hand, the disturbances in the cavitation zone cannot propagate in a short time far from their source because of small a_m. This is why the flow velocity and the pressure at a given cross-section in the cavitation zone are nearly constant. These properties are almost the same as those, which appear in the liquid when the pressure at the considered cross-section decreases to the saturated vapour pressure. Therefore, one can assume the cavitation zone to possess both constant pressure equal to the saturated vapour pressure and variable flow velocity, which depends only on the x-coordinate.

Direct determination of the lines of motion (in the x, t plane) of the boundaries of the cavitation zone involves considerable complication of numerical procedures as well as longer calculation time. In this paper an indirect procedure is applied, in which the temporary locations of the cavitation boundaries are determined from pressure control in the characteristic nodes (Ref. 14).

5. RESULTS AND DISCUSSION [1]

In Fig. 4 - 6 the traces of the pressure in the vicinity of the valve and of the position of the valve plate are given. The calculated pressure traces are compared with the experimental ones. From these traces it may be concluded that both the pressure and the plates position vary periodically with amplitude up to 5 p_o and frequency up to 120 Hz. However, the traces in each cycle differ somewhat one from another. The reason for these differences can be explained as follows. When the plate is close to the valve seat, then its relatively small displacement causes a proportionally strong change of $\emptyset$. Thereby, it causes a strong change of amplitude of the reflection wave for a given incident wave. A relatively small deviation of the trace of plate position from the exactly periodical one causes strong deviation of pressure trace when the plate is close to the valve seat. This influences the traces of the pressure and of the plate position in the next cycle. The differences in the calculated pressure traces for each cycle are much stronger than the differences in the measured case. One can suppose, that it is influenced by the initial conditions as well as by the numerical discretyzation with finite Δx. Nevertheless, the differences between calculated and measured pressure traces do not exceed 12% for frequency and for the mean amplitude.

---

[1] In the calculations the constant discharge coefficient β = 0.35 measured as a mean value for steady liquid flow through the valve, has been assumed.

In the general case the frequency and the amplitude of the investigated pulsating flow depend upon applied power, inertia of the liquid and of the valve, as well as on the elasticity of the plate support and the dimension of the system (see sections 3 and 4). One can express these parameters by the following non-dimensional variables: $\bar{p}_o$, $\bar{L}$, ϕ_*, k, $\overline{\Delta l}$, $\overline{M}$. In Figs. 7 - 9 the functions of mean pressure of impulse p_W and frequency of the pulsating flow versus $\bar{p}_o$, $\overline{\Delta l}$ and $\bar{L}$ are shown. As was mentioned above, adjacent cycles are not identical. Therefore, in Figs. 7 - 9 the ranges of $\bar{p}_W$ and f marked by open columns for theory and by filled columns for experiment are given. They are evaluated from some cycles of calculated and measured traces, respectively.

From Fig. 7 one can conclude that the mean pressure of impulse (p_W) and frequency f increase with an increase of supplied pressure ($\bar{p}_o$). At the same time, the ratio of the mean pressure to the supply pressure remains almost constant. If the increase of the mean pressure results directly from an increase of the supplied pressure, then, an increase of frequency will follow from a shortening of the time of flow acceleration due to the greater $\bar{p}_o$. The p_W increases and f decreases with growth of $\overline{\Delta l}$ (see Fig. 8.), while with greater displacement of the plate, the time in which the valve is opened increases. That causes simultaneously an increase of flow velocity.
For longer tubes (see Fig. 9) frequency decreases, and this arises from an increase of the time of acceleration of a longer water column by the same supplied pressure.

Calculated pressure distributions in plane x,t and measured pressure traces at some cross-sections of the tube show,that cavitation emerges several times during one cycle of pulsation. The cavitation region is relatively long and momentarily occupies almost the whole length of the tube. The boundary of this region is visible in the flow photograph as shown in Fig.10.

The calculated and measured results justify the wave pattern described in the introduction and shown in Fig. 11a. The strength of each wave is illustrated in Fig. 11b by thin continuous lines. In that figure the thick curve correspond to the boundary conditions at the tube ends: the continuous curve at the buffer tank and the broken curves at the valve. Each dashed curve corresponds to one phase of a valve opening (ϕ).
The curves over and under the abscissa correspond to the outflow and to the inflow of the liquid through the valve, respectively. The period of pulsation is composed of the time interval t_1, in which the valve is closed and by interval t_2, in which it is opened. Time t_1 depends on the length of the tube as well as the velocity of disturbances in the liquid (a_r). Time t_2 depends, moreover, on supplied pressure and spring deflection Δl. This time is proportional to the number of waves travelling in the tube and causing an increase of the flow velocity to the value required for closure of the valve. For example, pressure traces in Fig.12 , show three distinguishable disturbances in one cycle, corresponding to three return travels of waves.

6. CONDITIONS OF TRANSITION FROM STEADY FLOW TO PULSATING FLOW

In the hydraulic system considered in this paper steady flow occurs for low supply pressures (p_o). When p_o increases a transition from steady into pulsating flow is observed. In order to determine the conditions associated with this transition process of valve closure due to a slow pressure increase is investigated. It seems that the steady flow changes into pulsating one in this case when a compression wave strong enough emerges in the liquid, at the end connected with the valve, in the moment of valve closure or during its closing. In this connection, the process of valve closure for the following two extremal cases is considered: (1) assuming quasisteady liquid flow in the tube, (2) assuming nonsteady liquid flow accompanied by a simple compression wave.

The first case corresponds to the short tubes, in which the disturbance travels many times during valve closure. The second one corresponds to the long tubes, in which the return transition time of the disturbance is lower or at least equal to the time of valve closure.

The transition considered in this section is controlled by the interaction force between the fluid and the plate as well as by a spring force. The interaction force see Eqs.(10 - 12) can be expressed by nondimensional variables as follows

$$\bar{P} = \bar{p}(1 - \phi_*) + (\bar{v} + \bar{u})^2 \left[1 - \frac{\phi_*}{2}(1 + 1/\phi_*^2 \beta^2)\right]. \tag{14}$$

6.1. The short tube

Substituting $\bar{v} = \phi \beta \sqrt{2\bar{p}_o}$, $\bar{p} = \bar{p}_o(1 - \phi^2\beta^2)$ and $\bar{u} = 0$ into Eq. (14) the following formula for the force of the liquid on the quasi-motionless valve plate is obtained.

$$\bar{P}_s/\bar{p}_o = (1 - \phi^2\beta^2)(1 - \phi_*) + 2\phi^2\beta^2\left[1 - \frac{1}{2}(\phi_* + 1/\phi_*\beta^2)\right]. \tag{15}$$

The non-dimensional force of the spring referred to the supplied pressure is as follows:

$$\bar{S}/\bar{p}_o = \bar{k}\ (\Delta\bar{l} - \bar{y})/\bar{p}_o \tag{16}$$

Relations between these two forces during valve closure determine the motion of the plate. This motion is quasisteady when the liquid and spring forces are equal to each other and it is accelerated, when the liquid force is greater than the spring one. Dependence of the liquid force during valve closure results from the function $\phi(y)$ - Eq.(7). For $y > y_*$, ϕ = const and P_s/p_o = const. For $y < y_*$, ϕ linearly depends on y and P_s/p_o changes in a parabolic manner. Dependence of $\bar{S}/\bar{p}_o(y)$ is linear. Gradient of $\bar{S}/\bar{p}_o$ for $\bar{p}_o = 0$ is infinite and decreases with growth of $\bar{p}_o$.

Functions of liquid force and the spring force during valve closure due to slow increase of supplied pressure are shown in Fig. 13a,b,c. In the case (a) these forces are equal to each other at each position of the valve plate. In this case the plate is displaced until valve closure in a quasi-steady manner and no compression wave appears. In the case (b) equilibrium between the liquid force and the spring force occurs for $\bar{y} > \bar{y}_*$ and for $\bar{y} < \bar{y}_A$. In the remaining range of $\bar{y}$ the liquid force is greater than the spring one. In case (c) such a relation appears for $\bar{y} < \bar{y}_*$. In this case the motion of the plate becomes accelerated when it reaches $\bar{y} = \bar{y}_*$. Therefore, at the moment of valve closure both the plate and the liquid have certain velocities. At the same time the velocity of the liquid is greater or at least equal to the velocity of the plate. This means that in case (c) the compression wave emerges at the moment of valve closure. One can suppose that the compression wave can also appear in the case (b) when $\bar{y}_A$ is relatively small while in this case the plate due to its inertia can have a certain velocity at the moment of valve closure. It arises from considerations given above that due to the slow increase of supply pressure the compression wave emerges at some $\bar{p}_o$ when $\Delta \bar{l} > \Delta \bar{l}_2$ (see Fig. 13d). For $\Delta \bar{l}$ in the range $\Delta \bar{l}_1 < \Delta \bar{l} < \Delta \bar{l}_2$ the compression wave is possible whereas for $\Delta \bar{l} < \Delta \bar{l}_1$ it is impossible. The values of $\Delta \bar{l}_1$ and $(\bar{k}/\bar{p}_o)_1$ are determined by the tangent to the curve $\bar{P}_s/\bar{p}_o(\bar{y})$ at $\bar{y} = \bar{y}_*$ as well as the values of $\Delta \bar{l}_2$ and $(\bar{k}/p_o)_2$ by the secant to the curve $\bar{P}_s/\bar{p}_o(\bar{y})$ at $\bar{y} = \bar{y}_*$ and $\bar{y} = 0$.

For example

$$(\bar{k}/\bar{p}_o)_2 = 4(1 - \phi_* \beta)^2 \sqrt{1 - \phi_*}. \tag{17}$$

6.2. The long tube

The simple compression wave can appear in the long tube for $\bar{y} < \bar{y}_*$, i.e. when flow is chocked during slow valve closure.

The flow properties in the simple wave change according to the following equation — see Eq.(4):

$$\bar{v} - \bar{v}_c = -(\bar{p} - \bar{p}_c), \tag{18}$$

where $\bar{v}_c$ and $\bar{p}_c$ are the flow velocity and the pressure corresponding to the plate position $\bar{y}_c$ at which the compression wave emerges. From continuity and energy equations for the flux through the valve we have

$$\bar{p} + \bar{v}^2/2 = \bar{v}^2/(2\phi^2 \beta^2). \tag{19}$$

Solving Eqs. (17) and (18) are

$$\bar{v} = -N + \sqrt{N[N + 2(\bar{v}_c + \bar{p}_c)]} \,, \qquad \bar{p} = \bar{p}_c + \bar{v}_c - \bar{v} \,, \tag{20}$$

where $N = \phi^2 \beta^2/(1 - \phi^2 \beta^2)$.

Substituting Eqs. (20) and $\bar{u} = 0$ into Eq.(14) one obtains the following expression for the liquid force.

$$\bar{P}_d/\bar{p}_o = \left\{ [\bar{v}_c + \bar{p}_c - \bar{v}](1 - \phi_*) + \bar{v}^2[1 - \tfrac{1}{2}(\phi_* + 1/\phi_* \beta^2)]\right\}/\bar{p}_o \tag{21}$$

Farther on, this force is called the rapidly-increasing liquid force as opposed to the liquid force determined by Eq.(15).

Functions of $\bar{P}_d/\bar{p}_o(y)$ for some $\bar{p}_o$ are given in Fig. 14. One can see that curves $\bar{P}_d/\bar{p}_o(y)$ are lightly convex in relation to the abscissa. Therefore, it can be assumed, that the rapidly-increasing liquid force is greater than the spring force under the following conditions (see Fig. 15):

$$\left[(\bar{P}_d/\bar{p}_o)_{\phi=0} - (\bar{P}_d/\bar{p}_o)_{\phi=\phi_c} \right]/\bar{y}_c \geqslant \bar{k}/\bar{p}_o , \tag{22}$$

$$\Delta\bar{l} \geqslant \bar{y}_c + (\bar{P}_d/\bar{p}_o)_{\phi=\phi_c}/(\bar{k}/\bar{p}_o) , \tag{23}$$

where

$$(\bar{P}_d/\bar{p}_o)_{\phi=0} = \left[\sqrt{\frac{2}{\bar{p}_o}}\, \phi_c\, \beta + (1 - \phi_c^2\, \beta^2) \right] (1 - \phi_*) \tag{24}$$

from Eq. (21),

$$(\bar{P}_d/\bar{p}_o)_{\phi=\phi_c} = (1 - \phi_c^2\, \beta^2)(1 - \phi_*) + 2\phi_c^2\, \beta^2 \left[1 - \tfrac{1}{2}(\phi_c + 1/\phi_*\, \beta^2) \right] \tag{25}$$

from Eq. (15).

The curves $\bar{k}(\bar{p}_o)_{\phi_c=\text{const}}$ and $\bar{k}(\bar{p}_o)_{\Delta\bar{l}=\text{const}}$ shown in Fig. 16 correspond to Eqs. (22) and (23). All of the curves coincide at the point $\bar{k} = \bar{p}_o = 0$. The curves $\bar{k}(\bar{p}_o)_{\Delta\bar{l}=\text{const}}$ for all of $\Delta\bar{l}$ cross the curve for $\phi_c = 0$, but only the ones for $\Delta\bar{l} > \bar{y}_*$ cross the curve $\phi = \phi_*$. The cross points of these curves signified by "1" and "2" in Fig. 16a respectively, correspond to certain conditions which have essential meaning in the process at valve closure.

It results from Fig. 16 b, that the conditions (22), (23) can be fulfilled for each value of $\Delta\bar{l}$, if $\bar{k} < \bar{k}_1 (\Delta\bar{l})$. It means that in this case, as opposed to the short tube, the liquid force can be greater than the spring force also for $\Delta\bar{l} < \Delta\bar{l}_1$ (see Fig. 13 d) including $\Delta\bar{l} < \bar{y}_*$.

The process of valve closure in the case of $\Delta\bar{l} < \bar{y}_*$ due to the increase of $\bar{p}_o$ is illustrated in Fig. 17. In the first phase of this process gradient of the $\bar{P}_d/\bar{p}_o$ is lower than the gradient of $\bar{S}/\bar{p}_o$. Therefore, the plate initially moves is a quasisteady manner. At the position of the plate, namely at $\bar{y} = \bar{y}_c$ these gradients became equal to each other. A certain value of $\bar{p}_s = \bar{p}_{oc}$, which depends on $\bar{k}$ and $\Delta\bar{l}$ (see Fig. 16 b), corresponds to this position. When this value is exceeded the motion of the plate becomes accelerated.

In the case of $\Delta\bar{l} > y_*$ the motion of the valve plate, due to increase of $\bar{p}_o$, up to position $\bar{y} = \bar{y}_*$ is a quasisteady one. For $\bar{y} > \bar{y}_*$ the motion can remain quasisteady or accelerated one depending on $\bar{k}$ see Fig. 18 . For $\bar{k} < \bar{k}_2$ (see Fig. 18a) it is accelerated while in this case the gradient of $\bar{P}_d/\bar{p}_o$ at $\bar{y} = \bar{y}_*$ is greater than the gradient of $\bar{S}/\bar{p}_o$. For $\bar{k} > \bar{k}_2$ (see Fig. 18.b) it is still quasisteady up to such a position when the gradients mentioned above become equal to each other. Further motion is of the accelerated type.

In all of the cases, in which accelerated motion of the plate occurs, the compression wave during valve closure appears.

7. THE LIMITING CASE OF PULSATING FLOW

It was mentioned is section 5, that the frequency of pulsating flow increases with growth of supply pressure. It was explained that this phenomenon is connected with the decrease of the time of acceleration of the flow, due to the increase of pressure difference at the ends of the water column. Decrease of the acceleration time corresponds to the reduction of the number of the waves contributing to the theoretical cycle, Fig. 11 . It means that with increase p_o the theoretical cycle tends to a certain asymptotic form comprising four waves Fig. 19 . A square tooth trace of the plate position corresponds to such a cycle Fig. 19b . Flow properties for each phase of the asymptotic cycle can be found by means of Eqs.(5). From these equations result the following non-dimensional equations corresponding to the four waves of the cycle.

$$\bar{p}_2 - \bar{p}_o = -\bar{v}_3, \qquad \bar{p}_o - \bar{p}_4 = -\bar{v}_3 + \bar{v}_4,$$

$$\bar{p}_1 - \bar{p}_4 = \bar{v}_1 - \bar{v}_4, \qquad \bar{p}_2 - \bar{p}_1 = \bar{v}_1. \qquad (26)$$

The set of Eqs.(26) is completed by the two following arising from the boundary conditions,

$$\bar{p}_o = \bar{p}_1 + \bar{v}_1^2/2, \qquad \bar{p}_4 = 0.5 \cdot \bar{v}_4^2 \left(1/\phi^2 \beta^2 - 1 \right) \qquad (27)$$

where ϕ is a relative area of minimum cross-section of the valve slit for phase "4".

Eqs. (26),(27) can be reduced to the following one

$$\frac{\bar{v}_1^2}{8} \left(1/\phi^2 \beta^2 - 1 \right) + \bar{v}_1 - \bar{p}_o = 0. \qquad (28)$$

Moreover, the following relations between liquid force P and spring force S ought to be fulfilled: in the phase "4" $P_4 = S_4$, in the phase "2" $P_2 \geqslant S_2$, in the phase "5" $P_5 \leqslant S_5$ (see Fig. 19c).
On the basis of these relations and of Eqs.(14)(where $p = p_4$, $v = v_4$) and (16) the following equation and inequalities can be written, respectively

$$\bar{p}_4(1 - \phi_*) + \bar{v}_4^2 \left[1 - \frac{1}{2} \left(\phi_* + 1/\phi_* \beta^2 \right) \right] = \bar{k}(\Delta\bar{l} - \phi/4 \sqrt{1-\phi_*}), (29)$$

$$\bar{p}_2 (1 - \phi_*) \geqslant \bar{k} \; \Delta\bar{l} \geqslant \bar{p}_5 (1 - \phi_*). \qquad (30)$$

The flow properties at the phases "2", "4", "5" can be determined from Eqs.(26),(27).
On the basis of Eqs.(28),(29) and inequalities (30) one can find the value of $\bar{p}_o$, corresponding to the emergence of the extremal cycle, for given $\bar{k}$ and $\Delta\bar{l}$. In Fig. 20 the values of $\bar{p}_o$ versus $\bar{k}$ for three values of $\Delta\bar{l}$ are shown (dashed lines). It can be seen from this figure, that $\bar{p}_o$ increases, achieves a certain maximum value and next de-

creases as $\bar{k}$ increases. The maximum value of $\bar{p}_o$ increases with growth of $\Delta\bar{l}$. Simultaneously, one can prove by means of Eqs.(28),(29), that

$$\bar{p}_o \longrightarrow \bar{k}\ \Delta\bar{l}/(1 - \phi_{\varkappa})\ \text{for}\ \phi \longrightarrow O(\bar{k} \longrightarrow \bar{k}_1).$$

It should be said, that the asymptotic cycle considered above in practice does not appear, because the valve opens and closes in a finite time. This cycle can be considered only as a certain theoretical asymptotic one, i.e. when the supply pressure, corresponding to this cycle is exceeded then the liquid flow stops.

8.   RANGES OF PULSATING FLOW

It arises from sections 6 and 7 that the pulsating flow is possible in the range determinated by the following lines shown in Fig. 20.

a)   At the side of low supplied pressure by the lines $\bar{k}(\bar{p}_o)_{\Delta\bar{l}=const}$ determined from Eq.(22) - long tube, or by the line $\bar{k}(\bar{p}_o)_{\Delta\bar{l}=const}$ determined from Eq.(17) - short tube.

b)   At the side of high supplied pressure by the lines $\bar{k}(\bar{p}_o)_{\Delta\bar{l}=const}$ determined from Eqs. (28-30).

All of these lines and the lines $\bar{k}(\bar{p}_o)_{\phi_c=const}$ coincide in the point $\bar{k} = \bar{p}_o = O$. For the conditions corresponding to the field on the left side of the line "a" the liquid flow is a steady one (opened valve). In the field on the right side of the lines "b" the liquid is stationary (closed valve). For $k > k_1$, which corresponds to the intersection point of the line "a" and "b" for given $\Delta\bar{l}$ the steady flow changes directly into a stationary state due to the increase of $\bar{p}_o$. The values of $\bar{p}_o$ corresponding to this transition are determined by the line $\bar{k}(\bar{p}_o)_{\phi_c=O}$.

Examples of numerical calculations Fig. 20 performed by the method described in chapter 4 and observations of the operation of hydraulic systems considered here prove that the transition from steady to pulsating flow for tubes with L greater than about 1 m occurs when $\bar{p}_o$ has a value close to that corresponding to long tubes.

These examples show, simultaneously, that the stopping of the pulsating flow occurs for supplied pressures, which are much lower than these resulting from the line "b" corresponding to given $\Delta\bar{l}$.

9.   CONCLUSIONS

In a hydraulic system containing a tank, a tube and a valve with an elastically supported plate, in certain circumstances a pulsating flow appears. The flow is accompanied by cavitation. From calculations it arises that during one cycle of pulsation a cavitation zone emerges several times. Frequency and amplitudes of the flow properties depend on the following parameters: $\bar{p}_o$, $\bar{M}$, $\bar{k}$, $\Delta\bar{l}$, $\phi_{\varkappa}$, L, which express supplied energy, inertia of the liquid and of the valve as well as elasticity of the plate support and the dimensions of the system.

The calculated and measured results show that the amplitude of the pressure can several times exceed the supplied pressure.

The pulsating flow appears only upon conditions which are determined by $\bar{k}$, $\Delta\bar{l}$, $\bar{p}_o$, $\varnothing_*$.

When plate support is rigid enough (large $\bar{k}$) the pulsating flow does not occur.

10. REFERENCES

1. Klein, F., Mueller C.: "Encyklopaedie der Mathematischen Wissenschaften, Band IV Mechanic, Teil 3, Teubner Verlag Leipzig, 1908.

2. Erhardt, G.: "Zur Optimierung des hydraulischen Widders", Abh. Aerodyn. Inst. RWTH Aachen, 21, 1979.

3. Iversen, H.W.: "An Analysis of the Hydraulic Ram", Journal of Fluid Engineering, Trans. ASME, June 1975, pp. 191-196.

4. Kraemer, K.: "Bemerkung zum hydraulischen Widder", Max-Planck-Institut für Strömungsforschung, Göttingen, Bericht 118/1978.

5. Schweitzer, P.H., Szebehely, V.G.: "Gas Evolution in Liquids and Cavitation", Journ. of Applied Physics, Vol. 21, No 12, 1950, pp. 1218-1224.

6. Baltzer, R.A.: " Column Separation Accompanying Liquid Transients in Pipes", Journal of Basic Engineering, Trans. ASME, Vol 89, Dec. 1967, pp. 837-846.

7. Safwat, H.H.: "Photographic Study of Water Column Separation", Proceedings of ASCE, Vol 98, No HY4, 1972, pp. 739-746.

8. Safwat, H.H., van den Polder J.: "Experimental and Analytic Data Correlation Study of Water Column Separation", Journal of Fluids Engineering, Trans. ASME, March 1973, pp. 91-97.

9. Driels, M.R.: "Investigation of Pressure Transients in a System Containing a Liquid Capable of Air Absorbtion", Journal of Fluid Engineering, Trans. ASME, Sept. 1973, pp. 408-414.

10. Martin, C.S., Padmanabahan M.: "The Effect of Free Gases on Pressure Transients", L´Energia Elettrica, No 5, 1975, pp. 262-267.

11. Wiggert, D.C., Sundquist, M.J.: "The Effect of Gaseous Cavitation on Fluid Transients", Journal of Fluid Engineering, Trans. ASME, March 1979, pp. 79-86.

12. Feldman, C.L., Nydick, S.E., Kokennak, R.P.: "The Speed of Sound in Single-Component Two-Phase Fluids: Theoretical and Experimental", Progress in Heat and Mass Transfer, Vol 6, Pergamon Press, 1972, pp. 671-700.

13. van Wijngaarden, L.: "Sound and Shock Waves in Bubbly Liquids", Cavitation and Inhomogeneities in Underwater Acoustics, Proceedings of the First International Conference, Göttingen, July 9-11, 1979, Springer Verlag, 1980, pp. 127-140.

14. Szumowski, A.P.: "Investigations of Wave Phenomena in Waterhammer", Warsaw Technical University Publications, z. 61, 1980, in polish .

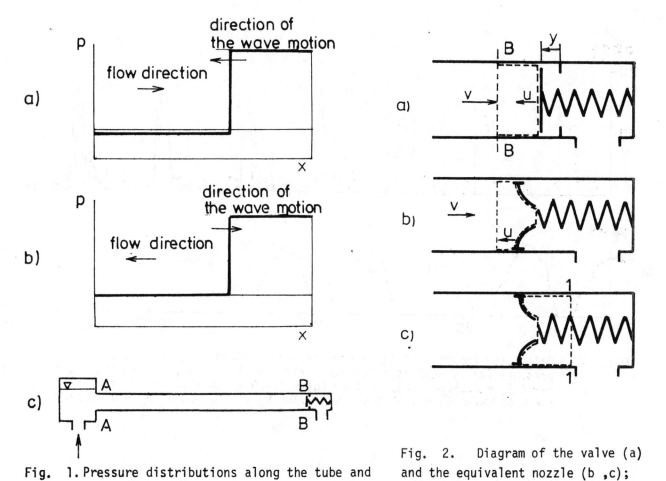

a)

direction of
the wave motion

flow direction

b)

direction of
the wave motion

flow direction

c)

Fig. 1. Pressure distributions along the tube and
the diagram of the hydraulic system considered.

Fig. 2. Diagram of the valve (a)
and the equivalent nozzle (b ,c);
broken lines mark control surfaces.

Fig. 3. The net of characteristics.

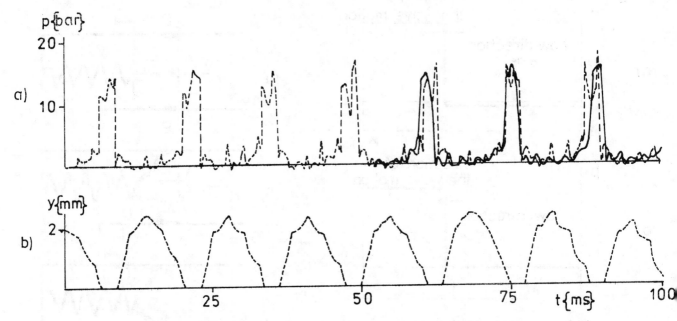

Fig. 4. Calculated (continuous line) and measured (dashed line) traces of the pressure (a) as well as calculated trace of the plate position (b) . $p_o= 3$ bar ($\bar{p}_o = 0.15.10^{-3}$), $\Delta l = 2$ mm ($\overline{\Delta l} = 0,02512$), $L = 1.6$ m ($\bar{L} = 20$), $K = 0.66$ kN/mm ($\bar{k} = 0.005366$), $M = 0.358$ kg ($\bar{M} = 0.9038$), $\emptyset_x = 0.03733$.

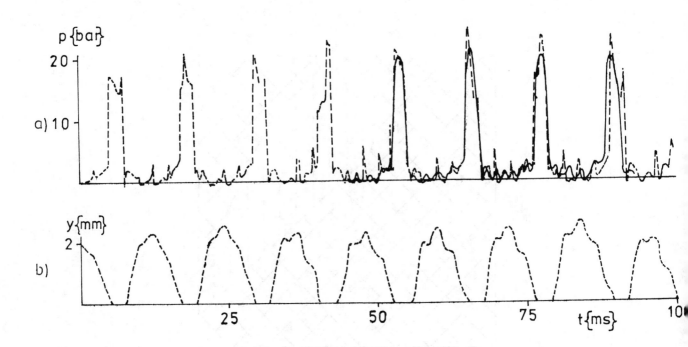

Fig. 5. $p_o= 4$ bar ($\bar{p}_o= 0.2.10^{-3}$). Explanation and the other dates in Fig. 4.

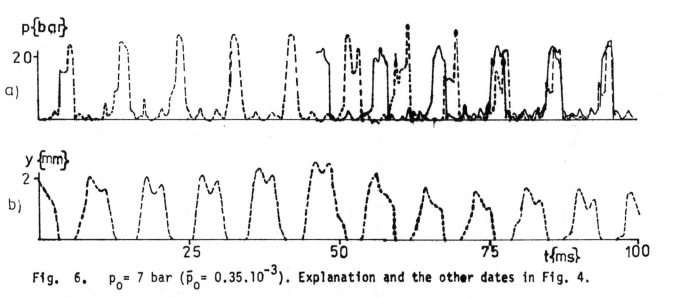

Fig. 6. $p_0 = 7$ bar ($\bar{p}_0 = 0.35.10^{-3}$). Explanation and the other dates in Fig. 4.

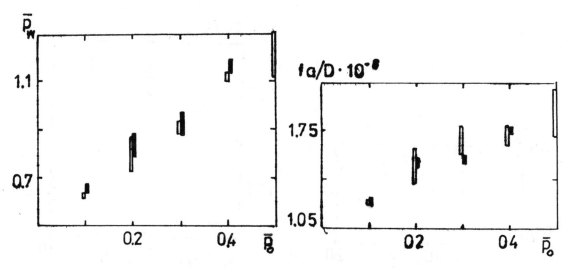

Fig. 7. Influence of the supplied pressure on the mean pressure of impulse (p_w) and on the frequency (f) . L = 1.6 m ($\bar{L}$ = 20), Δl = 2 mm ($\bar{\Delta l}$ = 0.02512), k = 0.66 kN/mm ($\bar{k}$ = 0.005366), M = 0.358 kg ($\bar{M}$ = 0.9038), $\emptyset_\star$ = 0.03733.

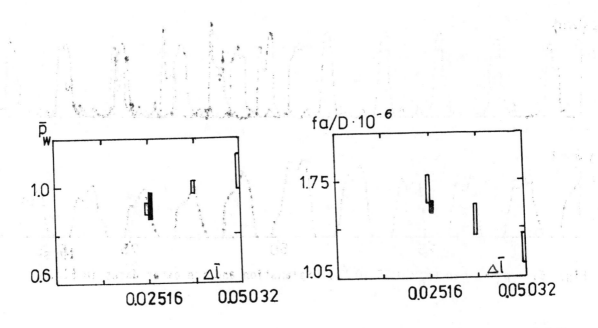

Fig. 8. Influence of the nominal displacement of the valve plate on p_w and f for $\bar{p}_o = 0.25 \cdot 10^{-3}$; $\bar{L}$, $\bar{k}$, $\bar{M}$, $\emptyset_*$ as in Fig. 7.

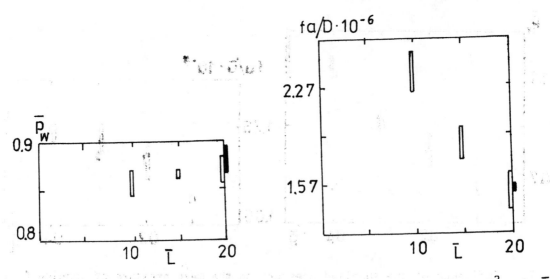

Fig. 9. Influence of the length of the tube on p_w and f for $\bar{p} = 0.2 \cdot 10^{-3}$; Δl, $\bar{k}$, $\bar{M}$, $\emptyset_*$ as in Fig. 7.

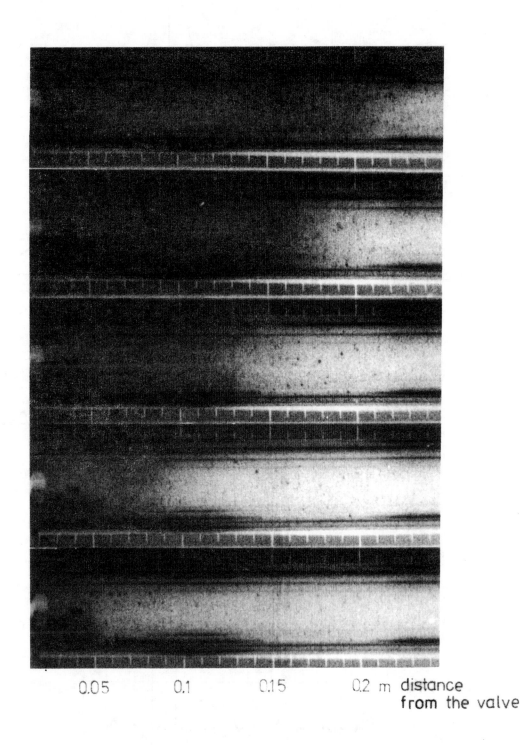

0.05   0.1   0.15   0.2 m distance
                                        from the valve

Fig. 10. Photographs showing the cavitation zone (dark portions) in vicinity of the
valve. Time distances between successive photographs 0,125 ms.

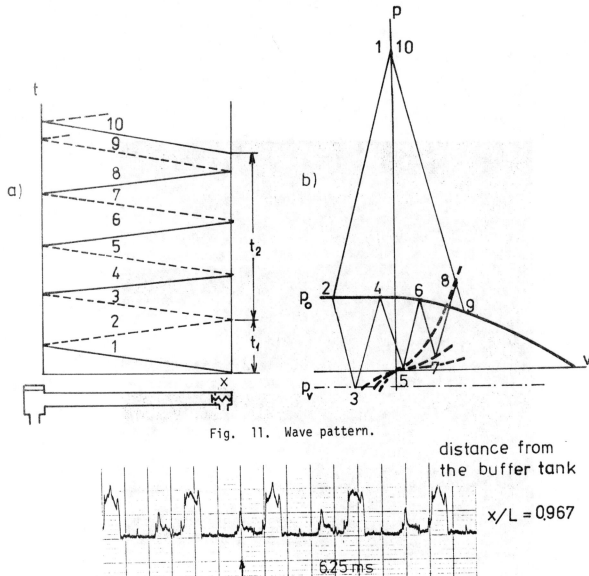

Fig. 11. Wave pattern.

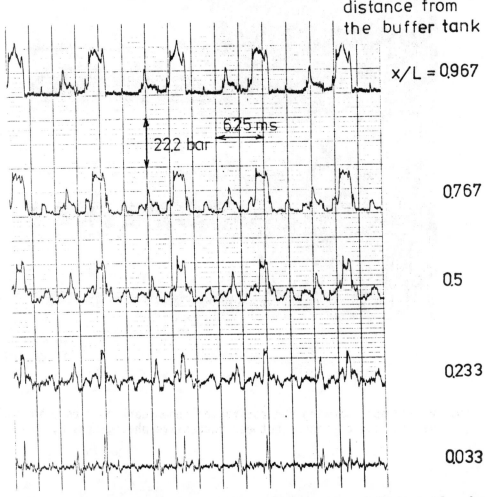

distance from
the buffer tank

x/L = 0.967

22.2 bar   6.25 ms

0.767

0.5

0.233

0.033

Fig. 12. Measured pressure traces at some cross-sections, p_o = 2.33 bar, Δl = 3 mm, L = 1.5 m, k = 45 N/mm, M = 0,042 kg, D = 50 mm, D_p^o = 44 mm.

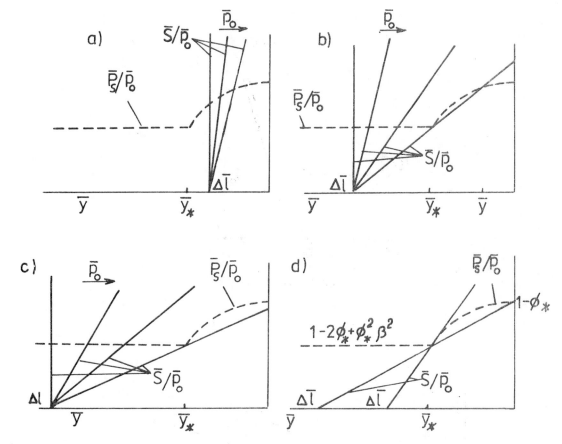

Fig. 13. Relations between the liquid force and the spring force during valve closure due to slow increase of supply pressure.

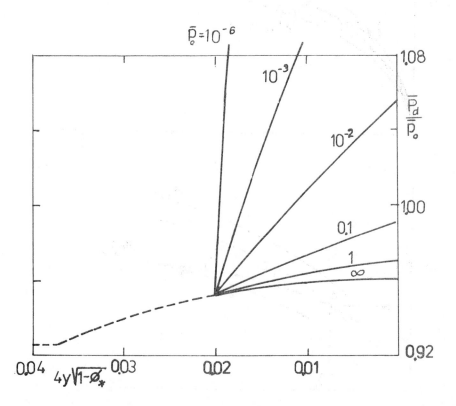

Fig. 14. Traces of the rapidly-increasing liquid force.
$\phi_* = 0.03733$, $\phi_c = 0.02$, $\beta = 0.35$.

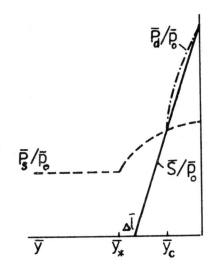

Fig. 15. For criteria (22), (23).

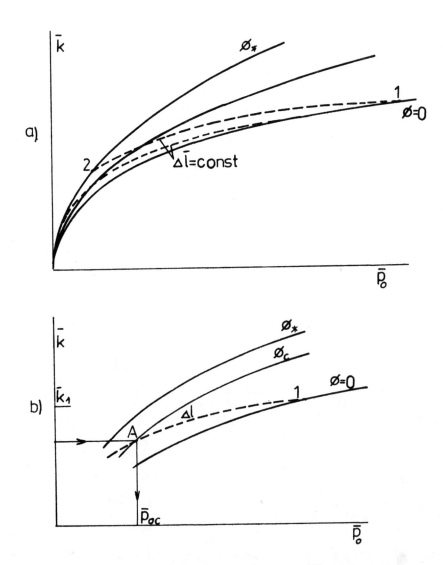

Fig. 16. Functions of spring constant versus supplied pressure for ϕ_c = const. (continuous lines) and for $_\Delta\bar{l}$ = const. (dashed lines).

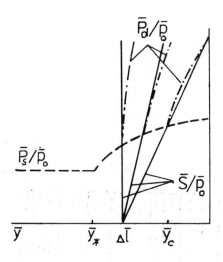

Fig. 17. Relations between the rapidly-increasing liquid force and the spring force during valve closure due to slow increase of $\bar{p}_0$ in the case $_\Delta\tilde{l} < \tilde{y}_*$.

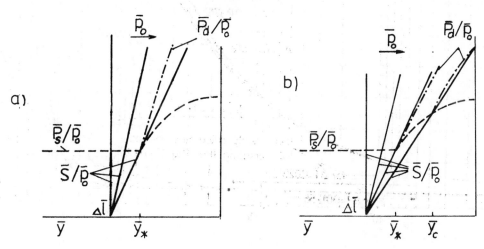

a)

b)

Fig. 18. Relations between the rapdily-increasing liquid force and spring force during valve closure due to the slow increase of $\bar{p}_0$ in the case $_\Delta\tilde{l} > \tilde{y}_*$.

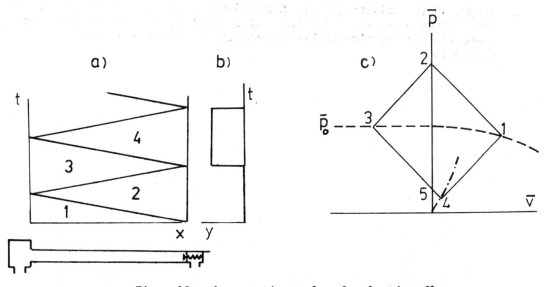

a)   b)   c)

Fig. 19. Asymptotic cycle of pulsating flow.

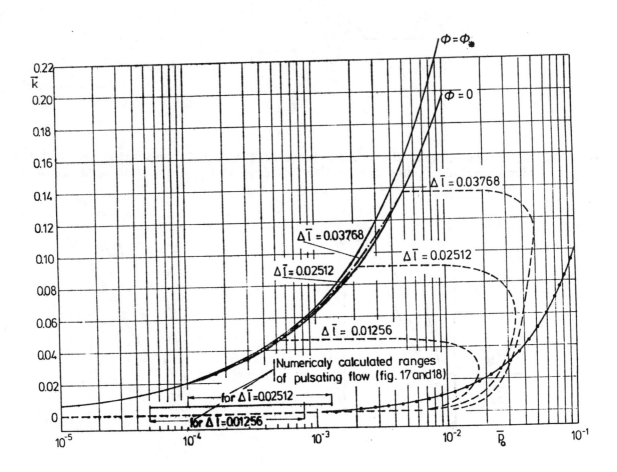

Fig. 20. Ranges of the pulsating flow. Boundaries at the side of low supply pressure are: lines —·—·—·— for long tubes and line —•—•—•— for short tubes. Boundaries at the side of high pressures are lines — — —. $\phi_* = 0.03733$. $\beta = 0.35$.

International Conference on

FLOW INDUCED VIBRATIONS
IN FLUID ENGINEERING

Reading, England: September 14-16, 1982

PAPER G2

UNSTABLE FLOW IN PIPELINES: TWO CASE STUDIES

P. Dawson

Hydraulic Analysis Limited, U.K.

Summary

This paper describes two cases where extreme vibration induced by the pulsation of flow in liquid-carrying pipe networks has threatened the shutdown of vital equipment. These cases demonstrate different methods of identifying the source of the problem and also show contrasting means of ensuring stable flow.

Organised and sponsored by
BHRA Fluid Engineering, Cranfield, Bedford MK43 0AJ, England

NOMENCLATURE

p   =   pressure generated by liquid

w   =   specific weight of liquid

ρ =   mass density of liquid

K   =   bulk modulus of liquid

v   =   mean flow velocity of liquid

d   =   diameter of pipe

t   =   wall thickness of pipe

E   =   Young's modulus of elasticity for pipe

z   =   elevation of pipe

A   =   area of flow in pipe

T   =   viscous shear stress

P   =   wetted perimeter of pipe

dp   =   pressure increment

$d\rho$ =   density increment

dA   =   area increment

dv   =   velocity increment

dx   =   distance increment

dz   =   elevation increment

dt   =   time increment

g   =   gravitational constant

1.0 CASE ONE: FIRE FIGHTING SCHEME

1.1 Introduction

The first case study gives a general description of an unsteady flow problem which is often experienced in fire fighting systems and other similar water supply networks on both oil platforms and onshore petrochemical installations. For this paper a fire fighting scheme is discussed but the problem is common to any scheme where a recirculation or dump line is installed to provide a temporary discharge of liquid from a high pressure system.

1.2 Piping System

The fire fighting system on an oil production platform or a refinery typically comprises a pump area containing two or more high capacity pumps, delivering water into a ring main which supplies hydrants and deluges (see Figure 1). To ensure adequate pressure at the hoses the ring main must be maintained above 5.5 bar, although pressures of over 8.0 bar are frequently required to supply high level deluges, on upper decks of the platform, for example. Hence, the fire system is designed so that the fire pumps are started automatically if the ring main pressure falls significantly below this supply level, with jockey pumps to accommodate any leakage from hydrants, etc.

In contrast, the fire pumps are switched off manually when the emergency is passed to avoid the possibility of an accidental instrument shutdown. However, it is common for the maximum pressure generated by the fire pumps to be close to the pipeline rating so that the full capacity of the system can be utilised. Hence an automatic dump line is also installed in the pump area to discharge flow overboard during the period between hydrant closure and pump shutdown. This line usually includes an automatic control valve, (arranged to open if the ring main pressure exceeds a preset value) which maintains a low flow from the pumps and so prevents piping overpressure or relief valves lifting (see Figure 2). Additionally, this prevents the pumps running under no flow conditions until they are switched off.

1.3 Vibration

The typical piping configuration described above has been adopted for many supply systems and, during normal fire fighting operation when the dump line valve is closed, steady flow and pressures are maintained. However, extreme vibration of the control valve and pump discharge pipework occurs as the hydrants are shutdown and flow is diverted from the ring main through the dump line. In more than one instance, the severity of these factors threatened a total shutdown of the fire fighting system and hence the entire plant.

During commissioning tests and subsequent operation, engineers have reported physical movement of the valve body, together with hunting of the valve positioner and the control mechanism; a high level of noise exceeding 105 dB and unstable action of the nearby relief valves have also been observed. Additionally, the pattern of flow leaving the discharge of the dump line has been described as resembling a Christmas tree (see Figure 3).

1.4 Mechanism

This description of unsteady flow and excessive vibration contains two features which are characteristic of flow instability induced by cavitation of the dump line control valve, namely the high noise level and the discharge flow pattern. However, this effect is compounded by the control functions of the valve, as well as cavitation and flow rotation in the dump line itself.

Consideration of the energy changes occurring in the body of the control valve demonstrates the primary reason for cavitation in the dump line. A globe valve dissipates energy by a sudden increase in pipe area and the corresponding turbulence as flow expands; hence the aim of the valve design is to accelerate flow through the valve inlet and cage and therefore allow this expansion in the outlet section.

$$\frac{p}{w} + \frac{v}{2g} + z = \text{constant}$$

When flow is converging energy losses are low but, as defined by Bernoulli's equation (above), the pressure (or potential energy) decreases as the velocity increases. Subsequently, a pressure rise occurs when flow decelerates in the discharge section (as shown on Figure 4) but energy is also lost in this phase. Hence the final (outlet) pressure is lower than the inlet pressure, with the pressure in the vena contracta (or minimum flow area) being lower than both values.

In the case of fire fighting systems, the pressure in the ring main may be only about 8.0 barg but, as there is no risk of contamination, flow is dumped overboard into free air; hence a total of 8 bar must be dissipated with the dump line discharging at atmospheric pressure. To prevent the problem described above, this valve discharge pressure must be avoided; the piping designer, therefore, refers to the valve manufacturer's catalogue to determine the minimum pressure allowed and an orifice plate is often sized to generate the necessary back pressure. But, in the case study described, the significance of the vena contracta at the orifice plate itself was underestimated. Thus, gassing occurred at the plate which also reduced the back pressure supplied to the control valve so that vapour still formed in the valve body.

The first sign of this problem is the characteristic noise of cavitation as the vapour bubbles, generated in the valve body, collapse against the wall of the pipe where pressure recovery occurs some metres downstream of the valve; but the continued evolution of gasses and vapour from the valve and orifice plate also causes a pocket to collect at a downstream bend, thus restricting flow. In turn, the upstream pressure rises and, as the control valve starts to open to overcome this trend, the vapour/gas pocket is displaced. This leads to a pulse of flow which gives the pattern of a Christmas tree as the jet drops vertically from the discharge pipe.

Ideally, the action of the control valve should overcome this oscillating pressure by opening as flow is baulked (and the pressure rises) and then closing partially as the restriction is removed (and the pressure at the sensor falls). But in practice, the stroke time of a large control valve, together with its inertia, the response of the controller etc., is too slow to follow the pulsation, which has been measured at about 0.5 - 1.0Hz. Moreover, the internal forces on the valve plug constantly change as the valve body is filled alternately with liquid and vapour. Therefore the valve cannot maintain a stable position which, in turn, leads to oscillating flow and a corresponding pressure variation at the valve control sensor.

The effects of vapour blocking and control instability can each cause serious vibration, but in this fire fighting system the development of flow through the relief valves caused a further flow pulsation. Spring operated safety/relief valves were installed to ensure the pipe system could not be overpressurised if the control valve in the dump line failed to open. Normally, therefore, these valves would be closed but, in this case, the fluctuation in flow in the ring main was sufficient to cause the pressure to exceed the relief setting. The valves lifted, diverting such a high flow rate that the local pressure dropped below the blowdown pressure and the valves closed. This, in turn, created a surge pressure rise as the relieved flow was stopped, which added to the existing flow pulsation in the ring main and caused a further pressure rise. The relief valves again lifted and hence another mechanism of unstable flow was developed.

1.5   Investigation

The hypothesis presented above is supported by simple calculations defining the pressure changes occurring through the control valve and the dump line; however, a more detailed investigation has never been deemed necessary. These calculations, together with experience of the same problem in several installations, are sufficient to confirm that the components causing unstable flow and excessive vibration all derive from cavitation in the valve body and the subsequent vapour blocking. Thereafter, a typical resonance problem ensues where the flow disturbance is not attenuated but is amplified by the characteristics of the piping configuration, the action of the control valve and the instability of the relief valves. The energy provided by the pump is accumulated, causing pressure and flow oscillations of increasing amplitude and vibration of an unacceptable magnitude.

1.6   Solution

The prime objective in eliminating the vibration was, therefore, to prevent the initial development of cavitation; thus all attention was transferred to redesigning the dump line using the two basic alternatives of:-

1.   retaining the globe valve but installing further head loss devices in the dump line to increase the back pressure on the valve.

2.   replacing the control valve by a drag valve (or similar) which divides flow into small streams with tortuous flow paths and dissipates head by friction losses and jet impact; this eliminates the low pressure vena contracta.

In practice, both options have been adopted on different schemes and in all cases the discharge system has subsequently maintained a stable upstream pressure by diverting a steady flow and so relief action, excessive vibration and noise have been avoided. However, the former option is preferred as this requires less capital expenditure although, as shown by this case study, restriction orifices are not always the best solution. The following factors were therefore considered in the redesign of the fire fighting system dump line (see Figure 5).

1.   a high velocity taper (nozzle) on the discharge of the line can dissipate most of the loss required downstream of the control valve.

2.   the diameter of the dump line downstream of the control valve should be the same as the exit flange of the valve with, preferably, only one 90° bend included. This avoids the possibility of high velocity flow rotation and cavitation being initiated at pipe expansions and elbows.

3.   the control valve should be located at least ten pipe diameters from the tee to the ring main. This allows flow to straighten and the instability of boundary layer separation at the tee to be dissipated.

4.   the back pressure at a control valve (or orifice plate if used) should be higher than the minimum acceptable value calculated

theoretically. This accommodates the significant variations in flow from the mean velocity, accentuated by random, fast moving vortices.

1.7   Summary Of First Case Study

When unsteady or pulsating flow is discussed during the design of a scheme, the problems are usually grouped in two categories, namely surge and resonance. The first typically entails a single incident, such as pump start, valve closure etc., and is usually investigated using surge analysis techniques. The second encompasses cyclic problems, which are mainly considered for high energy systems (particularly in hydro electric schemes where governor hunting, guide-vane vibration, cavitation and vortex shedding have all caused resonance) or when the prime mover pulsates (pumps).

In contrast, this case study highlights additional areas (i.e. all dump lines and pump low-flow recirculation lines) where resonant problems can develop. The provision of a dump line on these schemes sought to overcome any surge problems following shutdown of flow in the ring main and slow control functions were adopted for the control valve to avoid unstable action. However, the vibration caused by valve cavitation negated the design aims of the line and a complete redesign (as described above) was required to avoid shutdown of a major system.

2.0   CASE TWO : LUBRICATION SYSTEM

2.1   Introduction

The second case study describes a large oil lubricating system that showed such an excessive amount of piping vibration, during the start up tests for the booster pump, that total shutdown was initiated within 5s. Pressure traces were taken during the test and so the first requirement of an investigation was to reproduce the pressure variation analytically, thus validating the techniques utilised and supporting the recommended modifications.

2.2   System Description

The system being tested was a combined network, distributing lube oil from a central reservoir to provide hydraulic actuation for the governor of turbo compressors, as well as fluid for hydraulic seals and lubrication for bearings (see Figure 6). Thus the scheme was unusual in terms of a lube oil plant by covering a large plan area, with some pipe runs exceeding 30m in length, and utilising relatively large bore pipework, i.e. 40mm - 200mm.

Typically flows of about 0.76 l/s were required for each seal with 3.15 l/s for governor control. However, up to 8 l/s would be required within 0.25s during a sudden demand for pressure at the governor servo motor; hence two parallel centrifugal primary pumps were installed, in series with two rotary positive displacement booster pumps (each set with a standby) to deliver a maximum flow of 40 l/s at a pressure of about 31 bar.

From the common manifold downstream of the booster pumps a 200/150mm line led through filters to a header from which 75mm or 50mm lines supplied each seal, bearing and governor; at these locations control valves were arranged to maintain a constant pressure at each unit (i.e. with the controller sensing on the downstream side of the valve). Equal percentage characteristic 25mm globe valves with pneumatic actuators, a 3.0s stroke and two term controllers were used for the seal oil control valves. However, quick acting characteristic valves with 0.25s stroke were adopted for the governor control (25mm, hydraulic actuator, direct control) to avoid large speed variations in the turbo compressors.

Additionally, a recirculation (kickback) system was included downstream of the booster pumps to accommodate fluctuations in the required flow, by maintaining a constant pressure at the header. This system comprised a 150mm line with a 50mm equal percentage pressure control valve (PCV) with an actuator capable of achieving end-to-end valve movement in 1.0s and a two term, proportional and integral controller.

2.3   Vibration

The possibility of on-site vibration was first identified when the pumps were tested by the manufacturer in a piping configuration resembling the proposed installation and including the control valves. During start up of the first booster pump, excessive vibration occurred in the kickback line and the pump had to be shutdown. Subsequently, pressure traces were taken, as presented on Figure 7, and the following observations were made:-

1.   Steady flow and pressure conditions could be achieved when the primary pumps alone were supplying oil at low pressure to the seals etc. However, start up of a booster pump (run up to speed over 1s) caused the kickback valve to open and a pressure oscillation immediately developed with an amplitude of about 6.5 bar and a frequency of about 20.5 Hz.

2.   The pipework vibration adjacent to the pumps and kickback valve was severe and necessitated shutdown of the booster pump.

3.   Resonant vibration did not occur if two booster pumps were running. This was determined by starting the two pumps in rapid succession.

4.   Resonant effects also occurred after a rapid flow demand by the governor servo motor. Minor resonance was observed as the governor control valve opened but this decayed within 0.3s. However, a major pressure oscillation developed at a frequency of about 16 Hz as the governor valve closed and most of the flow was re-established to the kickback line.

2.4   Mechanism

From the results of this testing it was apparent that the pressure oscillation was primarily forced on the system by the actuation of the control valve in the kickback line. Hence the basic mechanism causing the pressure oscillation was formulated from an assessment of the transient conditions occurring between the header and the valve (see Figure 8).

As the booster pump is started, oil is discharged into the manifold, compressing the liquid already contained in that pipe section (1) and hence causing a local pressure rise. This imposes a positive pressure gradient across the next element of liquid (2), accelerating the column and in turn pressurising the next element (3). Hence a rising pressure wave is transmitted along the pipeline to the header at the acoustic velocity, typically 1000 m/s. However, an 18mm pipe leads from the header to the control valve, acting as the sensor line for the controller; thus the pressure wave also travels along this line (4) and, after a transmission delay, the diaphragm pressure in the actuator rises and the kickback control valve starts to open. Then, as the valve moves, liquid accelerates from the upstream line (5) and the local pressure falls. A drop in pressure, therefore, travels back along the main line to the header and subsequently the pressure at the valve sensor position falls; but the signal to partially close the valve is again delayed by the transmission time of the sensor line.

It is apparent from the above that the transient effects occurring in both the sensor line and the kickback line induce a cyclic stroking of the valve. Moreover, this is exacerbated by the delay inherent in the controller itself, together with the time taken to accelerate the valve plug and the subsequent tendency of the plug to overshoot the required position due to its acquired momentum. Hence, the frequency of the pressure fluctuation forced on the test rig was a complex function of:-

1. the length of main pipe and controller sensor line.

2. acoustic velocity.

3. controller delay.

4. valve plug acceleration rate.

5. valve stroke time.

6. valve plug momentum.

7. valve pressure/flow characteristic.

Based on this information obtained from the test rig it might have been possible to overcome similar problems occurring on-site by extending the stroke time of the kickback valve and/or changing the sensor location to a point immediately upstream of the control valve. However, there was a danger in this action of limiting the flow to the governor when a rapid response was required. Moreover, it was difficult to assess the significance of the on-site piping lengths being different from those in the test rig and so it was necessary to analyse the possible modifications more closely.

2.5   Analysis

The operators of the system insisted that any computerised analysis techniques utilised should first reproduce the sequence of events following the start up of the booster pump on the test rig; this would confirm the hypothesis of the mechanism causing vibration as well as predicting the possibility of on-site vibration and ensuring that any proposed modifications would not delay the system response. A modelling method was, therefore, adopted (based on the Method Of Characteristics) as the results from such a study are in the same format as the output from the transducer measurements on the actual system.

Subsequently, a frequency response or free vibration analysis could have been undertaken (using impedance or transfer function methods) to determine the best way of reducing the resonant features of the system. But the likelihood of the kickback and governor control valves forcing flow pulsations of different frequencies, together with the complex interaction of transient effects in the pipework made such an approach difficult; hence, Characteristic Modelling techniques were used throughout the study.

2.6   Computer Modelling By Method Of Characteristics

In contrast to a resonant study (which defines the gain or damping inherent in a system subjected to a preselected cyclic disturbance) the aim of computer modelling (based on the Method Of Characteristics) is to simulate a real incident, such as pump start up/shutdown. Thus the pressure and flow effects in the system and the subsequent action of any control valves, relief facilities etc., can be determined. This topic is covered extensively in text books but, in brief, the analysis is based on two fundamental laws of physics, namely that in any pipe section:-

1. the acceleration of the enclosed liquid column depends on the mass and the applied forces (pressure, self weight, friction) i.e. Newton's second law.

2. the difference between inflow and outflow is stored liquid, accommodated by radial expansion of the pipe and compression of the enclosed liquid (n.b. this storage capacity dictates the acoustic velocity).

In algebraic form, the dynamic and continuity equations based on these laws are:-

1. $$pA - \left(p + \frac{\delta p}{\delta x}.\delta x\right)\left(A + \frac{\delta A}{\delta x}.\delta x\right) + \left(p + \frac{1}{2}\frac{\delta p}{\delta x}.\delta x\right)\left(\frac{\delta A}{\delta x}.\delta x\right)$$
$$- TP\,\delta x - wA\,\delta x.\frac{dz}{dx} = \rho A\,\delta x.\frac{dv}{dt}$$

2. $$\rho\,Av\,\delta t - \left(\rho + \frac{\delta \rho}{\delta x}.\delta x\right)\left(A + \frac{\delta A}{\delta x}.\delta x\right)\left(v + \frac{\delta v}{\delta x}.\delta x\right)\delta t$$
$$= \rho A\,\delta x\,\frac{\delta p}{\delta t}.\delta t\left(\frac{1}{K} + \frac{d}{tE}\right)$$

These equations can be solved using either resonant techniques or the Method Of Characteristics; but computer modelling using the latter approach has the advantage of implicitly incorporating transient variations in both wavespeed and friction while calculating the interaction of flow and pressure changes at any point in a system. A disturbance can be initiated (i.e. pump start up) and the accompanying effects are calculated throughout the system. These calculations are repeated a short time after start up and again at regular time intervals until a complete picture of the pressure changes at the pump, for example, is obtained as well as a realistic assessment of the transmission of the effects through the network.

2.7   Investigation

Although the basic computer program used in this investigation is fully verified, providing numerous correlations with independently measured field results, some unusual features had to be included. Firstly, the aim was to reproduce an incident which led to a high frequency pressure oscillation, dependent on the piping lengths and pipework configuration. Hence, it was necessary to model the scheme in unusual detail, ensuring piping accuracy of ± 0.25m. However, the simulations showed that this alone was not sufficient to provide accurate correlation and so attention was focused on the analytic model of the valve.

In most transient simulations any control valves can be assumed to adopt a series of quasi-steady state positions (being delayed only by the stroke rate) as the signal to the controller changes. But, in this case, the exact response of the valve was essential and hence it was necessary to include an algorithm defining the phase shift and gain of the pressure controller (based on the manufacturer's data, Figure 9). Additionally, the mass of the valve stem and plug was modelled, allowing the acceleration and momentum to be calculated; this, together with both stiction and hysterisis effects, proved important, particularly as the end-to-end stroke of the kickback valve was produced by only 30mm stem lift.

2.8   Results

Using these modelling techniques, the start up of the first booster pump and the sudden demand of a governor were again simulated and the general shape of the predicted pressure waves was very similar to the values measured on the test rig. It was also confirmed that the amplitude and frequency of the pressure oscillation were highly dependent on stiction, hysterisis and momentum effects in the control valves and that it would prove difficult to "tune" the theoretic values on four valves to obtain perfect correlation. However, frequency and amplitude predictions of within 95% and 90% (respectively) of the on-site measurements could be obtained, sufficiently accurate to verify the modelling techniques adopted.

The simulations also affirmed that the start up of the second booster pump would not cause valve oscillation, the probable reasons being:-

1.   at the increased flow provided by two pumps, friction losses were higher which dissipated energy and hence offered some damping.

2.   at the higher flow, the kickback control valve was further open to maintain a constant header pressure and was therefore operating at a more stable stroke position.

2.9   Solution

Computer simulations of start up and governor demand were then undertaken on a model of the complete lube oil scheme but these showed that the proposed system would also be subject to serious vibration. As stated earlier, the basic design criteria of the scheme allowed only minor modifications to the control valve stroke times and control sensor lines and, as anticipated, further simulations confirmed that these changes did not reduce the pressure oscillation to an acceptable level. Hence the installation of gas filled accumulators was proposed (see Figure 10), being a standard method of suppressing resonant vibration but this, in itself, posed a further problem.

Although the level of pressure oscillation (and accompanying vibration) in the system was excessive it was apparent from the simulations that the forcing frequency was not in fact occurring at the dominant natural frequency of the piping. The inclusion of any accumulation in the system would slightly lower the natural resonant frequency but at the same time the forced frequency could also be reduced, possibly enough to lead to higher amplification of the pressure variation. It was, therefore, necessary to eliminate this potential problem by modelling the system with a different number of accumulators.

In the final phase of the investigation, pump start up and a sudden demand of governor flow were therefore resimulated to confirm the optimum location and capacity of the suppression. As shown by the predicted trace (see Figure 11) the recommended accumulators reduced the pressure oscillation significantly and after their installation the lube oil system runs smoothly, with no vibration. However, it is interesting to note that further verification of the accuracy of the modelling of the complete installation was obtained when the accumulators were isolated for a short period; excessive vibration occurred when a booster pump was started and the system had to be shutdown.

2.10   Summary Of Second Case Study

This case study shows another problem of vibration in a piping system which is induced by pulsating flow but, unlike the previous example, a complex computerised analysis was required to fully identify the source of the problem and subsequently ensure stable flow. An analysis

based on the Method Of Characteristics is not ideal if a general
scan of the resonant response of a system is required under various
imposed oscillating frequencies; but as shown here, the method can
be used with great accuracy to verify the mechanisms causing resonance
and to ensure that the proposed suppression facilities do not induce
a higher degree of vibration.

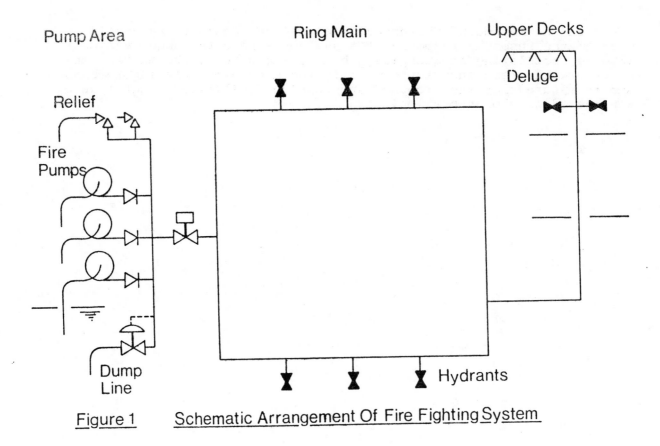

Pump Area   Ring Main   Upper Decks

Relief

Fire Pumps

Dump Line

Hydrants

Deluge

Figure 1   Schematic Arrangement Of Fire Fighting System

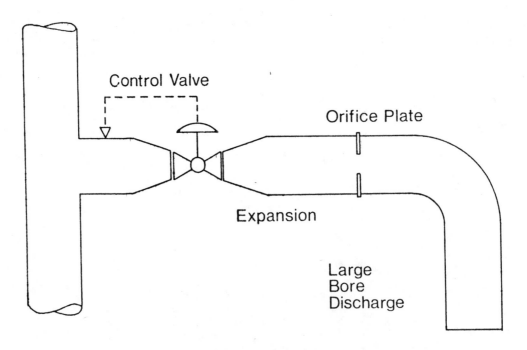

Control Valve

Orifice Plate

Expansion

Large Bore Discharge

Figure 2   Typical Dump Line Prone To Vapour Blocking

Figure 3

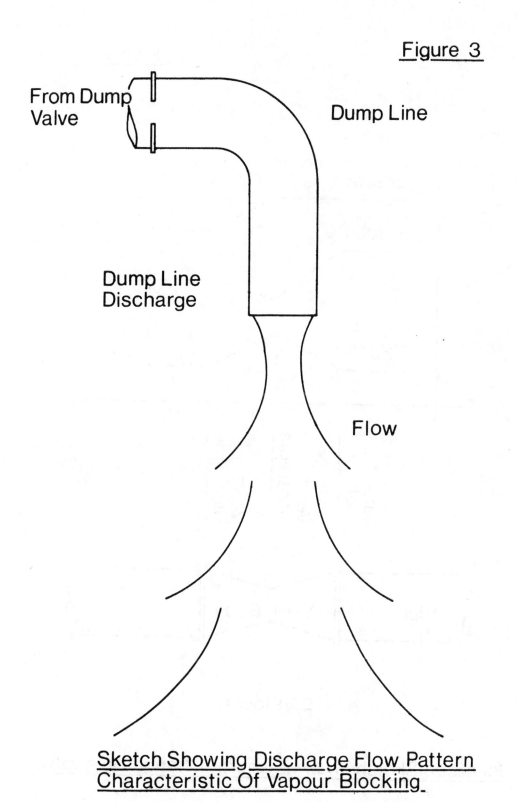

From Dump Valve

Dump Line

Dump Line Discharge

Flow

Sketch Showing Discharge Flow Pattern Characteristic Of Vapour Blocking

313

Figure 4

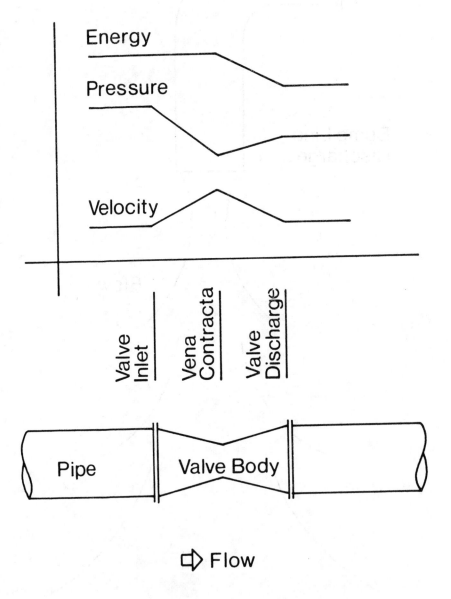

Idealised Sketch Showing Energy Changes In A Valve

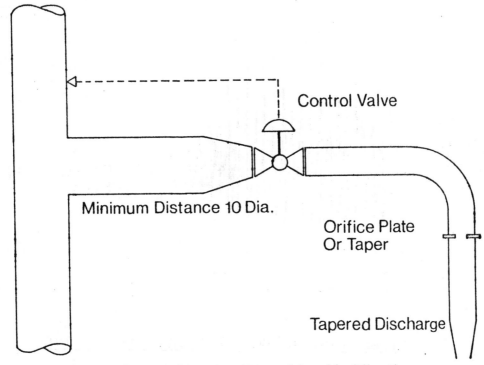

Figure 5   Sketch Showing Dump Line Modifications

Control Valve

Minimum Distance 10 Dia.

Orifice Plate
Or Taper

Tapered Discharge

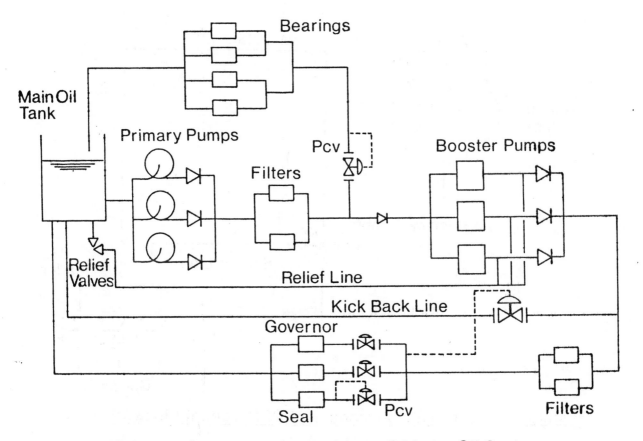

Figure 6   Schematic Arrangement Of Lube Oil System

Bearings

Main Oil
Tank

Primary Pumps

Pcv

Booster Pumps

Filters

Relief
Valves

Relief Line

Kick Back Line

Governor

Seal

Pcv

Filters

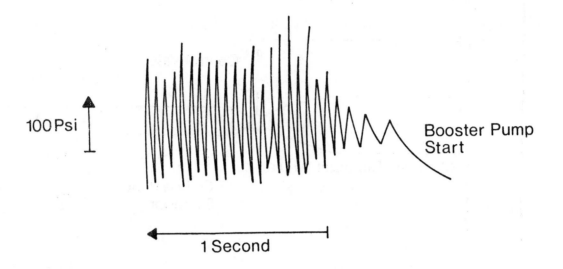

100 Psi

Booster Pump
Start

1 Second

Figure 7   Pressure Trace Taken Downstream Of Governor

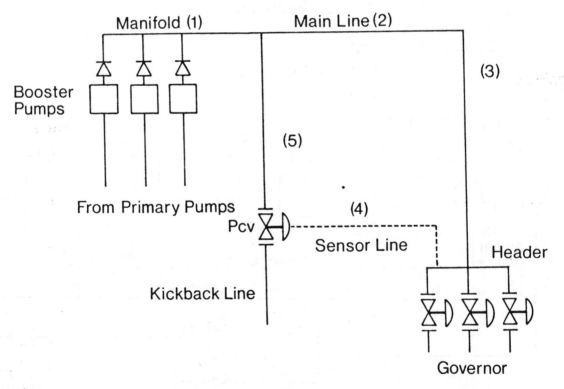

Manifold (1)   Main Line (2)

(3)

Booster
Pumps

(5)

From Primary Pumps

(4)

Pcv

Sensor Line

Header

Kickback Line

Governor

Figure 8   Schematic Arrangement Of Resonant Pipework

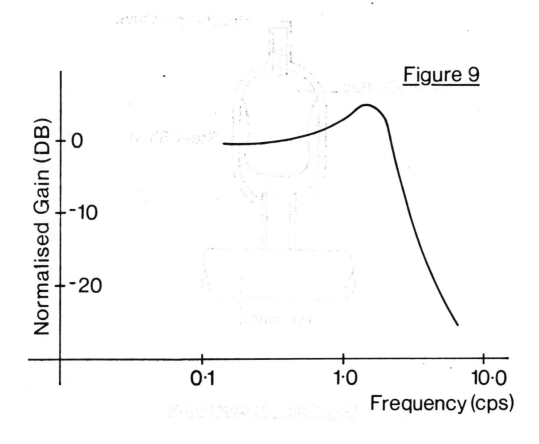

Figure 9

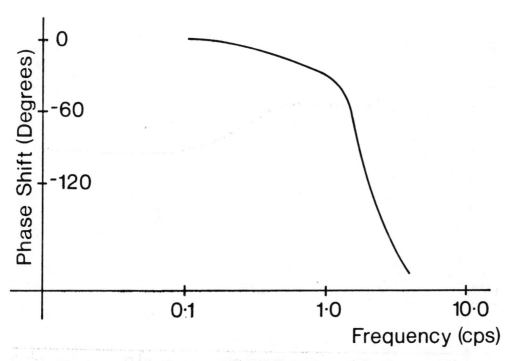

Frequency Response Of Pressure Controller

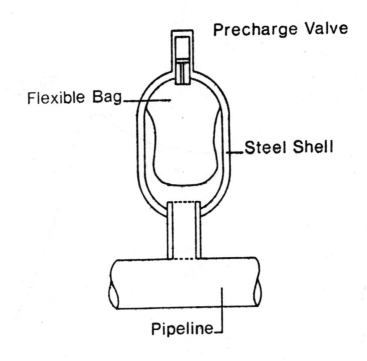

Precharge Valve

Flexible Bag

Steel Shell

Pipeline

Figure 10   Gas Filled Accumulator

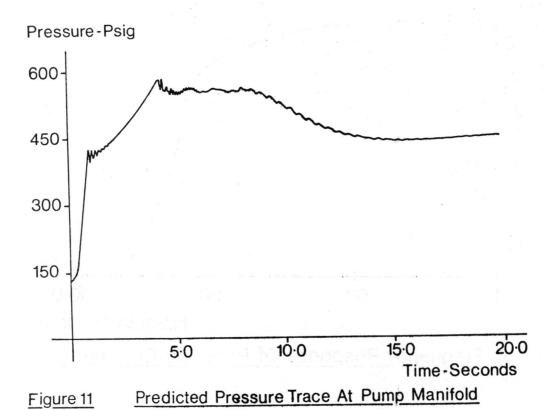

Figure 11   Predicted Pressure Trace At Pump Manifold

FLOW INDUCED VIBRATIONS IN A VALVE-WASHER COMBINATION

R.F. Boucher

Sheffield University, U.K.

Summary

Three conical valves, each weighing about 2 tonnes, seat against the flow direction to control pressure in a blast furnace. Water droplets in the annular channels provide gas washing. The shafts, 7.5 m between bearings, were found to be vibrating laterally during commissioning of the furnace. A vibration mechanism, involving small divergence of the channel height between the cone and its outer skirt in the flow direction, was postulated. Experiments were conducted on a 1/30 scale model and observations confirmed the hypothesis. The plant was subsequently modified to accommodate new cones with slightly converging flow channels. The blast furnace has subsequently been commissioned without further shaft vibration problems.

Organised and sponsored by
BHRA Fluid Engineering, Cranfield, Bedford MK43 0AJ, England

NOMENCLATURE

a, b	lengths from shaft supports to c.g. of load (m)
D	cone diameter (m)
E	modulus of elasticity (N/m^2)
F	force (N)
f	frequency (Hz)
I	second moment of area (m^4)
j	see text
L, ℓ	length of shaft between supports (m)
M	mass of cone (kg)
p	pressure (N/m^2, bar)
P	tension force (N)
U	see text
V	fluid velocity (m/s)
w	weight of cone (N)
y	channel height (m)
δ	deflection (m)
ρ	fluid density (kg/m^3)

Suffices

1, 2	small and large ends of cone
M	model
A	actual plant

1. INTRODUCTION

Modern blast furnaces, like the British Steel Corporation Plant at Redcar, operates under high pressure, typically up to 3.5 bar. Besides its supply of "wind", the furnace produces its own gases and is operated in batch mode. It is therefore essential to provide automatic regulation of the furnace pressure, and this is done by a valve or valves on the discharge from the furnace. Before such gases are discharged to the atmosphere, they must be cleaned. A common technique is wet scrubbing. In a wet scrubber the dirty gas is accelerated through a nozzle or venturi into which water droplets have been sprayed. The water droplets, being much larger than the dust particles, accelerate more slowly. The dust particles therefore collide with the water droplets and these large, dirty droplets are then easily removed by centrifugal action in a cyclone-type of device.

The functions of control valve and scrubber have been conveniently combined into one device on the Redcar furnace, a novel concept due to Gottfried Bischoff GMBH of West Germany. The arrangement is shown schematically in Fig. 1. The valve cone is actuated hydraulically and water is sprayed from special atomising nozzles situated upstream of the cone. On the Redcar furnace a typical gas flow is 720,000 Nm^3/h and one litre of water is added by the spray system to each normal cubic metre of gas. To accommodate such large flows, three valve-scrubber combinations are used in parallel. These are large structures, each cone weighing approximately 2,000 kg, and the distance between the upper and lower bearings being 7.5 m. At the lower bearing, the shaft passes outside the structure to attach to the hydraulic actuator. This length of shaft is thus visible to the operators.

During the commissioning of the Redcar furnace, this visible length of shaft was seen to be vibrating laterally when the furnace had only been worked up to less than half its final operating pressure. Fatigue fracture of the shaft was feared. Commissioning could not proceed and even the down-rated operation was threatened until the problem was solved.

2. DIAGNOSIS OF PROBLEM

To help understand the possible sources of the vibration, the layout of the system is shown in Fig. 1. The three valves are distributed symmetrically to discharge into the large downstream chamber through which the shafts pass. This chamber is 7 m diameter, and the distance from the bottom of the skirt to the bottom of the chamber is approximately 5 m.

The oscillation frequency was found to be centred around 18 Hz. Although the shaft executed lateral vibrations, with a precessing motion, the first possibility considered was that this lateral vibration was excited by an essentially longitudinal forcing mechanism. This forcing mechanism would basically be pressure fluctuations arising at the valve inlet, due to "noise" from the furnace which might be promoting duct work resonance or exciting a Helmholtz resonance in which the annular valve channels would be the "neck" and the large downstream chamber would be the vessel. However, no flow resonances were in evidence. The vibration appeared to be confined to the shaft alone, with little or no effect on the flow. It seemed therefore, that the logical sequence of the investigation was as follows. Firstly, to determine that the observed frequency did indeed correspond to the shaft-cone natural frequency; secondly, to identify the lateral forcing mechanism; thirdly, to verify this experimentally; and finally, and most important, to implement a solution.

2.1 Frequency of Vibration

The system can be simply modelled as a beam 7.5 m long, tubular section (136.5 mm outer radius, 124.0 inner). The cone constitutes the main inertia load and is assumed to have its c.g. 1.5 m from the top end bearing in the working position. Since the lower shaft end outside the casing exhibits large vibrations, the fixing is plainly 'simply supported'. In this case the flexibility δ/w (δ = deflection for load w) is

$$\frac{\delta}{w} = \frac{4a^3}{3EI} \left(1 - \frac{a}{L}\right)^3 \tag{1}$$

where the load is at distance a from one end and δ is measured at a.

Using the well known Rayleigh approximation, the frequency is given by

$$f = \frac{1}{2\pi} \sqrt{\frac{W}{M\delta}} \qquad (2)$$

where M is the mass of the cone, concentrated at a.

Using the following data:

a/L = 1500/7500 = 0.2 ; R = 136.5 mm

a = 1.5 m   ; r = 124 mm

E = 210 x 10^9 N/m^2

M = 2062 kg   ; $I = \frac{\pi}{4} (R^4 - r^4)$

the frequency f is 10 Hz (600 cycles/min), which is almost half the observed frequency.

2.2   Forcing Mechanism

A lateral forcing mechanism was sought, arising from pressure forces on the cone. When the cone is central, the geometry and flow pattern are axisymmetric, and static equilibrium can be expected. However, the stability of the cone when disturbed from its central position must be examined. If the resulting flow asymmetry gives rise to pressure forces tending to restore the cone to the centre, it will be stable. If however the forces are such as to displace the cone even farther off centre, then they are a potential source of instability.

It will be shown that such decentralizing forces may arise from two sources - pressure changes due to friction effects and due to inertia effects in the fluid if the channel is not perfectly parallel. The former are considered of little relevance in the present problem, where fluid velocities are so high that inertia forces are far more significant than friction losses. However, for viscous fluids used in oil hydraulic systems with very small annular gaps, the reverse is true and decentralizing forces arise giving rise to the phenomenon of 'hydraulic lock'. It is instructive to consider this phenomenon first before the second 'Bernoulli Induced' instability.

2.2.1   Hydraulic Lock Instability

This problem arises usually in nominally parallel annular gaps which diverge, however little, in the direction of the flow. It occurs because of changes in the rate of pressure loss due to friction in the direction of the flow, and is usually associated with spool valves in oil hydraulics (Ref. 1).

If the centre body is moved slightly off centre the rate of divergence of area is increased on the narrow side and decreased on the wide side. The rate of pressure loss due to viscous friction varies as $1/y^3$ where y is the channel height so the pressure falls more rapidly on the narrow side (Fig. 2). The centre body is thus unstable and is forced hard to one side.

This phenomenon only arises if the channel diverges. If the channel converges, the configuration is stable, the net force always being towards the centre. Because of the low velocities in viscous flow, diffuser action in the diverging channels is negligible.

2.2.2   Bernoulli Induced Instability

Where velocities are high, pressure changes arising from velocity changes through Bernoulli's equation can be appreciable. These velocity changes are brought about by changes of area. For the washer, the area from inlet to outlet increases by a factor of 1.8 when the gap is 30 mm (the factor reduces slightly with cone opening). This has some very important effects.

Because it is a diffuser, the pressure rises in the direction of flow to 1.1 bar at exit (Fig. 3). The pressure at the smallest diameter of the channel is well below the exit pressure (i.e. very high vacuum) so that the flow is nearly choked. (Despite the high pressures involved, the flow in the actual washer never reaches sonic velocity due to the high density of the two-phase mixture of gas and water).

322

The vacuum in the channel results in an upward force on the cone and shaft. This tensioning raises the natural frequency (see below) above that of the unloaded shaft.

The most relevant effect of the diffuser action is on providing an instability mechanism. Although the mechanism is different, the effects of various geometrical changes are the same as for hydraulic lock. Consider that the channel height diverges slightly in the direction of flow and that the cone has moved slightly off centre (Fig. 3).

The area ratio of the narrower diffuser section A is greater than the wider section B. This results in the pressure distribution shown which is destabilizing forcing the cone further off centre.

2.2.3 Misalignment

If the cone is mounted parallel to the centre line but displaced slightly from it then so long as there is no channel divergence, it is stable i.e. eccentricity alone is not destabilizing. If the cone is at an angle to the centre line then, even though the channel would otherwise be parallel, one side now diverges and the other converges. This would again lead to instability. To avoid this, the angle of the inner cone should be greater by an amount which can absorb such misalignment without producing a diverging channel anywhere around the cone. Even with no misalignment, it is plainly essential that tolerances be specified such that there will be no channel divergence.

2.2.4 The Oscillatory Cycle

The above arguments demonstrate that the cone will be subject to a decentralizing force unless the cone included angle is greater than that of its outer skirt, producing converging channels (although the total annulus area is always increasing in the flow direction). However, the stiffness of the shaft produces a centralizing force, as shown in Fig. 4. When the cone has made contact with the skirt this force increases very rapidly. The line xox represents a hypothetical 'fluid spring' force such as might be generated by the effects described above. The fluid spring must be stiffer than the shaft to decentralize it and equilibrium would be established at points x with cone-skirt contact. However, the linear fluid spring characteristic shown is in practice, impossible for two reasons. Firstly, the pressure difference between two opposite sides of the cone will tend to relieve sideways (peripherally) around the cone. Secondly, the area ratio of the two-dimensional channel diffuser tends to infinity as the cone approaches the skirt. A high area ratio diffuser will produce flow separation and substantial loss of diffuser action, resulting probably in a hysteretic characteristic, shown by the broken lines in Fig. 4. Due to its inertia, the cone will overshoot the regions of very fast change in fluid force (hysteretic or not) and oscillate about them or through the central position.

2.3 Effect of Cone Force on Frequency

Due to the Bernoulli effect, the net cone force is upward. Avoiding unnecessary complication, consider a 30 mm gap and assume the gas flows from the throat at 250 m/s (V_1). Then the exit velocity V_2 is $250/1.8 = 140$ m/s. Crudely, we apply Bernoulli. The force over the tapering cone is

$$F = \int_{D_1}^{D_2} p \, \pi D \, dD \tag{3}$$

where D is the diameter. Using Bernoulli with pressures relative to outlet pressure

$$p = \tfrac{1}{2}\rho V_2{}^2 \left(1 - \frac{D_2{}^2}{D^2} \right) \tag{4}$$

This gives

$$F = \frac{\pi}{2} \rho V_2{}^2 \left[\frac{D_2{}^2 - D_1{}^2}{2} - D_2{}^2 \ln \left(\frac{D_2}{D_1} \right) \right] \tag{5}$$

323

which, on substituting, is 16.5 tons upwards. The downward force is due to 1.6 bar above outlet pressure acting on 0.9 m diameter i.e. 10.5 tonnes. Net upward force is 6 tonnes approximately.

For a beam with a point load distance a from one end, b from the other (a + b = ℓ, length), and subjected to tension force P, the flexibility is

$$\frac{\delta}{w} = \frac{j}{P} \left[\frac{\text{Sinh } b/j \text{ Sinh } a/j}{\text{Sinh } U} - \frac{ab}{\ell j} \right] \qquad (6)$$

where $j = (EI/P)^{\frac{1}{2}}$ and $U = \ell/j$

Using P = 6 tonnes, this predicts a shaft flexibility δ/w of 2×10^{-7} m/N against 3.15×10^{-8} for the untensioned shaft. This is a factor of 6.3 which raises the natural frequency 2.5 times. If in addition one allows for the shaft weight of 0.6 tonnes this reduces the untensioned frequency to about 7 Hz and thus the tensioned case to about 18 Hz, which was the frequency observed.

2.4   Effect of Bearings on Vibration

All the mechanisms for feeding a vibration depend on the <u>ratio</u> (y_2/y_1) of channel height at outlet to inlet <u>increasing</u> as the cone moves <u>towards</u> it. This is achieved when

1) the cone angle is less than the skirt angle and the cone moves sideways

2) cone and skirt angles are equal but there is axial misalignment and the cone moves sideways

3) Cone and skirt angles are equal (or possibly even the cone angle is slightly more than the skirt angle), the two are axially aligned but the top bearing slack allows the top of the cone to move sideways more than the bottom does. This results in a narrower inlet than outlet height and produces instability from an otherwise stable situation.

3.   EXPERIMENTAL INVESTIGATION

3.1   Design of Model

The problem posed modelling difficulties. Since the flow annulus constituted a diffuser, the possibility of flow separation existed and therefore Reynolds number scaling was desirable. The actual flow was two-phase (gas and water droplets) but the time scale prohibited the use of such elaborate modelling. Despite the high ultimate working pressure of the furnace (3.5 bar) the manufacturers indicated that the flow was not choked due to the high density of the mixture. It was thus decided to use a model working with lower pressure air only. Reynolds number scaling was therefore impossible. This was not felt to be a serious discrepancy since flow separation was in any case more likely at the lower model Reynolds numbers, although it would probably occur earlier than in the actual case.

For instability to occur, the fluid spring rate must be greater than the shaft spring rate. Thus the relationship between the two on the plant and the model should be the same. With geometrical similarity, and using steel for the model shaft, the shaft stiffness is proportional to I/L^3 for fixed a/L. The fluid spring stiffness will be proportional to $(pL^2)/L$, using L to simply represent size. Thus the group pL^4/I should be the same on model and plant. Using model pressures approximately ¼ of those on the plant, with a 1/30 scale model, indicates that $I_M = 3 \times 10^{-7} I_A$. This is conveniently achieved with 4.8 mm (3/16 in) steel rod. The distance between the shaft bearings will be (7.5/30)m i.e. 250 mm.

Thus a 1/30 scale model was constructed, and is shown in Fig. 5. A number of cones were manufactured with angles between 22° and 18° with a skirt angle of 20°. The cruciform pieces had central holes and acted as the top and bottom bearing supports. Some cruciform pieces were made with eccentrically drilled holes to simulate shaft misalignment. The skirt was held inside a short length of steel tube with a Perspex tube extension to permit visual observation of the shaft. A pressure Tapping was drilled in the middle of the skirt and connected to a pressure transducer.

The model was mounted vertically with an air supply connected to the top. The lower (Perspex tube) end was open to atmosphere. The shaft extended approximately 5 cm below the perspex tube so that the cone position could be altered by inserting packing between the end of the shaft and the laboratory floor.

3.2   Experimental Observations

The experiments were largely qualitative in nature. Any vibrations of the shaft could be felt by hand, viewed stroboscopically and directly, often heard and sometimes viewed as pressure fluctuations on the oscilloscope.

With a skirt angle of 20°, an 18° cone exhibited oscillations over the whole pressure range (0-1 bar) at various openings. In contrast, the 22° cone showed no sign of instability over the whole range of pressure and openings. The tests were then repeated with the upper cruciform bearing hole enlarged to 5.5 mm diameter. The 22° cone was again stable whilst the 18° cone exhibited extremely strong vibrations with large amplitude shaft deflections clearly visible. The upper bearing integrity was clearly important. On the other hand, a tight upper bearing and very loose lower bearing produced no observable change in behaviour.

To ensure that the stability of the 22° cone in the 20° skirt was not simply associated with a specific flow geometry a 20° cone and 18° skirt angle combination were also tested and found to be equally stable throughout the pressure range.

Since the actual plant cones and skirts were both nominally 20°, the question arose of how little divergence in the channel (due to small imperfections in manufacture) was capable of causing instability. Due to the less than perfect scaling from plant to model, the question could not be answered accurately. However, 3 cones (20°, $20\frac{1}{2}^{\circ}$, $19\frac{1}{2}^{\circ}$) were tested in the 20° skirt and the $19\frac{1}{2}^{\circ}$ cone was again found to have unstable operating regions, whilst the other two cones were stable.

To test the effect of misalignment, the $20\frac{1}{2}^{\circ}$ cone was installed with an eccentric upper bearing, producing about 1° slope from the centre line. The effect was to cause oscillations of this otherwise stable configuration to return over parts of its opening range.

4.   CONCLUSIONS

The experiments conducted produced evidence to support the hypothesis explaining the basic oscillatory forcing mechanism. This is associated with a cone angle which is slightly smaller than the skirt angle producing a decentralizing 'spring' force whose stiffness exceeds the shaft stiffness. The problem is cured by ensuring that tolerances never allow the cone angle to become equal to or smaller than that of the skirt. This ensures that the fluid spring force is a centralizing force, as indicated by line Y-Y in Fig. 4. Although the problem could also be cured by stiffening the shaft, it may in practice be difficult to estimate the necessary stiffness, and more cumbersome and expensive to implement.

During the early part of 1980, new cones having angles 1.5° greater than the skirt angles were installed in the Redcar furnace. The furnace was commissioned up to its full working pressure with no further problems of valve-washer assembly vibration.

5.   REFERENCES

Blackburn, J. F., Reethof, G., and Shearer, J. L.: "Fluid Power Control". New York, MIT Press, 1960.

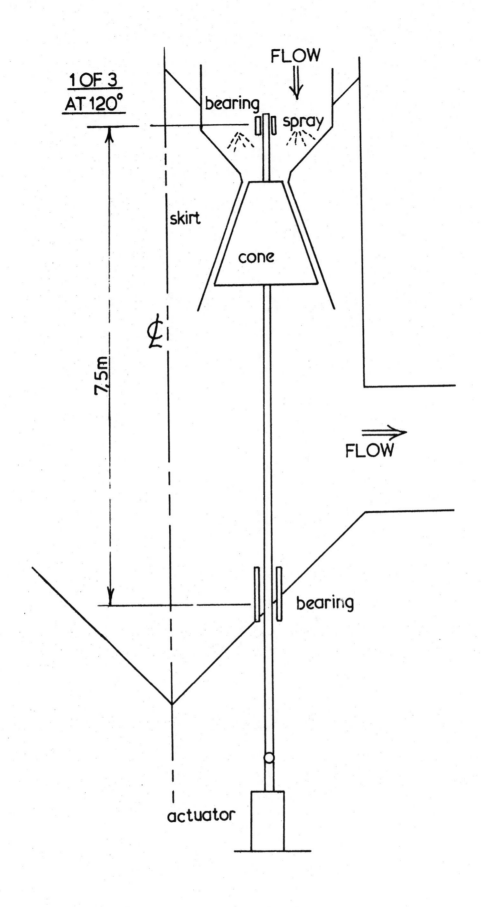

Fig. 1   General layout of valve-washer assembly

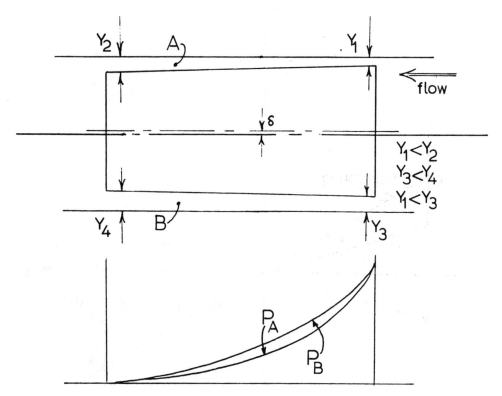

Fig. 2   Effect of lateral displacement in viscous flow producing
hydraulic lock in diverging channels

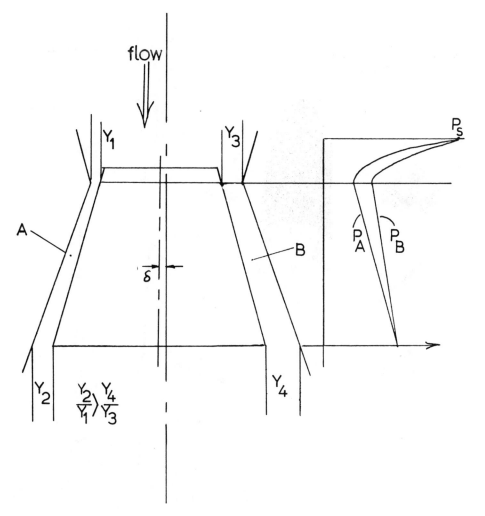

Fig. 3   Effect of lateral displacement in annular-diffuser cone
producing decentralizing force in diverging channels

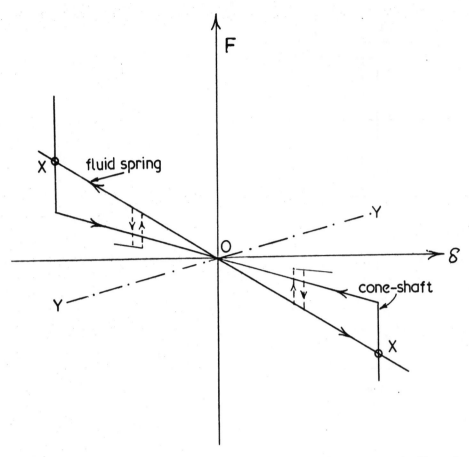

Fig. 4   Force-displacement characteristics due to shaft and
          fluid springs

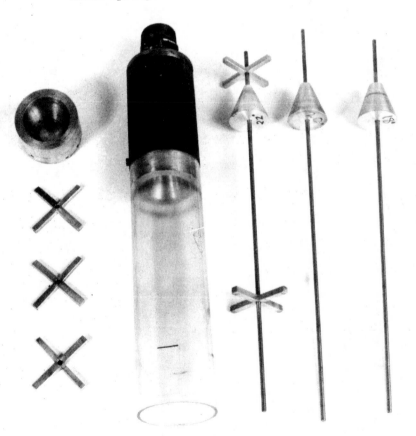

Fig. 5   Photograph showing 1/30 scale model with skirt, cones
          and cruciform bearing pieces

328

**FLOW INDUCED VIBRATIONS
IN FLUID ENGINEERING**

Reading, England: September 14-16, 1982

CONTRIBUTIONS TO SELF-EXCITING VIBRATIONS OF SOME
FLOW REGULATING UNITS – CONICAL POPPET VALVE TYPE –
IN DOUBLE PHASE FLOW

D. Milan and G. Berthollon

Neyrpic Company, France

Summary

To illustrate the problem of structural hydroelastic vibrations in double-phase flow involving fluid jet inertia, the lateral vibrations of a conical poppet valve are considered. The results of a series of measurements on the dynamic behaviour of the device in air, water and air/water mixture are presented. Interpretation of the results is given from theoretical considerations based on the application of the momentum theorem.

Organised and sponsored by
BHRA Fluid Engineering, Cranfield, Bedford MK43 0AJ, England

b   =   width of back-spring blade (m)

c   =   damping factor (kg s^{-1})

D   =   supply tube diameter (m)

e   =   thickness of back-spring blade (m)

E   =   modulus of elasticity (N m^{-2})

f   =   frequency (Hz)

F_y =   lateral hydrodynamic force (N)

F_p =   resultant of pressure forces (N)

G   =   mass discharge per unity section (kg s^{-1} m^{-2})

h   =   poppet valve lift (m)

ΔH =   headloss (N m^{-2})

K   =   headloss factor

k   =   valve restoring factor (N m^{-1})

l   =   length of back-spring blade (m)

L   =   damping length of fluid jet pulsations (m)

m   =   mass (kg)

R   =   supply tube radius (m)

t   =   time (s)

V   =   flow velocity (m s^{-1})

$\frac{1}{2}\Sigma\rho v^2$ =   kinetic energy of fluid volume unity (N m^{-2})

x   =   mass quality

y   =   displacement perpendicular to the flow (m)

Re   =   Reynolds' number

α =   half vertical angle of the poppet valve (°)

μ =   dynamic viscosity (N s m^{-2})

ρ =   specific mass (kg m^{-3})

ζ =   damping ratio

1. INTRODUCTION

While the recent years have seen significant developments in the control of structural vibrations induced by single-phase flows, little changes have occurred in the treatment of vibrations in double-phase flows which generally remain badly known. However, the simultaneous transport of liquid and gas phases constitutes a particularly dangerous potential excitation source such as illustrated, in particular by pumps and gates subjected to air entrainment and by circuits and heat exchangers in presence of water and water steam.

The analysis of the bibliography shows that the present topics most often studied mainly concern cylindrical tubes and cover three different fields : internal axial flows of gas-liquid mixtures in pipelines, Hara (Ref.1, 2 and 3) ; external axial flows in parallel heat exchangers and nuclear fuel pins, Gorman (Ref.4 and 8), Haris (Ref.5), Cedolin (Ref.6), Dellavalle (Ref.7), Pettigrew (Ref.9), Hara (Ref.10 and 13), Degli Espinosa (Ref.11), Fokin (Ref.12); external perpendicular flows in heat exchangers and steam generators, Pettigrew (Ref.14 and 15), Heilker (Ref.16), Remy (Ref.17).

Dynamic interaction phenomena between the fluid phases and the structure appear to become very complex according to the double-phase flow pattern which itself depends on numerous parameters still badly mastered, even for simple geometric configurations. Many investigations are therefore needed for a better understanding of such phenomena.

It may be interesting to determine whether self-exciting vibrations can occur like in single-phase flow. The example of a conical poppet valve is considered below. This is a typical flow regulating unit which is subjected, under some conditions, to strong self-oscillations. When the liquid jet pulses, the applied hydrodynamic force and the motion are out-of-phase due to jet inertia. Energy transfer occurs between the fluid and the structure. This coupling maintains the initial vibration while amplifying it. This instability occurs above a critical velocity which should be accurately known in all cases. What happens in multi-phase flows ?

2. TESTED DEVICE AND EXPERIMENTAL PROCEDURE

For the study of self-exciting lateral vibrations in a hydraulic conical poppet valve, the experimental device is sketched-out in Fig.1 . Table I gives the main characteristics selected. Air (up to 0.15 kg/s) and/or water (up to 4 kg/s) discharges may pass through the vertical supply tube designed with a straight length of 100 x diameter ensuring a well-steady inlet flow, Fig.2. The dynamic valve motion, perpendicular to the flow, is sensed by means of calibrated strain gauges the signal of which is recorded and processed by SPECTRAL DYNAMICS SD360 analyzer.

Tests have been made for air and water single-phase increasing flow and then for increasing flow of the double-phase mixture with quality being kept constant. In each case, the vibratory motion of the poppet valve, with a maximum travel of ± 5 mm before reaching the end-stop, has been analysed.

3. EXPERIMENTAL RESULTS

3.1. Vibratory characteristics in fluid at rest

In a first stage, the lateral mode frequency of the poppet valve was measured. The values obtained - 7.5 Hz in air, 7.2 Hz in water - are quite consistent with the geometrical and mechanical characteristics of the experimental device.

../..

3.2. Results obtained under flow

57 tests were made : 14 in water, 12 in air and 31 in water and air including for the latter, 8 tests with the quality of 0.05, 13 tests with the quality of 0.1 and 10 tests with the quality of 0.2. The results obtained concern the frequency and r.m.s. amplitude of the induced motion as well as pressure differential between the upstream and downstream side of the poppet valve.

Since the tests were mainly intended to show the incipient instability, high flow velocities were used. As concerns double-phase flow, it can be demonstrated by reference to Spedding (Ref.18) that all points broadly correspond to an "annular" flow pattern in the supply tube, with a liquid film in the wall and entrainment, by the continous gas phase, of dispersed liquid micro-drops.

Velocity slip between phases, upstream of the experimental device, can be calculated using the Lockart-Martinelli correlation for an upward vertical flow according to Oshinowo (Ref.19). This slip, in the range of 10, is far from being negligible. However, it must be considered as an average, without differentiation between the slow liquid film and the liquid micro-drops entrained at high speed by the gas.

3.3. Poppet valve headloss factor

3.3.1. Single-phase flow

Fig.3 gives the measured values in air and water of single-phase headloss factor, $K_{single-phase}$, versus Reynolds' number, $Re = \rho_{upstream} \ V_{upstream} \ D/\mu_{upstream}$:

$$K_{single-phase} = 2 \ \Delta H \left[\rho_{upstream} \ V^2_{upstream} \right]^{-1} \quad \dots\dots\dots\dots\dots\dots\dots \quad (1)$$

ΔH : measured headloss; D : conduit diameter; $\rho_{upstream}$: specific fluid mass ;

$\mu_{upstream}$: dynamic viscosity; $V_{upstream}$: flow velocity.

The values show a rather small dispersion with an average of $K_{single-phase} = 8.5$ which is in close agreement with the value given by Idel'cik (Ref.20) for a similar configuration. However, it should be stressed that contrarily to what happens in water, the headloss increases at high air velocities associated to the large dynamic valve amplitudes. The cause for such a phenomenon has not been found.

3.3.2. Double-phase flow

The use of an homogeneous model based on equal phase velocities being recommended by many authors according to Bergles (Ref.21) for calculating the irreversible headloss of a singularity constituted by a contraction followed by an expansion, with this assumption the headloss factor can be written as :

$$K_{double-phase} = 2 \ \Delta H \left[\Sigma \ \rho V^2 \right]^{-1} \quad \dots\dots\dots\dots\dots\dots\dots\dots \quad (2)$$

where the kinetic energy of fluid volume unity is :

$$\tfrac{1}{2} \Sigma \ \rho V^2 = \tfrac{1}{2} \ G^2_{upstream} \left[x/\rho_{gas} + (1 - x)/\rho_{liquid} \right] \dots\dots\dots\dots\dots\dots\dots \quad (3)$$

$G_{upstream}$ is the total mass discharge per unity section of the supply tube ;

x is the mass quality of the mixture.

../..

332

$K_{double-phase}$ values, as determined from the above formula, are given on Fig.4 versus the homogenous model Reynolds'number $Re = G_{upstream} D/\mu_{double-phase}$ where $\mu_{double-phase} = \mu_{liquid} \mu_{gas} [(1 - x)\mu_{gas} + x\mu_{liquid}]^{-1}$. The values show of course , a higher dispersion than in single-phase flow and are slightly smaller: 20% as an average. If velocity slip between phases were considered in the calculation of volume unity kinetic energy, this value would be conveniently reduced but too much since the $K_{double-phase}$ value with the estimated overall slip would thus be equal to 40, as an average, instead of 8.5

It can be therefore seen from the above that calculation of headloss using an homogeneous double-phase model is rather satisfactory and that a quasi-homogeneity of the velocities of the two phases seems to be achieved in the singularity.

3.4. Response analysis and hydroelasticity definition

3.4.1. Single-phase flow

Figs.5 and 6 give the evolutions of the frequency and r.m.s. amplitude of poppet valve vibratory motion induced by the air and water flows. If frequency decreases steadily versus the kinetic energy of the fluid volume unity, the amplitude evolution slope, which is approximately equal in a first step to 1, shows a sharp rise above a threshold. By definition, this threshold, characterized by the change in the slope of the evolution curve of poppet valve motion r.m.s.amplitude, will be considered as the hydroelastic instability point. Below this threshold, the poppet valve responds to the turbulent flow pressure fluctuations existing in the annular space between the poppet valve and its seat. The smaller damping ratio can account for the slightly higher amplitude level observed for the air flow.

3.4.2. Double-phase flow

Similar evolution of valve induced-motion characteristics is observed in water/air mixture flow. On Figs.7 and 8 the evolutions of frequency and amplitude are plotted versus the volume unity kinetic energy, calculated on homogeneous model. The instability threshold, characterized by the change in the slope of the amplitude curve appears in double-phase flow. Table II summarizes the instability conditions observed. It can be noted that before instability occurs, vibratory levels are significantly higher in double-phase flow due, in this case, to a much higher level of pressure fluctuations.

4. THEORETICAL CONSIDERATIONS AND COMPARISON WITH THE EXPERIMENT

4.1. Equationing principle

The lateral vibratory motion of a conical poppet valve was studied by Maeda (Ref.22) through the application of the momentum theorem, in non-permanent flow. The Maeda's analysis, partially used below, is completed according to experimental results. The lateral hydrodynamic force, F_y, acting on the valve shifted by a distance y and calculated in the control volume Ω (ABCDEFGHI) of surface S, shown on Fig.1, is written as :

$$ F_y = \left[\vec{F_p} - \iint_S \rho \, v_n \vec{v} dS - \iiint_\Omega \frac{\partial(\rho \vec{v})}{\partial t} \, dv \right]_y \quad \dots\dots\dots\dots\dots\dots\dots\dots\dots\dots\dots\dots\dots \quad (4) $$

F_p = resultant of pressure forces on control surface S

V_n = velocity normal to control surface S.

As a first approximation, the hydrodynamic force F_y can be expressed in the form of :

$$ F_y = - (c_{fluid} \, \dot{y} + k_{fluid} \, y) \quad \dots\dots\dots\dots\dots\dots\dots\dots\dots\dots\dots\dots\dots\dots\dots\dots \quad (5) $$

so that the equation of poppet valve motion is :

$$ m \, \ddot{y} + (c + c_{fluid}) \, \dot{y} + (k + k_{fluid}) \, y = \text{turbulent exciting force} \quad \dots\dots\dots \quad (6) $$

../..

where :

$m = m_{valve} + 0.23\,\rho_{structure}\ b\ l\ e +$ added fluid mass

$k = E\ b\ e^3/_4\ \ l^3$: valve restoring factor

c : specific valve damping factor.

The expressions of k_{fluid} and c_{fluid}, in terms of flow velocity, are given herein-after, the fluid compressibility and viscosity being neglected.

4.2. Evolution of induced-motion frequency

k_{fluid} factor, in terms of fluid kinetic energy per volume unity, is written as :

$$k_{fluid} = \frac{\rho\ V^2_{upstream}}{2}\ \pi\,R^2 \left[\frac{R\ cotg\alpha\ (1-\ 3h\ tg\alpha\ /R)}{4h^2\ \sin^2\alpha\ (1-h\ tg\alpha/R)} - \frac{K-1}{l} - \frac{\cos\alpha}{h}\ \sqrt{K} \right]\ \ \dots\dots\ (7)$$

The first two terms come from the pressure resultant, F_p, without neglecting the torque due to the vertical force ; the last term is the permanent term of the momentum equation, calculated in the contracted outlet section under the following assumptions: $r_{contracted} = R + h$ and contracted velocity $= \sqrt{K}\ V_{upstream}$ (Fig.1): control volume : ABC'D'EF'G'HI

4.3. Determination of instability threshold

4.3.1. Calculation of hydrodynamic damping factor

The non-permanent term of momentum theorem leads on the one hand to the calculation of the added fluid mass which is dependent of flow velocity and, on the other hand to the calculation of the hydrodynamic damping factor which varies with flow velocity.

If the section normal to the flow is A (η) at point η (Fig.1), velocity is $V = d\ q_\theta\ /dA$ and volume element is $dv = dA\ dy$, hence :

$$\left[\iiint_\Omega \frac{\partial\,(\rho\,V)\ dv}{\partial t} \right]_y \# 2\ \frac{d}{dt} \int_o^\pi \int_o^{\eta_c} \rho\,\sin\frac{\alpha}{2}\ \cos\theta\ d\ q_\theta\ d\eta$$

$$+ 2\ \frac{d}{dt} \int_o^\pi \int_c^{\eta_c+L} \sin\alpha\ \cos\theta\ d\ q_\theta\ d\eta\ \ \dots\dots\dots\dots\ (8)$$

assuming that mean flow direction, for η comprised between 0 and η_c in the contracted section is $\alpha/2$ (Fig.1). L is the damping length of annular jet pulsations.

Before the contracted flow section, for $0<\eta<\eta_c$, we have $d\ q_\theta = V_{upstream}\ h_\theta$. $R^2\ d\theta/2\ h\ \sin\alpha$. In the annular wall jet emerging from the contracted flow section, it can be assumed by analogy with the radial wall jet that the considered flow is proportional to η due to the entrained air (cf.Rajaratnam, Ref.23). Or $d\ q_\theta = \eta\,V_{upstream}\ h_\theta\ R^2\ d\theta/2\eta_c\ h\ \sin\alpha$. With $h_\theta = h\ \sin\alpha - y\ \cos\alpha\ \cos\theta$ it comes out :

$$c_{fluid} = -\ \frac{\rho\ V_{upstream}\ \pi\,R^2\ \cos\alpha}{2\ h\ \sin\alpha} \left[\sin\frac{\alpha}{2}\ \eta_c + \sin\alpha\,(\frac{L}{2\eta_c} + 1)\ L \right]\ \ \dots\dots\dots\ (9)$$

where $\eta_c = R + h\ (1 - tg\alpha)/\sin\alpha$.

This expression of the hydrodynamic damping factor, c_{fluid}, shows that this term increases with flow velocity since the damping length, L, cannot decrease. Factors c and c_{fluid} being of opposite signs, the response y of equation (6) tends to the infinity when velocity is such that the sum $c + c_{fluid}$ tends to zero. In fact, second-order terms are involved and limit the amplitude. When the increase to infinity of the response is assimilated to the instability threshold defined above, the condition $c + c_{fluid} = 0$ gives the critical instability velocity. Consequently, the latter varies on the one hand with the specific valve damping,c,and on the other hand with the damping length, L, of the pulsations of the annular jet emerging from the poppet valve.

../..

4.3.2. Determination of specific valve damping

Such as indicated above, the poppet valve is subjected to turbulent flow fluctuations, below the critical velocity. The frequency shape of the measured curve of valve response spectral density permits then to determine the effective values of specific damping factor, c . In actual fact, in case of a large band random excitation near the structural frequency, the power spectral density of the response is proportional to $\left\{ \left[1 - (f/f_{structure})^2 \right]^2 + (2\xi f/f_{structure})^2 \right\}^{-1}$. By adjustment on the measured curves, the values of specific damping factor, c, are derived and shown on Table III.

4.3.3. Numerical application

In order to determine the order of magnitude of jet pulsation damping length, L, the condition $c + c_{fluid} = 0$ will be used with the value c_{fluid} given by formula (9) and the measured instability conditions, given in Table II. Table IV shows that damping length L nearly varies like $V_{upstream}$.

The damping length determined as above is worth 0.5 to 10 times the poppet valve annular space thickness. This result is to be compared with that obtained by Kolkman (Ref.24). This author shows in fact that length of the fluid volume to be considered for calculating the inertia term, in the case of vertical self-exciting vibrations in a sluice gate with small opening, is of the order of such opening. Anyhow, thorough investigation of this point should be of interest as it is of major importance in the type of problems of interaction between a fluid and a structure where the inertia of a fluid jet is involved.

6. CONCLUSIONS

The lateral vibrations of a poppet valve, for a given lift , are studied in water, air and double-phase mixture flows. Considering the discharges used, the latter correspond to annular flow patterns, with liquid wall film. Experimental results on the evolution of the frequencies and amplitudes of the dynamic device motion are given versus the discharges.

- the general shapes of the dynamic behaviour are similar for all the qualities considered : frequency decrease and amplitude increase permitting to define an hydroelastic instability threshold before the poppet valve reaches the end-stop.

- the results show that double-phase mixtures can be treated, with a good approximation through the singularity, as homogeneous fluids characterized by equal velocities of both phases.

7. BIBLIOGRAPHY

1. <u>Hara, F., Shigeta, T. and Shibita, H.</u> : "Two phase flow induced random vibrations" in Symposium on Flow-Induced Structural Vibrations - IUTAM/IAHR Karlsruhe, Germany ; Aug. 14-16 1972, Paper G5, 5 pp.

2. <u>Hara, F.</u> : "A theory on the two-phase flow induced vibrations in piping systems" in : 2nd Int. Conf. on Structural Mechanics in Reactor Technology Berlin ; 10-14 Sept. 1973, Paper F 5/1.

3. <u>Hara, F.</u> : "Two phase flow induced vibrations in a horizontal piping system" Bull. JSME, Vol 20 n° 142 ; April 1977, pp. 419-427.

4. <u>Gorman, D.J.</u> : "An analytical and experimental investigation of the vibration of cylindrical reactor fuel elements in two phase parallel flow". Nuclear Science Eng. ; 44, 1971, pp. 277-290.

5. <u>Harris, R.W. and Holland, P.G.</u> : "Response of a cylindrical cantilever to axial air water flow". In : First Int. Conf. on Structural Mechanics in Reactor Technology - Berlin 1971, Paper E 3/6.

6. <u>Cedelin, L., Hassid, A., Rossini, T. and Solieri, R.</u> : "Vibrations induced by the two phase coolant flow in the power channels of a pressure type nuclear reactor". In : First Int. Conf. on Structural Mechanics in Reactor Technology Berlin 1971, Paper E 4/5.

7. <u>Dellavalle, F., Rossini, T. and Vaneli, E.</u> : "Vibrations induced by a two phase parallel flow to a rod bundle - Preliminary experiments on the surface tension and gas density effect". In : Proc. of BNES Int. Conf. on Vibration Problems in Industry - Keswick U.K. 1973, Paper 525.

8. <u>Gorman, D.J.</u> : "Experimental and analytical study of liquid and two phase flow-induced vibration in reactor fuel bundles". ASME Paper 75 - PVP 52 - 2nd Nat. Congress on Pressure Vessels and Piping ; San Francisco 23/27 June 1975.

9. <u>Pettigrew, M.J. and Païdoussis, M.P.</u> : "Dynamics and stability of flexible cylinders subjected to liquid and two phase axial flow in confined annuli". In : Trans. of the 3rd Int. Conf. on Structural Mechanics in Reactor Technology London U.K. 1-5 Sept. 1975.

10. <u>Hara, F. and Yameshita, T.</u> : "Parallel two phase flow-induced vibrations in fuel pin model". Journal of Nuclear Science and Technology - 15 [5] 1978, pp. 346-354.

11. <u>Degli Espinosa, P., Lelli, G., Possa, G. and Vanoli, G.</u> : "Fuel sting dynamics and pressure tube fretting corrosion in the CIRERE power channel". In : Prof. of BNES Int. Conf. on Vibration in Nuclear Plant. Keswick U.K. 1978, Paper 1-5.

12. <u>Fokin, B.S. and Gol'dberg, E.N.</u> : "Association of vibration of tubular elements of generating equipment with the dynamic characteristics of the sweeping two phase flow". Thermal Engineering 26 (7) 1979, pp. 419-422.

13. Hara, F. : "Vibration experiences with a fuel pin model system in a parallel two phase flow". In : Symposium on Practical Experiences with Flow-Induced Vibrations. IUTAM/IAHR, Karlsruhe, Germany ; 3-6 Sept. 1979, 4 pp.

14. Pettigrew, M.J. and Gorman, D.J. : "Experimental studies on flow-induced vibration to support steam generator design - Vibration of small tube bundlles in liquid and two-phase cross-flow". In : Proc. of BNES Int. Conf. on Vibration Problems in Industry- Keswick U.K. 1973, Paper 424.

15. Pettigrew, M.J. and Gorman, D.J. : " Vibration of heat exchanger components in liquid and two phase cross-flow". In : Proc. of BNES Int. Conf. on Vibration in Nuclear Plant. Keswick U.K. 1978, Paper 2.3.

16. Heilker, W.J. and Vincent, R.Q. : "Vibration in nuclear heat exchangers due to liquid and two-phase flow". Trans. ASME - Journal of Eng. for Power, Vol. 103, April 1981 pp. 358-366.

17. Remy, F.N. : "Etudes expérimentales sur l'instabilité aéro-hydroélastique de faisceaux tubulaires en écoulement diphasique". In : Réunion Thermohydroélasticité - SHF - 25/26 Mars 1980.

18. Spedding, P.L. and Van Thang Nguyen : "Régime map for air water two-phase flow". Chem. Eng. Science - Vol. 35, n° 4-B 1980, pp. 779-793.

19. Oshinowo, T. and Charles, M.E. : "Vertical two-phase flow". The Canadian Journal of Chem. Eng. Vol. 52, August 1974, pp. 438-448.

20. Idel'cik, I.E. : "Memento des pertes de charge". Eyrolles Paris 1969.

21. Bergles, A.E., Collier, J.G. , Delhaye, J.M. , Hewitt, G.F. and Mayinger, F. "Two phase flow and heat transfer in the power and progress industries" Hemisphere Publishing Corporation, 1981.

22. Maeda, T. : "Studies on the dynamic characteristics of a poppet valve" Bull. of the JMSE, Vol. 13 n° 56 1970, pp. 281-289.

23. Rajaratnam, N. : "Turbulent jets". Elsevier Scientific Publishing Cy 1976.

24. Kolkman, P.A. and Vrijer, A. : "Gate edge section as a cause of self-exciting vertical vibrations". 17th IAHR Congress - Baden Baden - August 1977, pp. 395-402.

TABLE I - GEOMETRICAL AND MECHANICAL CHARASTERISTICS

R	$=$	$19,1 \ 10^{-3}$ m	m_{valve}	$=$	$1,3$ kg
$R_{downstream}$	$=$	$73 \ 10^{-3}$ m	$\rho_{structure}$	$=$	$7 \ 800$ kg/m^3
R_{seat}	$=$	$29,1 \ 10^{-3}$ m	ℓ	$=$	$0,238$ m
$H_{cylinder}$	$=$	$20 \ 10^{-3}$ m	b	$=$	$30 \ 10^{-3}$ m
H_{cone}	$=$	$40 \ 10^{-3}$ m	e	$=$	$3 \ 10^{-3}$ m
h	$=$	$5 \ 10^{-3}$ m	E	$=$	$2 \ 10^{11}$ N.m^{-2}
α	$=$	$45°$			

TABLE II - OBSERVED INSTABILITY CONDITIONS

Fluid	Homogeneous model			Frequency	Amplitude
	Specific mass	Upstream velocity	Upstream specific kinetic energy $\rho v^2/2$		
	(kg m^{-3})	(m s^{-1})	(N m^{-2})	(Hz)	(mm)
Water	1 000	2,8	4 000	5,8	0,35
Air	1,65	68	3 800	5	0,44
Mixture x = 0,05	29	21	6 350	6,5	1,2
Mixture x = 0,1	15,4	29	6 550	6,3	1,0
Mixture x = 0,2	10,7	42	9 300	4,8	1,1

TABLE III - SPECIFIC DAMPING FACTOR c

Fluid	Quality x	c (kg s^{-1})	$f_{structure}$ (V = 0) (Hz)	m (V = 0) (kg)	ξ (%)
Water	0	4	7,2	1,46	3
Air	1	1,5	7,5	1,34	1,2
Mixture	0,05	2	7,5	1,34	1,6
Mixture	0,1	2	7,5	1,34	1,6
Mixture	0,2	2	7,5	1,34	1,6

with $\xi = c / 4\pi m f_{structure}$

TABLE IV - JET DAMPING LENGTH L

Fluid	Quality x	Instability conditions Homogeneous specific mass (kg m^{-3})	Homogeneous upstream velocity (m s^{-1})	Frequency (Hz)	Damping length L (m)	L/V$_{amont}$ 10^{-4}
Water	0	1 000	2,8	5,8	0,003	10,2
Air	1	1,65	68	5	0,067	9,9
Mixture	0,05	29	21	6,5	0,019	9,1
Mixture	0,1	15,4	29	6,3	0,027	9,3
Mixture	0,2	10,7	42	4,8	0,027	6,4

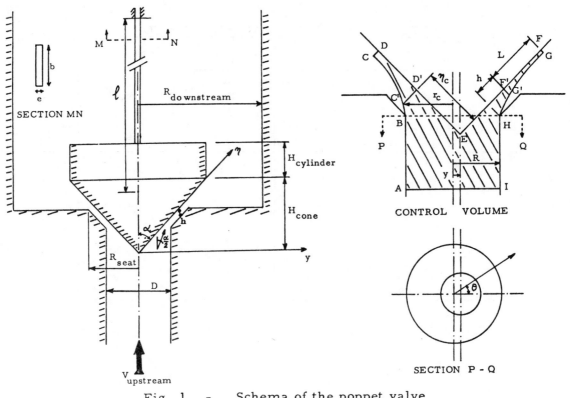

SECTION MN

Fig. 1 - Schema of the poppet valve

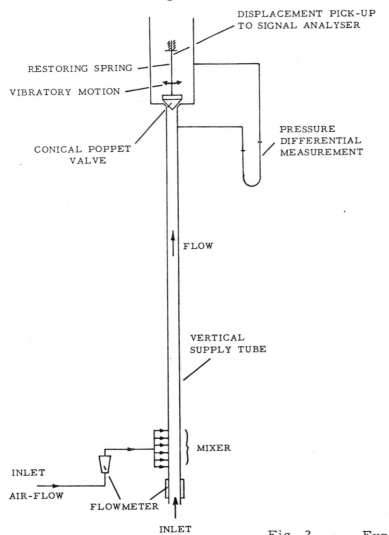

Fig. 2 - Experimental apparatus

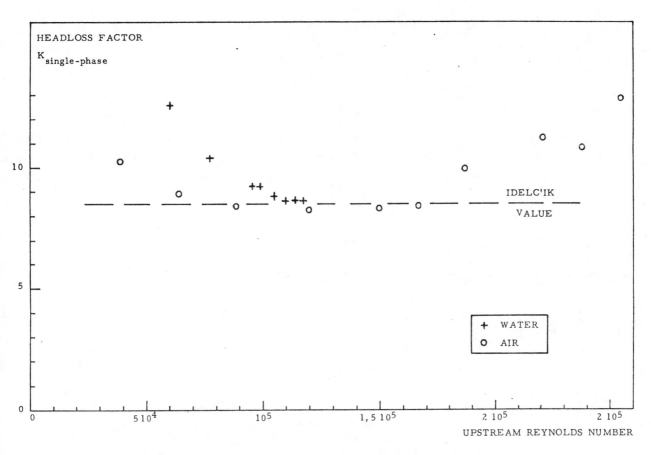

Fig. 3 - Single-phase headloss factor

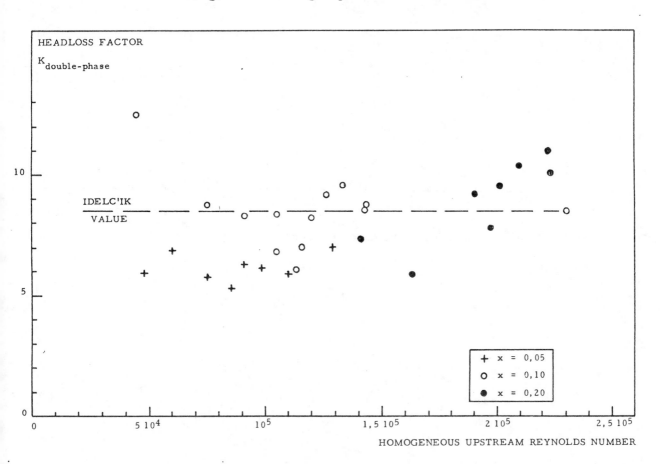

Fig. 4 - Double-phase headloss factor

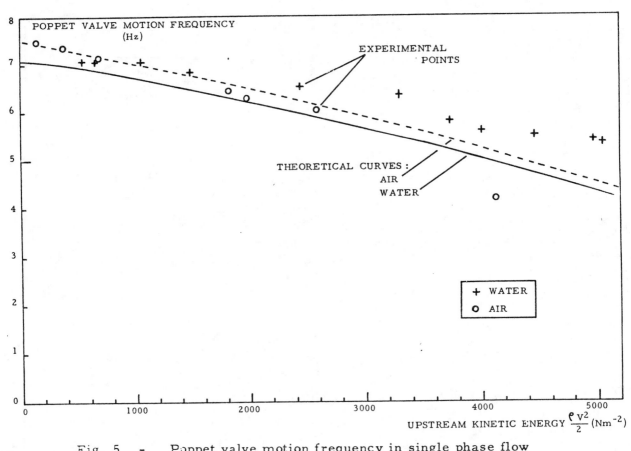

Fig. 5 – Poppet valve motion frequency in single phase flow

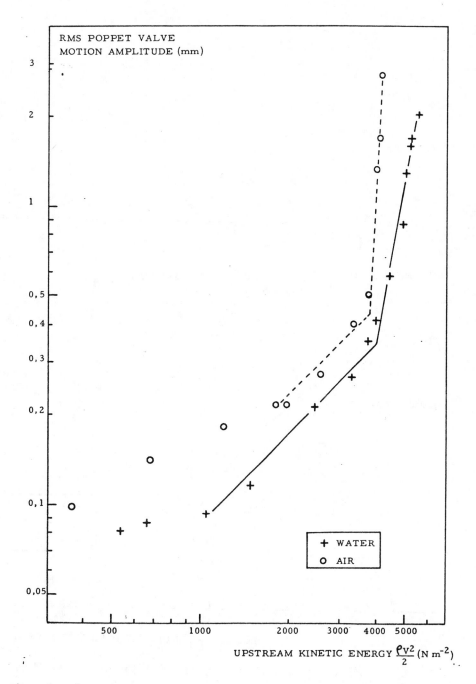

Fig. 6 - Poppet valve motion amplitude in single-phase flow

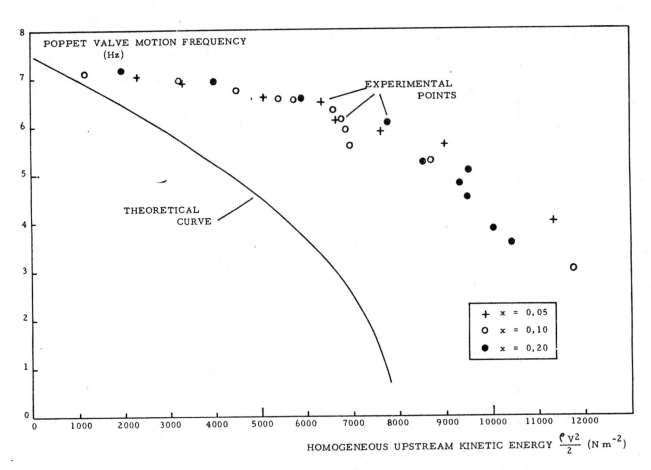

Fig. 7 — Poppet valve motion frequency in double-phase flow

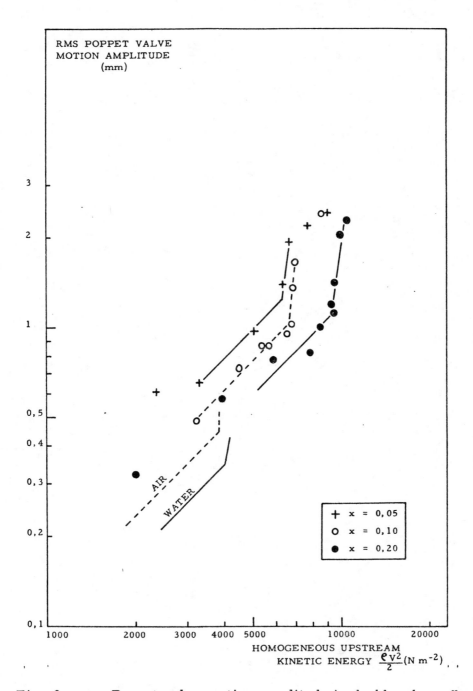

Fig. 8 – Poppet valve motion amplitude in double-phase flow

**FLOW INDUCED VIBRATIONS
IN FLUID ENGINEERING**

Reading, England: September 14-16, 1982

ON COINCIDENCE IN RELATION TO PREDICTION OF PIPE
WALL VIBRATION AND NOISE RADIATION DUE TO TURBULENT
PIPE FLOW DISTURBED BY PIPE FITTINGS

M. K. Bull

University of Adelaide, South Australia

M. P. Norton

University of Western Australia, Western Australia

Summary

A severe disturbance to fully-developed turbulent flow in a pipe results in the generation of intense broadband internal acoustic waves which can propagate throughout a piping system. The vibrational response of the pipe wall to this excitation, and hence the externally radiated acoustic power also, is predominantly determined by coincidence of higher-order acoustic modes with resonant bending modes of the pipe wall. Four principal wavenumber coincidences can be associated with each acoustic mode; and, because of asymmetry of the dispersion curves of the acoustic modes, which is a direct consequence of flow in the pipe, the principal coincidences occur at four different frequencies in the general vicinity of the cut-off frequency of a given acoustic mode. Corresponding peaks in the experimental vibration and acoustic spectra can be identified, and it is shown that the frequencies at which they occur can be accurately predicted.

The effects of structural damping and the relative bandwidths of structural response and joint acceptance functions are discussed, and it is concluded that for accurate prediction of vibration and acoustic radiation these effects, as well as the effects of asymmetry of acoustic mode dispersion curves and principal wave-number coincidences, must be taken into account.

It is shown that the total number of principal wavenumber coincidences in the frequency range of interest in practice will generally be large, but that this number, and hence the vibrational response and acoustic power radiation levels, can be reduced by increasing the ratio of wall thickness to pipe diameter.

Organised and sponsored by
BHRA Fluid Engineering, Cranfield, Bedford MK43 0AJ, England

NOMENCLATURE

A_{pq} = constant, Equ.(1)

$A_{\alpha}(\omega)$ = structural response function, Equ.(8)

a_i = internal radius of pipe

a_m = mean radius of pipe wall

B_{pq} = constant, Equ.(1)

c_e = speed of sound in external fluid

c_i = speed of sound in internal fluid

c_{LP} = longitudinal wave speed in plate of pipe wall material

E = Young's modulus of the pipe wall material

f = frequency, Hz

f_L = upper limiting frequency, Hz

h = pipe wall thickness

$j_{\alpha\alpha}^2$ = joint acceptance function, Equ. (12)

$(j_{\alpha\alpha}^2)_{MAX}$ = maximum value of $j_{\alpha\alpha}^2$

J_p = p-th order Bessel function of first kind

K_m = $m\pi/\Lambda$

K_x = $k_x a_m$

$\left.\begin{array}{c} k_{c_1} \\ k_{c_2} \\ k_{c_3} \\ k_{c_4} \end{array}\right\}$ = axial wavenumber components at principal wavenumber coincidences

k_x = axial wavenumber

ℓ = length of pipe

M = uniform flow Mach number = U/c_i

M_o = flow Mach number on pipe centre-line = U_o/c_i

M_{LPi} = c_{LP}/c_i

$\left.\begin{array}{c} m \\ \\ n \end{array}\right\}$ = axial and circumferential orders of resonant modes of pipe wall

p = circumferential order of acoustic mode

$p(r,\phi,x)$ = acoustic pressure

Q_{α} = damping quality factor of α-th structural mode

q	=	radial order of acoustic mode
q_o	=	$1/2\rho_i U_o^2$
(r,ϕ,x)	=	cylindrical coordinates
S	=	area of vibrating pipe surface
t	=	time
U	=	uniform flow velocity
U_o	=	flow velocity on pipe centre-line
Z_α	=	complex obstructance function, Equ.(9)
β	=	$h/2\sqrt{3}\ a_m$
ζ	=	displacement of pipe wall
$\dot{\zeta}$	=	velocity of pipe wall
$\ddot{\zeta}$	=	acceleration of pipe wall
$\langle \ddot{\zeta}^2 \rangle_{1/3}$	=	mean square acceleration of pipe wall in one-third octave band
$(\Delta_\nu)_A$	=	bandwidth of structural response function A_α
$(\Delta_\nu)_j$	=	bandwidth of joint acceptance function
δ_{np}	=	$\begin{cases} 1 \text{ for } n = p \\ 0 \text{ for } n \neq p \end{cases}$
κ_{pq}	=	radial wavenumber eigenvalue for (p,q)th acoustic mode
Λ	=	ℓ/a_m
μ	=	Poisson's ratio of pipe wall material
ν	=	non-dimensional frequency = ω/ω_r
ν_L	=	non-dimensional upper limiting frequency
$\left.\begin{array}{l} \nu_{c1s} \\ \nu_{c1s} \\ \nu_{c2s} \\ \nu_{c3s} \\ \nu_{c4s} \end{array}\right\}$	=	non-dimensional structural resonance frequencies at principal wave-number coincidences

$\left.\begin{array}{l} \nu_{c1a} \\ \nu_{c2a} \\ \nu_{c3a} \\ \nu_{c4a} \\ \\ \nu_{cia} \end{array}\right\}$ = non-dimensional acoustic frequencies at principal wavenumber coincidences

ν_c = theoretical non-dimensional complete coincidence frequency

ν_{c+} = theoretical non-dimensional complete coincidence frequency with associated positive axial wavenumber component

ν_{c-} = theoretical non-dimensional complete coincidence frequency with associated negative axial wavenumber component

ν_{co} = non-dimensional cut-off frequency

ν_{on} = cut-off frequency of structural modes of circumferential order n

ρ_e = density of external fluid

ρ_i = density of internal fluid

ρ_s = density of pipe wall material

ρ_{is} = ρ_i/ρ_s

σ_α = radiation ratio for α-th structural mode

Φ_p = $\phi_p U_o/q_o^2 a_i$

$\Phi_{\ddot\zeta}$ = $\phi_{\ddot\zeta}/\omega_r^3 a_m^2$

Φ_π = $\phi_\pi/\rho_e c_e^2 S a_m$

ϕ_p = power spectral density of wall pressure fluctuations

$\phi_{\dot\zeta}$ = power spectral density of pipe wall velocity

$\phi_{\dot\zeta\alpha}$ = contribution to power spectral density of pipe wall velocity due to α-th structural mode

$\phi_{\ddot\zeta}$ = power spectral density of pipe wall acceleration

ϕ_π = spectral density of radiated acoustic power

$\phi_{\pi\alpha}$ = contribution to spectral density of radiated acoustic power due to α-th structural mode

ψ = $(1-\mu^2)^{1/2}$

ω = frequency, rad/s

$(\omega_{co})_{pq}$ = cut-off frequency of (p,q)th acoustic mode

ω_r = ring frequency = c_{LP}/a_m

ω_α =   resonance frequency of α-th structural mode

$[\ ]$ =   average over S

1. INTRODUCTION

Following the disturbance of fully-developed turbulent pipe flow by a pipe fitting, undisturbed mean velocity and turbulence distributions become re-established at a downstream distance of some ten to fifteen pipe diameters. However the effect of the fitting is still felt, in that the flow has superimposed on it an acoustic field generated by and radiated upstream and downstream from the seat of the disturbance. Severe disturbances, such as those due to valves, junctions, sharp bends, and orifice plates, are accompanied by separation of the flow from the pipe walls and generate an intense acoustic field with pressure fluctuations very much larger than those due to the undisturbed flow turbulence. The acoustic field consists of plane waves and higher order modes which can propagate throughout a piping system. (Bull and Norton, Refs. 1, 2, 3). The latter in particular are a major source of pipe wall vibration and wall-generated external acoustic radiation, as a result of their coincidence with resonant vibrational bending modes of the pipe wall (Refs. 2, 4, 5). In this paper various aspects of coincidence as it relates to prediction of pipe wall vibration and external acoustic radiation will be considered.

A computational procedure has been set up, by means of which the response of the pipe wall to any given higher order acoustic mode as a function of frequency, theoretical complete coincidence (defined in a later section) frequencies and wavenumbers, acoustic and structural frequencies and wavenumbers associated with principal wavenumber coinicdences (also defined later), and a count of the number of principal wavenumber coincidences occurring up to any given fequency, can be obtained. The calculations use the method of Arnold and Warburton (Ref. 6) for evaluation of structural resonance frequencies.

Experimental results which are presented have been obtained from 2.92 m long test sections of 72.54 mm internal diameter steel pipe with wall thicknesses of 0.89 and 6.36 mm, mounted in an induced flow pipe rig 52.8 diameters downstream of the pipe fitting producing the flow disturbance. The rig has been previously described by Bull and Norton (Refs. 3, 4). The flow in it is choked downstream of the test section, so that the test section constitutes a resonant pipe section excited by travelling acoustic modes. Attention will be confined here to such a system, which is representative of a practical piping system with long runs of pipe made up from a series of relatively short sections but, apart from the fitting causing flow disturbance, uninterrupted internally.

2. THE ACOUSTIC FIELD IN A PIPE

The acoustic field inside a pipe of circular cross-section may consist of plane waves and higher order acoustic modes. In the absence of flow, the acoustic pressures associated with the (p, q)th such wave, of radian frequency ω and axial wavenumber k_x, propagating in the axial direction, can be represented by (Ref.7, p.509)

$$p(r,\phi,x) = (A_{pq}\cos p\phi + B_{pq}\sin p\phi)J_p(\kappa_{pq}r)e^{j(k_x x - \omega t)}, \qquad (1)$$

where r, ϕ, and x are radial, angular, and axial coordinates, J_p is the p-th order Bessel function of the first kind, κ_{pq} is the solution of the equation $J_p'(\kappa_{pq}a_i) = 0$, a_i is the internal radius of the pipe,

$$(\omega/c_i)^2 = \kappa_{pq}^2 + k_x^2, \qquad (2)$$

and c_i is the speed of sound in the internal fluid. It has p plane diametral nodal surfaces and q cylindrical nodal surfaces concentric with the pipe axis; it can propagate only at frequencies above its cut-off frequency $(\omega_{co})_{pq} = \kappa_{pq}c_i$.

In the presence of an idealised uniform flow with velocity U and Mach number $M = U/c_i$ parallel to the pipe axis at all (r,ϕ) the frequency, as seen by a stationary observer, of a wave with axial wavenumber component k_x is given by

$$\omega/c_i = \sqrt{\kappa_{pq}^2 + k_x^2} + Mk_x \qquad (3)$$

instead of Equ.(2), the additional term representing the Doppler frequency shift. The cut-off frequency is now reduced to

352

$$(\omega_{co})_{pq} = \kappa_{pq} c_i (1-M^2)^{1/2},\qquad(4)$$

and occurs at an axial wavenumber of

$$k_x = - (M/\sqrt{1-M^2})\kappa_{pq}\qquad(5)$$

instead of at $k_x = 0$ as in the no-flow case (Bull and Norton, Ref. 3). The dispersion curve (the variation of axial wavenumber with frequency) for these waves, which is symmetrical about the frequency axis in the case of stationary internal fluid, therefore becomes asymmetrical due to the influence of flow. It should be noted that the frequency of a wave with $k_x = 0$ (Equ. (3)) is $\kappa_{pq,1} c_i$ irrespective of the flow Mach number. These relationships are illustrated in Fig. 1. In general, replacement of M by M_o, the centre-line Mach number of the turbulent flow, will provide an adequate representation of the effect of the flow.

3. VIBRATIONAL RESPONSE OF THE PIPE WALL TO THE INTERNAL ACOUSTIC FIELD

We consider a section of pipe of length ℓ between supports, with a wall thickness h which is not too large in relation to the mean radius a_m ($h/a_m \ll 1$). In practice such lengths will be those between flanges or fittings, which represent discontinuities in the pipe structure. Each such length will constitute a resonant system with its own set of discrete natural frequencies, and as far as external acoustic radiation is concerned it is the resonant bending modes of the pipe wall which are of primary interest.

Let ζ, $\dot{\zeta}$, $\ddot{\zeta}$ be respectively the radial displacement, velocity, and acceleration of the pipe wall. The power spectral density $\phi_{\dot\zeta}$ of the velocity response, averaged over the area S of the vibrating surface, to a random pressure field of power spectral density ϕ_p, can be expressed as

$$[\![\phi_{\dot\zeta}(\omega)]\!] = \frac{\omega^2 \phi_p(\omega) S^2}{2} \sum_\alpha \frac{j_{\alpha\alpha}^2(\omega)}{|Z_\alpha(\omega)|^2}\qquad(6)$$

(Bull and Rennison, Ref. 8), where $j_{\alpha\alpha}^2(\omega)$ is the joint acceptance function for the α-th resonant structural mode and the applied pressure field, Z_α is the complex obstructance function for the α-th mode, $[\![....]\!]$ indicates a spatial average over the surface S, and the summation is over all structural modes. The contribution of the α-th mode to the spectral density ϕ_π of the radiated acoustic power is given by

$$\phi_{\pi\alpha}(\omega) = \sigma_\alpha \rho_e c_e S [\![\phi_{\dot\zeta\alpha}(\omega)]\!],\qquad(7)$$

where $\phi_{\dot\zeta\alpha}$ is the corresponding modal contribution to the power spectral density of the wall velocity, ρ_e and c_e are respectively the density of and speed of sound in the external fluid, and σ_α is the radiation ratio for the α-th mode. If $A_\alpha(\omega)$ is defined as

$$A_\alpha(\omega) = (\omega/\omega_\alpha)^2 / [(1-\omega^2/\omega_\alpha^2)^2 + (\omega/\omega_\alpha Q_\alpha)^2],\qquad(8)$$

then

$$|Z_\alpha(\omega)|^2 = \omega^2 \omega_\alpha^2 (\rho_s S h/4)^2 / A_\alpha(\omega),\qquad(9)$$

where ρ_s is the density of the pipe wall material, and ω_α and Q_α are respectively the resonance frequency and damping quality factor of the α-th mode. Expressions can then be written for the non-dimensional spectral densities of the pipe wall acceleration $\Phi_{\ddot\zeta} = \phi_{\ddot\zeta}/\omega_r^3 a_m^2$ and acoustic power radiation $\Phi_\pi = \phi_\pi/\rho_e c_e^2 S a_m$, where $\omega_r = c_{LP}/a_i$ is the ring frequency of the pipe, $c_{LP} = \sqrt{E/\rho_s \psi^2}$, $\psi^2 = 1-\mu^2$, and E and μ are respectively the Young's modulus and Poisson's ratio of the pipe wall material.

These are

$$\Phi_\zeta = \frac{a_i}{a_m} \cdot \frac{\rho_{is}^2 M_o^3}{6\beta M_{LPi}^3} \cdot \Phi_p \cdot \sum_\alpha \nu^2 \cdot \frac{A_\alpha j_{\alpha\alpha}^2}{\nu_\alpha^2} \qquad (10)$$

and

$$\Phi_\pi = \frac{c_i a_i}{c_e a_m} \cdot \frac{\rho_{is}^2 M_o^3}{6\beta M_{LPi}^2} \cdot \Phi_p \cdot \sum_\alpha \sigma_\alpha \cdot \frac{A_\alpha j_{\alpha\alpha}^2}{\nu_\alpha^2} , \qquad (11)$$

where $\rho_{is} = \rho_c/\rho_s$, ρ_i is the density of the internal fluid, U_o is the flow velocity on the centre-line of the pipe (so that $M_o = U_o/c_i$), $\beta = h/2\sqrt{3}a_i$, $M_{LPi} = c_{LP}/c_i$, $\Phi_p = \phi_p U_o^2/q_o a_i$ is the non-dimensional power spectral density of the internal wall pressure field, $q_o = \frac{1}{2}\rho_i U_o^2$, and $\nu = \omega/\omega_r$.

For a section of pipe with simply-supported ends (a reasonable approximation except for the lowest frequency modes) the $\alpha \equiv (m,n)$th mode has n structural wave-lengths around the circumference of the pipe and m half-waves between supports. The joint acceptance function for the α-th structural mode subject to excitation by the (p,q)th acoustic mode with axial wavenumber k_x is in this case given by

$$j_{\alpha\alpha}^2(\omega) = \delta_{np} K_m^2 (1-\cos\Lambda K_m \cos\Lambda K_x)/2\Lambda^2 (K_m^2-K_x^2)^2 , \qquad (12)$$

where $\delta_{np} = 1$ for $n = p$ and 0 for $n \neq p$, $K_m = m\pi/\Lambda$,

$K_x = k_x a_m$, and $\Lambda = \ell/a_m$.
$j_{\alpha\alpha}^2$ has its maximum value of 1/16 when $K_x = K_m$.

4. COINCIDENCE

4.1 Definitions of various forms of coincidence.

The term coincidence refers to matching in wavelength and frequency between the modes of the acoustic excitation and the resonant bending modes of the pipe wall. In the case of an infinitely long pipe, standing structural waves can occur in the circumferential direction only and these can be associated with travelling waves of any axial wavenumber component; there can then be a continuous variation of axial wavenumber component with frequency. Since the travelling acoustic modes also exhibit continuous variation of axial wavenumber component with frequency, simultaneous matching of both axial wavenumber components and frequencies of the two types of wave is in principle possible. However, when a finite section of pipe is considered and only discrete values of axial wavenumber component, corresponding to the resonance frequencies, are allowed, the two conditions cannot in general be simultaneously satisfied, even for excitation by travelling acoustic waves. For this latter situation we adopt the following definitions.
Wave number coincidence is defined as the condition in which the structural and acoustic waves have equal wavenumbers at the pipe wall. This requires that the circumferential distance between diametral nodes of the (p.q)th acoustic mode is equal to a half wavelength of the (m,n)th structural mode in the circumferential direction (i.e. n = p), and that the structural and acoustic waves have equal axial wavenumber components (i.e. $K_x = K_m$).
Complete coincidence is defined as wavenumber coincidence with, in addition, equality of structural resonance and acoustic frequencies.

In general, because the pipe has a set of discrete resonance frequencies and not a continuum, only wavenumber coincidence will occur. This is illustrated diagrammatically in Fig. 1.

The figure also shows that coincidence can occur at both positive and negative values of axial wavenumber. Furthermore, because of the asymmetry of the acoustic mode curve about the ν-axis, due to the presence of flow in the pipe, the frequencies of the positive and negative wavenumber coincidences will not be the same.

For travelling acoustic waves, for which frequency varies continuously with K_x, wavenumber coincidence is possible between any resonant structural mode and a given acoustic mode, and in each case the joint acceptance will have its maximum value of 1/16. However, the greater the difference between the (acoustic) frequency of excitation and the resonance frequency of the structural mode the smaller will be the response function $A_\alpha(\omega)$ and the smaller the overall response. The greatest structural response will occur, in general, in those modes which are closest in frequency to the theoretical complete coincidence frequencies ν_{c+} and ν_{c-} (that is to the complete coincidence frequencies which would be obtained by considering the continuous variation of structural wavenumber with frequency which would apply in the case of an infinitely long pipe). There will in general be four such structural modes for any given acoustic mode, with resonance frequencies ν_{c1s} and $\nu_{c2s}(< \nu_{c1s})$ and corresponding positive axial wavenumbers k_{c1} and k_{c2}, and ν_{c3s} and ν_{c4s} ($< \nu_{c3s}$) and corresponding negative axial wavenumbers k_{c3} and k_{c4}. The corresponding acoustic mode frequencies at the same wavenumbers are denoted by ν_{c1a}, ν_{c2a}, ν_{c3a} and ν_{c4a}. These various parameters are identified on Fig. 1. The wavenumber coincidences identified in this way will be referred to as *principal wavenumber coincidences*.

4.2.   Form of pipe wall response at coincidence.

While the preceding considerations lead to identification of the structural modes associated with the principal wavenumber coincidences, the maximum structural response in these modes will not in all cases occur precisely in the condition of wavenumber coincidence (although it will occur in a condition very close to it). The precise condition, as determined by the $A_\alpha j_{\alpha\alpha}^2$ product (Equ.(10)), will depend critically on the relative dispositions of the $j_{\alpha\alpha}^2$ and A_α curves against frequency. The curve of $j_{\alpha\alpha}^2$ against frequency is multi-lobed: the overall maximum value of 1/16 occurs in the main lobe at $K_x = K_m$; and successive side lobes each have a maximum value which is less than that of the preceding lobe. (Its form can be seen in calculated response curves shortly to be introduced). The half-width of the main lobe and full width of each of the side lobes is given by

$$(\Delta\nu)_j \simeq 2\pi K_m / \Lambda\nu M_{LPi}^2. \qquad (13)$$

The structural resonance function A_α has a sharp peak at the resonance frequency and decreases monotonically with frequency on either side of the peak. Its half power bandwidth is

$$(\Delta\nu)_A = \nu_\alpha / Q_\alpha. \qquad (14)$$

Thus in the region of a coincidence

$$(\Delta\nu)_j / (\Delta\nu)_A \simeq 2\pi K_m Q_\alpha / \Lambda\nu_c^2 M_{LPi}^2. \qquad (15)$$

Numerical values show that for low damping (high Q_α), short pipes (small Λ) and low radial-order (low q) acoustic modes $(\Delta\nu)_j \gg (\Delta\nu)_A$, while for high damping, long pipes, and high-q acoustic modes $(\Delta\nu)_j \ll (\Delta\nu)_A$. In the case that $(\Delta\nu)_j \gg (\Delta\nu)_A$ the following forms of response, depending on the separation between the acoustic and structural resonance frequencies ν_{cia} and ν_{cis} (i=1,2,3,4), can occur:

(1) Acoustic and structural frequencies equal, $\nu_{cia} = \nu_{cis}$. This corresponds to complete coincidence and will occur only in a few isolated cases if at all. An example is shown in Fig.2. Maximum response will occur at the frequency $\nu = \nu_{cia} = \nu_{cis} = \nu_{c+}$.

(2) ν_{cis} falls within the bandwidth of the main lobe of the $j_{\alpha\alpha}^2$ curve. Maximum response will occur essentially at the structural resonance frequency ν_{cis}. A response of this type is shown in Fig. 3.

(3) ν_{cis} is equal to or very close to one of the frequencies at which $j_{\alpha\alpha}^2 = 0$. Maximum response will occur at the acoustic frequency ν_{cia}, or possibly at the

frequency of one of the maxima of the side lobes of the $j_{\alpha\alpha}^2$ curve.

(4) ν_{cis} falls within the first or a subsequent side lobe of the $j_{\alpha\alpha}^2$ curve. There will be local peaks in response at the structural frequency ν_{ics}, at the acoustic frequency ν_{cia}, and at frequencies between these two corresponding to peaks of intermediate side lobes of the $j_{\alpha\alpha}^2$ curve. Which of these is the greatest will depend on the value of the structural damping. An instance in which the structural resonance occurs in the third side lobe of the $j_{\alpha\alpha}^2$ curve, but still produces maximum response is shown in Fig. 4.

At the other extreme, when $(\Delta\nu)_A \gg (\Delta\nu)_j$, the response has the following possible forms:

(1) Acoustic and structural frequencies equal. In the isolated cases in which this occurs, maximum response will occur at frequency $\nu = \nu_{cia} = \nu_{cis} = \nu_{c\pm}$, as in the previous case.

(2) ν_{cia} falls within the bandwidth of the A_α curve. Maximum response will occur at the acoustic frequency ν_{cia}, and its value will depend on Q_α.

(3) ν_{cia} falls outside the bandwidth of the A_α curve. Maximum response, as shown in Fig. 5, will again occur at ν_{cia} and will in general be independent of Q_α. Both here and in (2) the response curves have essentially the same shape as the $j_{\alpha\alpha}^2$ curve.

Maximum response at ν_{cia} corresponds to wavenumber coincidence. When the frequency for maximum response is not ν_{cia}, but ν_{cis} or some intermediate value, the structural and acoustic wavenumbers will not be precisely equal in the maximum response condition. However, in all cases the frequency for maximum structural response to a given acoustic mode, and for maximum power radiation, will be within the small range $\nu_{cis} \leqslant \nu \leqslant \nu_{cia}$ or $\nu_{cia} \leqslant \nu \leqslant \nu_{cis}$.

4.3.   Comparison of calculated principal wavenumber coincidence frequencies with experimental results.

Narrowband spectra (100 Hz bandwidth) of wall acceleration response of and acoustic radiation from a test section of pipe with $\beta = 0.007$, resulting from the disturbances associated with a 90° mitred bend, for flow Mach numbers $M_o = 0.22$ and 0.50, are shown in Figs. 6 and 7 Calculated frequencies corresponding to the principal wavenumber coincidences for a number of the lower order acoustic modes are marked on the figures.

It can be seen that in the cases of the (1,0) and (2,0) acoustic modes the calculated frequencies coincide closely with major measured spectral peaks at both flow speeds. For higher order modes some cases of matching of experimental peaks with calculated frequencies can be seen, although in other cases, as a result of overlapping of the frequency ranges of the principal wavenumber coincidences, the matching becomes less certain. There is also a significant peak in spectral density (particularly of the wall acceleration at both flow speeds) between the (1,0) and (2,0) modes which is not accounted for. However, the general indication is that the frequencies of the various peaks in the spectral curves correspond to principal wavenumber coincidences, and that the calculation procedure adequately takes account of the effect of flow speed in producing acoustic mode dispersion curves which are not symmetrical about the frequency axis.

4.4.   Influence of pipe parameters on number of coincidences up to a given frequency.

In relation to the practical problems of prediction of the levels of sound radiation from a given piping system and choosing pipe parameters to minimise such radiation, the number of possible coincidences in a given frequency range and the possibility of avoiding coincidence are of interest.

It has previously been shown (Bull and Norton, Refs. 2,4) that the frequency at which coincidence occurs is very close to the cut-off frequency of the acoustic mode involved. This is basically a consequence of the fact that the group velocities $(\partial\omega/\partial k_x)$ of the structural waves are generally much greater than those of the acoustic higher order modes. At first sight this would imply that the total number of principal wavenumber coincidences below any given frequency would be simply four times the number of acoustic modes with cut-off frequencies below the given frequency.

In reality the number is less than this because (i) the difference between principal wavenumber coincidence frequencies and acoustic cut-off frequencies, even though small, results in a general translation of coincidences to frequencies above the acoustic cut-off frequencies, and (ii) coincidence does not occur if the cut-off frequency of structural modes of a given circumferential order n exceeds that of an acoustic mode of the same circumferential order.

These effects can be seen in Fig.8 which shows structural and acoustic dispersion curves for the n = p = 6 modes under no-flow conditions (M_o = 0). For convenience the structural curves have been based on Heckl's approximation (Ref. 9) for the resonance frequencies, namely

$$\nu_{mn}^2 = \beta^2 (K_m^2 + n^2)^2 + \psi^2 K_m^4 / (K_m^2 + n^2)^2 \qquad (16)$$

The cut-off frequency below which no resonant structural modes exist, corresponds to K_m = 0, that is to

$$\nu_{o,n} = \beta n^2 . \qquad (17)$$

Suppose that the upper frequency limit is set at ν = 2.25, which limits the acoustic modes to be considered to the $(6,0)$ to $(6,7)$ modes. It can be seen that for some pipe wall thicknesses, e.g. $\beta = 10^{-2}$, there is coincidence with every acoustic mode. As β is decreased the number of coincidences is decreased (see for example $\beta = 3 \times 10^{-3}$) and at very small β (e.g. 10^{-3}) no such coincidences occur. (It might be noted that because of the reversed curvature exhibited by the structural dispersion curves at low K_m values when β is very small, two additional coincidences with a given acoustic mode become possible for some values of β). On the other hand, as β is increased beyond 10^{-2}, there comes a stage where the number of coincidences decreases because the increase in cut-off frequency of the structural modes eliminates coincidence with the lower order acoustic modes (e.g. $\beta = 3 \times 10^{-2}$) and ultimately at large β values no coincidences occur within the set frequency range. Thus for a given value of n = p the number of coincidences will be low for small β, will reach a maximum as β increases and will fall to zero at high values of β. A similar variation with β might then be expected for the total number of coincidences, summed over all n = p values.

We expect the number of coincidences N_c with frequencies below a given frequency ν to be of the form

$$N_c = N_c(\nu, M_{LPi}, \beta, M_o), \qquad (18)$$

and independent of Λ. Calculated values of N_c, for steel pipes with internal air flow at speeds of M_o = 0 and M_o = 0.50, obtained by means of the programme referred to in Section 1, are presented as a function of ν with β as a parameter in Figs. 9 and 10. Except at the smallest value of β, the change in flow speed produces no significant change in the number of coincidences. The variation of N_c with β at a particular value of ν (namely ν = 0.88, which corresponds to f ~ 20 kHz and ~ 18.5 kHz for the experimental test pipes with h = 0.89 and 6.36 mm respectively) is shown in Fig. 11. Again there is no significant effect of flow speed except at very small values of β.

If in practice sound radiation at frequencies up to a limiting dimensional frequency f_L (e.g. the limit of the audio-frequency range) is the main consideration, the limiting non-dimensional frequency, $\nu_L = 2\pi f_L a_m / c_i M_{LPi}$, will increase with

increasing pipe radius. For f_L = 20 kHz say, ν_L ~ 0.7 for a pipe with a_m = 30 mm, and Figs. 9 and 10 indicate that the number of coincidences in the range of interest would be typically about 35. However, for a pipe with a_m = 300 mm, ν_L ~ 7 and the number of coincidences would be correspondingly larger. It is clear that for pipes of practical interest the number of coincidences in the audio-frequency range, which will contribute to the acoustic power radiation due to broadband internal acoustic excitation (such as that associated with internal flow disturbances due to pipe fittings), will in general be large.

It has been indicated above that for a given circumferential mode order (n = p) the number of coincidences can be reduced by increasing β and thereby raising the cut-off frequency of the structural modes above those of some of the lower radial-order acoustic modes. The effect is illustrated by Fig. 12, even though this figure is based on the approximate expression, Equ.(17), for the structural cut-off frequencies, and applies for $M_o = 0$. The ordinate used, $\nu_{co}(1 - \sqrt{3}\beta)$, arises because acoustic cut-off frequencies are related to a_i while the structural cut-off frequencies are related to a_m. It has been assumed that $a_m = a_i + h/2$. It can be seen that increasing β does not eliminate appreciable numbers of coincidences until β becomes large and n is significantly greater than unity. It can also be seen that at a given n there is a limit to the number of possible coincidences which can be so eliminated, set by $\beta = 1/2\sqrt{3}$ corresponding to $h = 2a_i$ (before which, of course, relations such as Equs.(16) and (17) which are based on the assumption of a thin pipe wall will have become inaccurate). The figure indicates for instance that coincidence with the (1,0) acoustic mode could be avoided by a sufficiently large β (in fact $\gtrsim 0.17$, corresponding to $h \gtrsim 0.42\ a_i$) but that coincidence with all other p = 1 modes is unavoidable. In contrast coincidence of all p = 8 modes up to the (8,5) could be avoided even with $\beta = 0.03$, although it would still be possible for (8,q) modes with q > 5. But the inescapable conclusion from Figs. 9-12 is that, for all practical cases of broadband internal acoustic excitation, coincidence cannot be eliminated entirely by choice of pipe wall thickness, and the number of possible coincidences will generally be quite large.

Despite this, there are still advantages in the use of thick pipe walls. The number of coincidences will be reduced (for β values to the right of that for maximum N_c in Fig.11) and in addition acoustic power radiation will be reduced as $1/\beta^2$ (Equ.(11)). Both of these effects can be seen by comparing experimental values of $\beta^2 \langle \ddot{\zeta}^2 \rangle_{1/3}$ (where $\langle \ddot{\zeta}^2 \rangle_{1/3}$ is the mean square wall acceleration in a one-third octave band), for test pipes with $\beta = 0.0070$ and 0.0465 respectively, excited by the disturbance due to a 90° mitred bend, which are presented in Fig.13. The values of this parameter are very similar in the range $0.10 \leqslant \nu \leqslant 0.25$ where coincidences with the (1,0) and (2,0) acoustic modes occur for both test pipes, indicating response proportional to $1/\beta^2$ (in this case the change in β from 0.0070 to 0.0465 results in a 16.5 dB decrease in response). In the range $0.25 \leqslant \nu \leqslant 0.50$ the thicker pipe shows additional reduction in response of about 10 dB. This is attributable to the elimination of coincidences with the (3,0) - (6,0) acoustic modes. Similarly, reductions of 6-17 dB over the frequency range $0.50 \leqslant \nu \leqslant 1.00$ are attributable to elimination of coincidences with the (7,0)-(10,0), (4,1)-(9,1) and (5,2)-(8,2) modes.

5.   PREDICTION OF PIPE WALL RESPONSE AND ACOUSTIC RADIATION

It is expected that a good estimate of the vibrational response and acoustic radiation of a given pipe to broadband higher order mode acoustic excitation will be obtained by confining the summations in Equs. (10) and (11) to those structural modes for which coincidence is possible. Even so, the discussion of the various possible forms of response in Section 4.2. indicates that this may require consideration of a range of conditions from $(\Delta\nu)_A \ll (\Delta\nu_j)$ to $(\Delta\nu)_A \gg (\Delta\nu)_j$.

In all cases the lowest frequency at which coincidence can occur is $\nu \sim 0.12$, set by the (1,0) acoustic mode; and this coincidence will occur for all pipe wall thicknesses and diameters of practical interest. On the other hand, if only audio-frequencies are considered, with $f_L = 20$ kHz, the maximum value of $\nu(=\nu_L)$ will depend on pipe radius. For example $a_m = 30$ mm gives $\nu_L \simeq 0.7$, and coincidence will not involve acoustic modes of circumferential order $q \gtrsim 3$. In contrast, for $a_m = 300$ mm, $\nu_L \simeq 7$, and coincidence can occur for acoustic modes of order q up to about 34. Thus the larger the pipe diameter, the larger will be the range of $(\Delta\nu)_j/(\Delta\nu)_A$ values to be considered.

For low q, low Λ, high Q_α, in which case $(\Delta\nu)_A \ll (\Delta\nu)_j$, the maximum response will in general correspond to $A_\alpha j_{\alpha\alpha}^2 = Q_\alpha^2 \cdot j_{\alpha\alpha}^2 (\nu_{cia}, \nu_{cis})$. Fig.14 , which shows response to the (9,0) acoustic mode for several values of $(\nu_{cia} - \nu_{cis})$ with ν_{cis} within the main lobe of the $j_{\alpha\alpha}$ curve, demonstrates the necessity to account for the difference between ν_{cia} and ν_{cis}: even small differences reduce the maximum response considerably below its maximum possible value of $Q_\alpha^2 (j_{\alpha\alpha}^2)_{MAX} = Q_\alpha^2/16$. In some cases the maximum may be of the form $A_\alpha(\nu).j^2_{\alpha\alpha}(\nu,\nu_{cis})$, with $A_\alpha < Q^2_\alpha$ and $j^2_{\alpha\alpha}\leqslant(j^2_{\alpha\alpha})_{MAX}$, and may be independent of Q_α. For high q, high Λ and low Q_α, $(\Delta\nu)_A \gg (\Delta\nu)_j$, and the maximum response will in general correspond to $A_\alpha j^2_{\alpha\alpha} \doteq (j^2_{\alpha\alpha})_{MAX}.A_\alpha(\nu_{cia}) = A_\alpha(\nu_{cia})/16$. When ν_{cia} and ν_{cis} are very close together $A_\alpha(\nu_{cia})$ will depend on Q_α but at larger separations of the two frequencies will be independent of Q_α. Thus the resultant spectral density, due to all coincidences, will have both Q_α-dependent and Q_α-independent contributions.

It has been shown previously (Bull and Norton, Ref. 10) that the broad assumption that all coincidences produce maximum response with $A_\alpha j^2_{\alpha\alpha} = Q^2_\alpha/16$ (as at complete coincidence) leads to a considerable overestimate of the measured response and acoustic radiation. But it was also found that a similarly broad assumption that at maximum response $A_\alpha j^2_{\alpha\alpha} = A_\alpha(\nu_{cia})/16$, independent of Q_α, leads to estimates in quite fair agreement with experimental values for the $\beta = 0.007$ test pipe at all flow speeds up to $M_0 = 0.50$. However, the present considerations indicate that this approximation will not necessarily lead to reliable estimates in all cases, and that to improve it the estimation procedure should also take account of structural damping and the relative frequency bandwidths of the $j^2_{\alpha\alpha}$ and A_α curves. Suitable procedures which do this and which will lead to adequate prediction of pipe response and acoustic radiation in specified frequency bands are currently being investigated.

It is clear from Equs. (10) and (11) that for any prediction scheme to be successful a knowledge of the spectral density Φ_p of the internal acoustic field is essential. To be more precise, what is required is a modal spectrum (as could be obtained by the methods of Bolleter and Crocker, Ref. 11, or Kerschen and Johnston, Ref. 12) for each of the acoustic modes involved in the coincidences, for any particular disturbance-producing fitting; this is especially important for the lower order acoustic modes where modal separations are so large that statistical approaches are not appropriate and consideration of individual modes is essential.

6. CONCLUSIONS

Experimental and theoretical results relating to the pipe wall response and external acoustic radiation resulting from internal excitation by the broadband acoustic field produced by a severe flow disturbance have been presented and discussed. The conclusions which can be drawn can be summarised as follows.

(1) Pipe wall response and external acoustic radiation are predominantly determined by the phenomenon of coincidence of structural and acoustic modes.

(2) In order to make accurate estimates of pipe wall response and external acoustic radiation, it is necessary not only to distinguish between complete coincidence and wavenumber coincidence but also to take account of the principal wavenumber coincidences associated with each theoretical complete coincidence.

(3) For any given internal acoustic mode there may be two theoretical complete coincidences, one corresponding to a positive and one to a negative axial wavenumber component. Because the effect of flow in the pipe is to produce asymmetry of the dispersion curves of the acoustic modes about the frequency axis, these two theoretical complete coincidences occur at different frequencies.

(4) Each theoretical complete coincidence has associated with it two principal wavenumber coincidences. Each acoustic mode therefore has four principal wavenumber coincidences associated with it, all occurring at slightly different frequencies.

(5) The four principal wavenumber coincidences associated with each acoustic mode in general give rise to four distinct spectral peaks in pipe wall response and acoustic radiation, within a narrow frequency range, and therefore have a major influence in determining the form of the spectra. It is essential that they be taken into account if the form of these spectra is to be predicted accurately. Theoretical values of the four frequencies (based on Arnold and Warburton's method, Ref. 6, for estimating pipe resonance frequencies) agree well with experimental values, particularly for the lower order acoustic modes.

(6) Examination of the various possible forms of pipe wall response at the principal wavenumber coincidences suggests that previous procedures for estimating pipe response and acoustic radiation (Ref. 10), although fairly successful, could be improved by additional consideration of structural damping of the pipe wall and the relative frequency bandwidths of the joint acceptance and structural resonance curves.

(7) In the case of broadband internal excitation it is not possible to avoid coincidence entirely and the number of principal wavenumber coincidences occurring in the range of frequencies of practical interest will generally be large.

(8) Pipe wall acceleration and acoustic power radiation can be reduced by increasing the ratio of pipe wall thickness to pipe diameter, i.e. by increasing β. This has a twofold effect: there is a direct effect in proportion to $1/\beta^2$ and an additional effect due to the reduction in the number of possible principal wavenumber coincidences.

7. ACKNOWLEDGEMENT

The support of the work reported by Australian Research Grants Committee is gratefully acknowledged.

8. REFERENCES

1. BULL, M.K. and NORTON, M.P.: "Effects of internal flow disturbances on acoustic radiation from pipes". In : Proc. Institution of Engineers, Australia, Vibration and Noise Control Engineering Conference, Sydney, 1976, pp. 61-65.

2. BULL, M.K. and NORTON, M.P.: "Higher order acoustic modes due to internal flow disturbances in relation to acoustic radiation from pipes". In : Proc. 9th International Congress on Acoustics, Madrid, 1977, Contributed Papers Vol. II, p.726.

3. BULL, M.K. and NORTON, M.P.: "On the hydrodynamic and acoustic wall pressure fluctuations in turbulent pipe flow due to a 90° mitred bend". J. Sound Vib. 76, 1981, pp.561-586.

4. BULL, M.K. and NORTON, M.P.: "The proximity of coincidence and acoustic cut-off frequencies in relation to acoustic radiation from pipes with disturbed internal turbulent flow". J. Sound Vib. 69, 1980, pp. 1-11.

5. WALTER, J.L., McDANIEL, O.H. and REETHOF, G.: "The coincidence of higher-order acoustic modes in pipes with pipe vibrational modes". J.Acoust. Soc. Am. 62, 1977, p.S84.

6. ARNOLD, R.N. and WARBURTON, G.B.: "The flexural vibration of thin cylinders". Proc. I.Mech.E. 167, 1953, pp. 62-74.

7. MORSE, P.M. and INGARD, K.U.: "Theoretical acoustics". New York, McGraw-Hill, 1968.

8. BULL, M.K. and RENNISON, D.C.: "Acoustic radiation from pipes with internal turbulent gas flows". In: Proc. Noise Shock and Vibration Conference, Monash University, Melbourne, 1974, pp. 393-405.

9. HECKL, M.: "Vibration of point-driven cylindrical shells". J. Acoust. Soc. Am. 34, 1962, pp. 1553-1557.

10. BULL, M.K. and NORTON, M.P.: "Prediction of noise radiation from pipes with disturbed internal turbulent flow". In : Proc. 10th International Congress on Acoustics Satellite Symposium, Adelaide, 1980, pp. KK1-KK10.

11. BOLLETER, U. and CROCKER, M.J.: "Theory and measurement of modal spectra in hard-walled cylindrical ducts". J. Acoust. Soc. Am. 51, 1972, pp.1439-1447.

12. KERSCHEN, E.J. and JOHNSTON, J.P.: "A modal separation technique for broadband noise propagating inside circular ducts". J.Sound Vib. 76, 1981, pp.499-515.

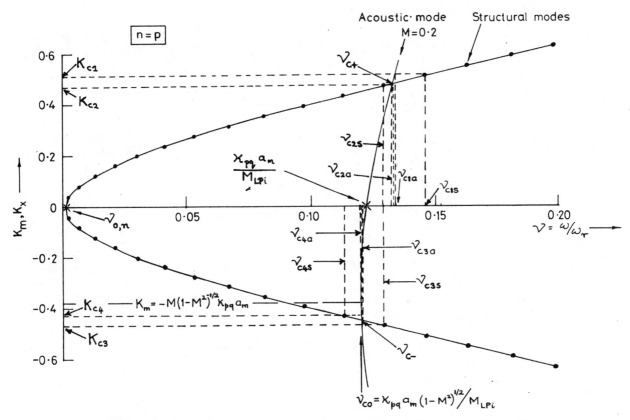

Figure 1   Example of structural and acoustic mode dispersion curves showing
(i) effect of flow in producing an asymmetrical acoustic dispersion curve,
(ii) theoretical complete coincidences, and (iii) principal wavenumber
coincidences and associated acoustic and structural resonance frequencies.

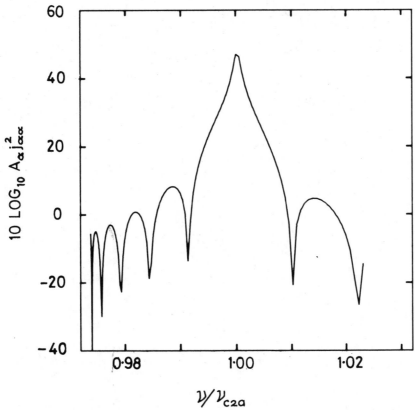

Figure 2   Form of structural response at complete coincidence when $(\Delta v)_j \gg (\Delta v)_A$:(1,0) acoustic mode, $m = 11$, $Q_\alpha = 1000$, $v_{c2a} = v_{c+} = v_{c2s}$.

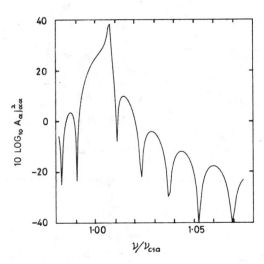

Figure 3   Form of structural response for $(\Delta\nu)_j \gg (\Delta\nu)_A$ and ν_{c1s} within main lobe of $j^2_{\alpha\alpha}$ curve: (1,0) acoustic mode, $m^j = 12$, $Q_\alpha^A = 1000$, $\nu_{c1s}/\nu_{c1a} = 1.007$.

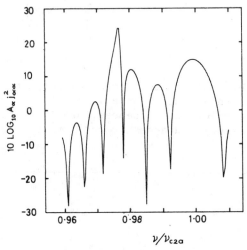

Figure 4   Form of structural response for $(\Delta\nu)_j \gg (\Delta\nu)_A$ and ν_{c2s} within a side lobe of $j^2_{\alpha\alpha}$ curve: (2,0) acoustic mode, $m^j = 28$, $Q_\alpha^A = 1000$, $\nu_{c2s}/\nu_{c2a} = 0.977$.

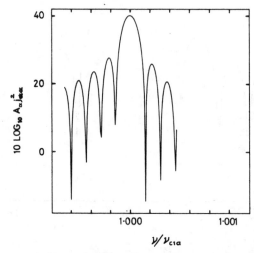

Figure 5   Form of structural response for $(\Delta\nu)_A \gg (\Delta\nu)_j$ and ν_{c1a} outside A_α bandwidth: (9, 36) acoustic mode, $m = 863$, $Q_\alpha = 500$, $\nu_{c1s}/\nu_{c1a} = 0.9993$.

362

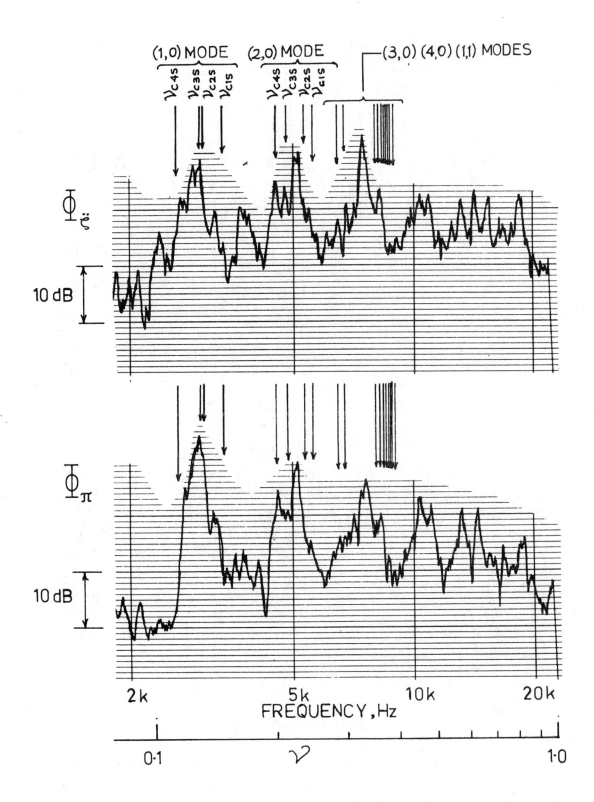

Figure 6 (a) Pipe wall acceleration spectrum, and (b) acoustic power radiation spectrum for $M_o = 0.22$: $h = 0.89$ mm, $\ell = 2.92$ m, $a_i = 36.27$ mm, $\beta = 0.0070$, $\Lambda = 79.5$. 90° mitred bend. Structural resonance frequencies for principal wavenumber coincidences are marked. Bandwidth = 100 Hz.

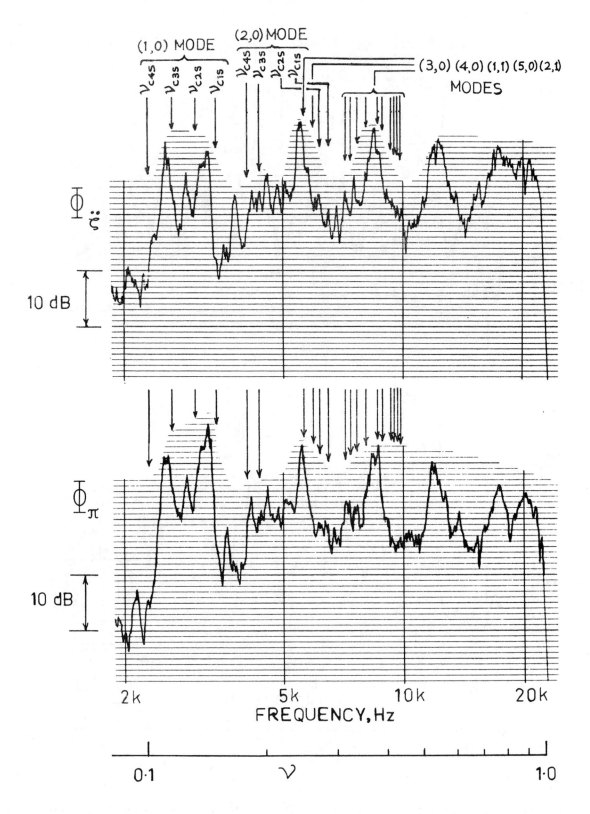

Figure 7   (a) Pipe wall acceleration spectrum, and (b) acoustic power radiation spectrum for $M_o = 0.50$.   Pipe details as for Fig. 6.

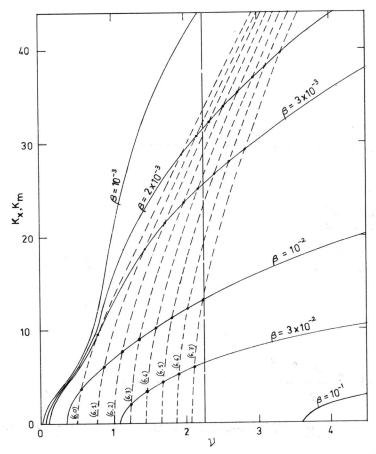

Figure 8   Acoustic and structural dispersion curves showing effect of structural mode
cut-off and shape of structural dispersion curves on number of coincidences
up to a set frequency limit : n = p = 6, M_O = 0.   Theoretical complete
coincidences indicated by ● .

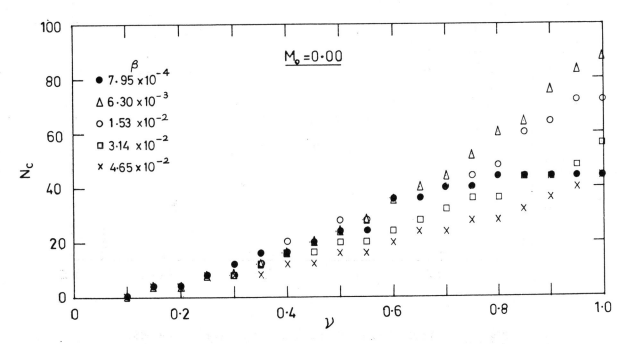

Figure 9   Variation of number of principal wavenumber coincidences N_c up to non-
dimensional frequency ν with ν and non-dimensional wall thickness β.
M_O = 0.00.

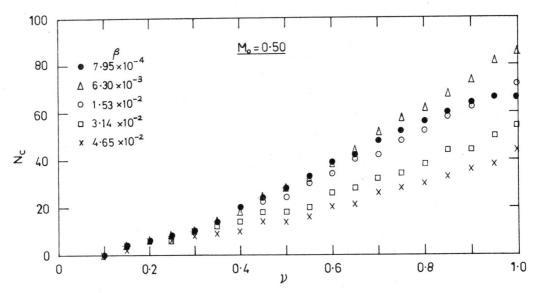

Figure 10   Variation of N_c with ν and β. M_o = 0.50.

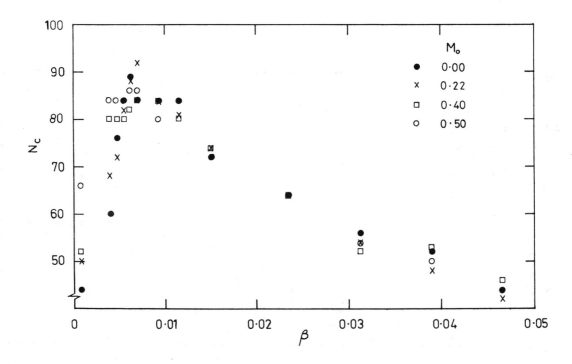

Figure 11   Variation of N_c with β and M_o for ν_L = 0.88.

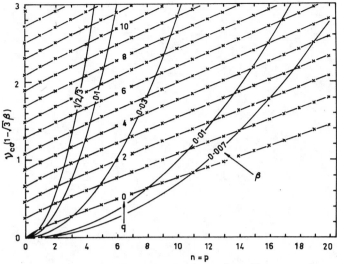

<u>Figure 12</u> Cut-off frequencies of structural (——) and acoustic modes (——x——) at
$M_o = 0$).   Coincidence can occur when the acoustic mode lies above the
structural curve for the appropriate value of β at a given value of circumferential mode n = p.

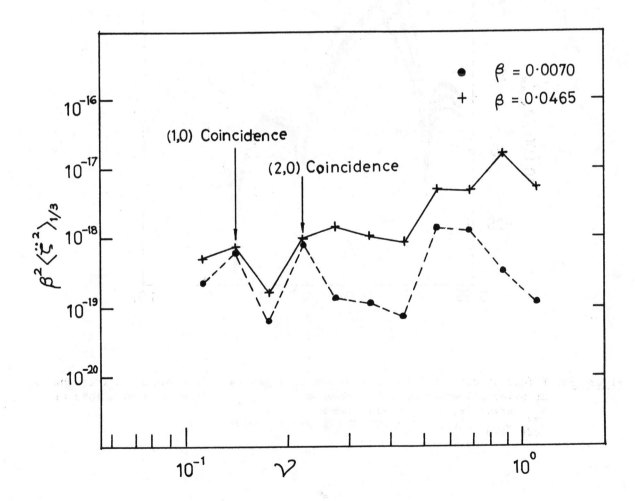

<u>Figure 13</u> Wall acceleration spectra for thin-walled (β = 0.0070) and thick-walled
(β = 0.0465) test pipes, downstream of a 90° mitred bend. $M_o = 0.50$.

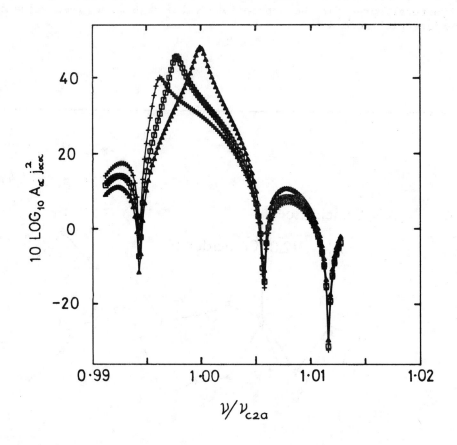

<u>Figure 14</u> Effect of difference between structural resonance and acoustic frequencies
at principal wavenumber coincidence when ν_{cis} falls within main lobe of
$j^2_{\alpha\alpha}$ curve: (9,0) acoustic mode, m = 66, Q_α = 1000.
ν_{c2s}/ν_{c2a} values : $\triangle$, 1.00; $\square$, 0.998; +, 0.996.

EFFECT OF A VISCOUS FLUID SUBSTRATE ON THE
FLOW-INDUCED VIBRATIONS OF A COMPLIANT COATING

P.W. Carpenter

University of Exeter, U.K.

A.D. Garrad

Wind Energy Group, Taylor Woodrow Construction, U.K.

Summary

Flow-induced vibrations, or instabilities, of a class of compliant coatings are studied theoretically. The theoretical model adopted corresponds closely to the original Kramer coatings. The coatings may be either inifinite or finite in length and consist of an elastic plate backed by a viscous fluid substrate and supported by an array of springs modelled by a continuous elastic foundation. The inifinite-length case corresponds to a travelling-wave flutter type of instability. Numerical solutions are presented showing how substrate viscosity and depth affect the development of travelling-wave flutter. The finite-length analysis deals with standing waves. It is shown that only a divergence-type instability is possible. Solutions are presented showing the effects of substrate viscosity and depth on the growth rate of divergence. For Kramer's optimum value of fluid substrate viscosity it is found that the growth rates for divergence and flutter are roughly the same. For lower values of viscosity the growth rate for divergence is larger and for higher values of viscosity the growth rates for travelling-wave flutter and boundary layer instabilities are larger.

Organised and sponsored by
BHRA Fluid Engineering, Cranfield, Bedford MK43 0AJ, England

NOMENCLATURE

A_i = integration constants, Eqn. (14)   (i = 1,2,3,4)

b = plate thickness

B = flexural rigidity, Eqn. (2)

B_{mn} = matrix coefficients, Eqn. (31)

c = complex wave speed

$\bar{c}$ = c/c_o

c_o = free wave speed

C_A = $\rho_e/(b\rho_m\alpha)$

C_B = $\pi^2 B/(b\rho_m g\ell^3)$

C_D = $\pi n C_s c_o/(g\ell)^{1/2}$

C_{KE} = $K_E\ell/(\pi^2 b\rho_m g)$

C_M = $\rho_e\ell/(b\rho_m)$

C_s = $12C_A\alpha\lambda_s\nu_s/\{c_o(\alpha H)^3\}$

D = $(\omega\ell^{1/2})/(\pi g^{1/2})$

E = elastic modulus

F = stream function amplitude, Eqn. (10)

Fr = $U/(g\ell)^{1/2}$

g = acceleration due to gravity

H = substrate depth

I_{inm} = integrals, Eqn. (34)   (i = 1,2,3)

K = spring stiffness

K_E = $K - g(\rho_e-\rho_s)$

ℓ = length of coating

m = mode number

M = $\mu_s/(b\rho_m c_o)$

n . = mode number

p = static pressure

$\hat{p}$ = amplitude of substrate pressure

t = time

u = velocity in x direction

U = main stream velocity

v	= velocity in y direction	
V	= U/c_0	
w	= surface deflection	
w_0	= amplitude of w	
x	= coordinate along surface	
y	= coordinate normal to surface	
α	= wave number	
α_m	= most unstable wave number	
β	= $\{\alpha^2 - i\alpha c/\nu_s\}^{1/2}$	
δp_1	= dynamic pressure fluctuation acting on upper surface	
δp_2	= dynamic pressure " " " lower surface	
λ_m	= most unstable wavelength	
λ_s	= ρ_s/ρ_e	
Λ_s	= $(1+e^{-2\alpha H})/(1-e^{-2\alpha H})$	
μ_s	= substrate viscosity	
ν_s	= μ_s/ρ_s	
ρ	= density	
ϕ	= phase shift of pressure fluctuation acting on upper surface	
ψ	= stream function for substrate flow	
ω	= complex frequency	

Subscripts.

$(\ )_D$ = divergence

$(\ )_e$ = external flow

$(\ )_F$ = flutter

$(\ )_I$ = imaginary part

$(\ )_m$ = plate material

$(\ )_R$ = real part

$(\ )_s$ = substrate

1. INTRODUCTION

Interest in compliant coatings was first aroused by Kramer's (Ref. 1) pioneering experiments. He found that a body of revolution, covered with a coating made to his design, experienced a substantial drag reduction compared to a rigid control model. The design of the compliant coating was based on the structure of dolphin skin. It consisted of a 2mm thick natural rubber plate supported on rubber stubs which were 1mm high. The stubs were attached to a 0.5 mm thick rubber base sheet which was backed by a rigid surface. The cavity between the outer rubber plate and the base sheet was filled with a highly viscous fluid. Such a compliant surface is shown schematically in Fig. 1.

For the practical application of such compliant coatings flow-induced vibrations, or instabilities, of the surface are an important consideration. These instabilities have been observed in many experimental investigations. See, for example, Puryear (Ref. 2), Ritter and Messum (Ref. 3) and Ritter and Porteous (Ref. 4). Moreover, it seems highly probable that such instabilities occurred in many other experiments but were not noticed. If such instabilities are present they may in extreme cases lead to damage or complete destruction of the coating. They can also play a role in prematurely triggering transition to turbulent flow in the boundary layer. In many ways this is even more important because Kramer was of the opinion that his compliant coatings tend to damp out the Tollmein-Schlichting instabilities in the boundary layer and thus achieve a drag reduction by delaying transition. This has been shown to be theoretically possible by Brooke Benjamin (Ref. 5) and Landahl (Ref. 6). Despite their importance, however, little appears to be known about these flow-induced surface instabilities, particularly of the original Kramer-type coating.

In the present paper the surface instabilities of a class of compliant coatings is studied theoretically. The theoretical model adopted (see Fig. 1) is intended to correspond closely to the original Kramer coating. The coatings consist of an elastic plate backed by a viscous fluid substrate having a density which may be different from the main flow. Provision is made for the plate to be supported by an elastic foundation thereby modelling the effects of the springy rubber stubs in the Kramer coating. Coatings of infinite and finite length are both investigated. Physically these two cases correspond to travelling-wave flutter and divergence respectively. The effects of the substrate fluid viscosity and depth on these two types of instability are studied.

2. THEORY FOR INFINITE LENGTH COATING

2.1 Equation of motion for plate.

Consider an elastic plate of thickness, b, placed over a fluid substrate of density, ρ_s, and dynamic viscosity, μ_s. The plate is supported by an array of springs the effect of which may be modelled by a continuous elastic foundation of spring stiffness K. For the motion of a two-dimensional plate the governing equation is given by:

$$b \, \rho_m \frac{\partial^2 w}{\partial t^2} + B \frac{\partial^4 w}{\partial x^4} + K_E \, w = \delta p_2 - \delta p_1 \qquad (1)$$

{see Kornecki (Ref. 7) for example} where ρ_m and ρ_e are respectively the densities of the plate material and the main stream, w is the surface displacement, B is the flexural rigidity of the plate, $K_E = K - g(\rho_e - \rho_s)$ is the equivalent spring stiffness, and δp_1 and δp_2 are respectively the perturbations in dynamic pressure acting on the plate from above and below. See Fig. 2 for definition of notation. The flexural rigidity is given by

$$B = E \, b^3 / \{12(1 - \nu^2)\} \qquad (2)$$

where E and ν are respectively the elastic modulus and Poisson's ratio of the plate material.

For simplicity the main flow will be taken as potential. It is possible to allow for the boundary layer effects by writing

$$\delta p_1 = C_{p_1} \exp(i\phi)(\delta p_1)_p \qquad (3)$$

where $(\delta p_1)_p$ is the dynamic pressure fluctuation in a potential flow, see Ref. 8 for further details. Herein we take $C_{p_1} = 1$ and the phase shift, $\phi = 0$.

Let the surface deflection, w, take the form:

$$w(x, t) = w_0 \exp\{i\,\alpha(x-ct)\} \tag{4}$$

where α is the wave number and c is the complex wave speed. The dynamic pressure perturbation due to the potential main flow is given by

$$\delta p_1 = -\rho_e \alpha(U-c)^2\, w \tag{5}$$

Substitute Eqns. (4) and (5) in Eqn. (1) and divide both sides by $b\rho_m\alpha^2 w$ to obtain

$$c_0^2 - c^2 = C_A(U-c)^2 + \delta p_2/(b\rho_m\alpha^2 w) \tag{6}$$

where c_0 is the free wave speed of the surface and is given by

$$c_0^2 = \{\alpha^4 B + K_E\}/(\alpha^2 \rho_m) \tag{7}$$

and

$$C_A = \rho_e/(b\rho_m\alpha)$$

Eqn. (6) is the characteristic equation for c but in order to solve it an expression for $\delta p_2/w$ must first be derived.

2.2   Substrate fluid motion and pressure.

Since the motion of the substrate fluid is due solely to the motion of the plate, low velocities can be assumed. This allows the Navier-Stokes equations to be linearised giving:

$$\rho_s \frac{\partial u_s}{\partial t} = -\frac{\partial p_s}{\partial x} + \mu_s \left(\frac{\partial^2 u_s}{\partial x^2} + \frac{\partial^2 u_s}{\partial y^2}\right) \tag{8}$$

$$\rho_s \frac{\partial v_s}{\partial t} = -\frac{\partial p_s}{\partial y} + \mu_s \left(\frac{\partial^2 v_s}{\partial x^2} + \frac{\partial^2 v_s}{\partial y^2}\right) \tag{9}$$

A stream function is introduced having the form:

$$\psi = F(y) \exp \{i\alpha(x-ct)\} \tag{10}$$

so that

$$u_s = \partial\psi/\partial y = F'(y) \exp \{i\alpha(x-ct)\} \tag{11}$$

and

$$v_s = -\partial\psi/\partial x = -i\alpha F(y) \exp \{i\alpha(x-ct)\} \tag{12}$$

Eqns. (11) and (12) are substituted into Eqns. (8) and (9). p_s is then eliminated by cross differentiation to obtain the following fourth-order ordinary differential equation for F:

$$\alpha c(F'' - \alpha^2 F) - i\,\nu_s(\alpha^4 F - 2\alpha^2 F'' + F'''') = 0 \tag{13}$$

where F'' and F'''' denote respectively the second and fourth derivatives of F. Eqn. (13) is, in fact, the well-known Orr-Sommerfeld equation with the main stream velocity set equal to zero. The general solution to Eqn. (13) is given by

$$F = A_1 e^{\alpha y} + A_2 e^{-\alpha y} + A_3 e^{\beta y} + A_4 e^{-\beta y} \tag{14}$$

$$\text{where} \quad \beta^2 = \alpha^2 - i\,\alpha\,c/\nu_s \tag{15}$$

The boundary conditions are:

$$u_s = 0 \quad \text{and} \quad v_s = \partial w/\partial t \quad \text{at} \quad y = 0$$
$$u_s = v_s = 0 \quad \text{at} \quad y = -H \tag{16}$$

The constants of integration A_1 etc. can be evaluated by applying the conditions (16) to Eqn. (14).

Let the pressure in the substrate fluid take the form:

$$p_s = \hat{p}(y) \exp\{i\alpha(x-ct)\} \tag{17}$$

Substitution of (11), (12) and (17) into Eqn. (8) and rearrangement gives

$$\hat{p} = (\rho_s c + i \mu_s \alpha)F' - i \mu_s F'''/\alpha \tag{18}$$

where F''' denotes the third derivative of F. With use of the boundary conditions (16) and Eqn. (14) it can be shown that at the surface, $y = 0$, Eqn. (18) reduces to

$$\hat{p} = \alpha \rho_s c(A_1 - A_2)$$

A_1 and A_2 may be evaluated by applying the boundary conditions (16) to give

$$\hat{p} = \alpha \rho_s c^2 w_0 \frac{\beta(\beta+\alpha)\left\{e^{(\beta-\alpha)H} - e^{-(\beta-\alpha)H}\right\} + \beta(\beta-\alpha)\left\{e^{(\beta+\alpha)H} - e^{-(\beta+\alpha)H}\right\}}{8\alpha\beta + (\beta-\alpha)^2\left\{e^{(\beta+\alpha)H} + e^{-(\beta+\alpha)H}\right\} - (\beta+\alpha)^2\left\{e^{(\beta-\alpha)H} + e^{-(\beta-\alpha)H}\right\}} \tag{19}$$

$\delta p_2/w$ in Eqn. (6) is equivalent to $\hat{p}/w_0$ and can be obtained as a function of c from Eqn. (19); it can be seen that in general this is not a simple relationship. There are two special cases, however.

If $\nu_s \to 0$ then Eqn. (19) reduces to

$$\hat{p} = \alpha \rho_s c^2 w_0 (1 + e^{-2\alpha H})/(1 - e^{-2\alpha H}) \tag{20}$$

This is the appropriate form for an inviscid substrate fluid.

If $H \to 0$ for fixed viscosity then Eqn. (19) reduces to

$$\hat{p} = 12 i w_0 \alpha^2 \mu_s c/(\alpha H)^3 \tag{21}$$

Eqn. (21) is approximately true for small substrate depths and moderate to high viscosity. It can be seen from the form of Eqn. (21) that under the small depth approximation viscous fluid damping is equivalent to adding a conventional damping term of the form d $\partial w/\partial t$ (where d is the damping coefficient) to the left-hand side of Eqn. (1) in that for both cases there would be an additional term proportional to c in Eqn. (6).

2.3   Solution of the characteristic equation.

In general, the characteristic equation (6) must be solved numerically when Eqn. (19) is substituted for $\delta p_2/w$.

For the simpler case of an inviscid substrate fluid Eqn. (6) becomes a quadratic in c with solutions:

$$c = \frac{C_A U \pm \left[C_A^2 U^2 - (C_A U^2 - c_0^2)\{1 + C_A(1 + \lambda_s \Lambda_s)\}\right]^{1/2}}{1 + C_A(1 + \lambda_s \Lambda_s)} \tag{22}$$

where $\lambda_s = \rho_s/\rho_e$ and $\Lambda_s = (1 + e^{-2\alpha H})/(1 - e^{-2\alpha H})$

Likewise Eqn. (6) also reduces to a quadratic in c for the small depth approximation with Eqn. (21) substituted for $\delta p_2/w$. In this case the solutions are:

$$c = \frac{C_A U}{(1 + C_A)}\left\{1 \mp \frac{C_s}{2\Delta}\right\} + i \frac{c_0}{(1 + C_A)}\left\{-\frac{C_s}{2} \pm \Delta\right\} \tag{23}$$

where

$$C_s = 12 C_A \alpha \lambda_s \nu_s / \{c_0(\alpha H)^3\} \tag{24}$$

$$\sqrt{2}\Delta = \left[\left\{(1 + C_A - C_A V^2 - C_s^2/4)^2 + (C_A C_s V)^2\right\}^{1/2} - 1 - C_A + C_A V^2 + C_s^2/4\right]^{1/2}$$

and $V = U/c_0$. (25)

2.4   Critical velocity and wave number.

Neutral stability occurs when $c = 0$. It can be seen from Eqns. (22) and (23) that this occurs when $U = U_D$ where

$$U_D = c_0/(C_A)^{\frac{1}{2}} \tag{26}$$

This result also holds when the full expression (19) is used for δp_2.

For an inviscid fluid substrate it can be shown from Eqn. (22) that neutral stability persists until U reaches U_F, the flutter velocity, at which $c_I = 0$ but $c_R \neq 0$. U_F is found by setting the argument of the square root on the right-hand side of Eqn. (22) equal to zero giving:

$$U_F = U_D \{1 + C_A/(1 + C_A \lambda_s \Lambda_s)\}^{1/2} \tag{27}$$

For a viscous substrate, on the other hand, flutter sets in immediately U_D is exceeded.

The minimum value of U_D and most unstable wave number, α_m, can be obtained by setting $\partial U_D/\partial \alpha$ equal to zero. In this way it is found that

$$\alpha_m = \{K_E/(3B)\}^{1/4} \tag{28}$$

3.   THEORY FOR FINITE LENGTH COATING

For a compliant coating of length, ℓ, standing-wave instabilities are investigated having the form:

$$w(x,t) = \exp(i\omega t) \sum_{n=1}^{\infty} \ell A_n \sin(n\pi x/\ell) \tag{29}$$

Strictly, the approximate result (21) for determining δp_2 only applies to compliant surfaces of infinite length. However, it would seem to be a reasonable approximation for finite coatings provided $H \ll \ell$. In order to render Eqn. (21) suitable for finite length surfaces α must be replaced by $\pi n/\ell$ and c by $\omega\ell/(\pi n)$. The main stream dynamic pressure perturbation is now obtained by thin-aerofoil theory[*] and Galerkin's method is followed to obtain a matrix equation for the coefficients A_n which takes the form:

$$B_{mn} A_n = 0 \qquad : \quad m,n = 1, \ldots, \infty \tag{30}$$

See Refs. 8 and 9 for the details of the derivation. In the present case the matrix coefficients B_{mn} are given by:

$$B_{mn} = \{m^4 C_B + C_{KE} + i C_D D - D^2\} \delta_{mn}$$
$$- (C_M/\pi)\{4i D Fr I_{1nm} + 2n Fr^2 I_{2nm} - \pi D^2 I_{3nm}\} \tag{31}$$

The non-dimensional coefficients are defined as follows:

$$C_B = \pi^2 B/(b \rho_m g \ell^3), \qquad C_{KE} = K_E \ell/(\pi^2 b \rho_m g) \tag{32}$$
$$C_D = \pi n C_s \{c_0^2/(g \ell)\}^{1/2} \quad \text{and} \quad C_M = \rho_e \ell/(b \rho_m)$$

The non-dimensional frequency and main-stream velocity are defined respectively as:

$$D = (\omega \ell^{1/2})/(\pi g^{1/2}) \quad \text{and} \quad Fr = U/(g \ell)^{1/2} \tag{33}$$

The latter is in the form of a Froude number hence the notation. I_{1nm}, I_{2nm} and I_{3nm} are integrals which are defined as:

$$I_{1nm} = \int_0^1 \int_0^1 \{\sin(n \pi \xi) \sin(m \pi \zeta)/(\zeta - \xi)\} d\xi \, d\zeta$$

$$I_{2nm} = \int_0^1 \int_0^1 \{\cos(n \pi \zeta) \sin(m \pi \zeta)/(\zeta - \xi)\} d\xi \, d\zeta$$

$$I_{3nm} = (2/n \pi)\left\{\left[1 + (-1)^{n+m}\right] \int_0^1 \ell n(\xi) \sin(m \pi \xi) d\xi - I_{2nm}\right\}$$

$$(\ldots)$$

---

[*]The main stream flow is modelled by a source distribution.

For Kramer-type coatings the mode numbers n and m are very high and it can be shown that in the limit as n and m tend to infinity $I_{1nm} = I_{3nm} = 0$ and $I_{2nm} = \pi/2$ if $n = m$ and $I_{1nm} = 1/(n-m)$ and $I_{2nm} = I_{3nm} = 0$ if $n \neq m$. Thus it follows that $B_{mn} = -B_{nm}$ for large m and n. This implies that modal coupling is very weak so that each mode may be considered separately. Under these conditions the characteristic equation is given by

$$B_{nn} = 0 \qquad (35)$$

Eqn. (35) is a quadratic in D and has the solutions:

$$D = i \, C_D/2 \pm \left\{n^4 \, C_B + C_{KE} - n \, C_M \, Fr^2/\pi - C_D^2/4\right\}^{1/2} \qquad (36)$$

The critical velocity at which D = 0 is given by

$$Fr_D = \left\{\pi(n^4 \, C_B + C_{KE})/(n \, C_M)\right\}^{1/2} \qquad (37)$$

When $Fr > Fr_D$ it can be seen from Eqn. (36) that D becomes purely imaginary which implies that the instability is divergence. A flutter-type instability would have been generated by modal interaction but as we have seen this is suppressed at high mode numbers.

The most unstable mode, i.e. the value of n for which Fr_D is a minimum, can be obtained from Eqn. (37) and takes the value:

$$n_m = \left\{C_{KE}/(3C_B)\right\}^{1/4} \qquad (38)$$

4. RESULTS AND DISCUSSION

Parameter values corresponding to the Kramer coating which achieved the greatest drag reduction, are used to illustrate the present theory. The elastic modulus, E, of the natural rubber used for such coatings is about 0.5×10^6 N m^{-2}. Taking the Poisson ratio equal to 0.5 gives $B = (8.9 \times 10^{-10})E$ N m. The spring stiffness can be related to E by assuming that the applied fluctuating pressure is supported by the springy stubs, the total cross-sectional area of which can be calculated as a fraction of the surface area. This gives $K = 230E$ N m^{-3}. Note that this quantity is not the same as the stiffness quoted by Kramer[*] who measured the coatings' response to a static point load rather than a sinusoidally distributed fluctuating pressure. Using Eqn. (28) we obtain $\alpha_m = 542$ or $\lambda_m = 2\pi/\alpha_m = 11.6$ mm for the most unstable wave. $C_A = 1.01$ and $c_0 = 16.8$ m/s for such coatings so Eqn. (26) gives a critical velocity of 16.7 m/s. This value is below Kramer's top operational speed of 18.0 m/s. Puryear (Ref. 2) reported wave-like instabilities on a Kramer-type coating. From his Fig. 7 it appears that the wavelength was about 14 to 18 mm which compares reasonably well with the present estimate.

The complex wave speeds corresponding to a fluid substrate viscosity of 200 cSt (Kramer's optimum value) are plotted against free stream velocity in Fig. 3. Both the unstable and stable solutions are shown. These were obtained by solving Eqn. (6) numerically with Eqn. (19) used to evaluate $\delta p_2/w$. Also shown in Fig. 3 are the corresponding solutions for the coating with no fluid substrate. These were obtained from Eqn. (22) with $\lambda_s = 0$.

The effects of varying the fluid substrate viscosity are shown in Figs. 4 and 5. Here, only the unstable solutions are shown; non-dimensional complex wave speed, $\bar{c} = c/c_0$ and free stream velocity, $V = U/c_0$ are used. Also shown in Fig. 4 is the solution for an inviscid fluid substrate obtained from Eqn. (22). Note that this solution is qualitatively similar to that for no substrate. It is apparent from Fig. 4 that substrate viscosity has a destabilising effect in that for free stream velocities up to $V \simeq 1.17$ the growth rates (i.e. $\bar{c}_I$) of the instabilities rise with an increase in viscosity. However, this effect does not continue indefinitely with an increase in viscosity and, in fact, the maximum growth rate occurs for a non-dimensional viscosity parameter of $M = \mu_s/(b \, \rho_m \, c_0) \simeq 0.012$ (corresponding to just over 400 cSt). The approximate theory for $\alpha H \ll 1$ {see Eqn. (23)} also yields the same qualitative behaviour but gives $C_S \simeq 1.50$ as the value for maximum growth rate. This value of C_S corresponds to $M \simeq 0.019$ in the present case of $\alpha H = 0.55$ so the approximate theory tends to over-

---

[*]For $E \simeq 0.5$ MNm^{-2} Kramer quoted a stiffness of 220 MNm^{-3}.

estimate M_{max}.

All the results discussed above correspond to $\alpha H = 0.55$ which is the value for the Kramer coating. The effect of varying substrate depth for a fixed viscosity of 200 cSt is shown in Fig. 6. Note that the effect on the growth rates is very marked. As would be expected M_{max} varies with αH. This is shown in Fig. 7. The trend shown in Fig. 7 of M_{max} rising with an increase in αH does not continue indefinitely since an inviscid substrate fluid gives the maximum growth rate for infinite depth. To obtain M_{max} the growth rates were evaluated at $V = V_F$ given by Eqn. (27). According to the approximate theory M_{max} corresponds to a fixed value of C_s which implies that M_{max} varies as $(\alpha H)^3$. The approximate theory tends to overestimate M_{max} as can be seen in Fig. 7. However, if it is assumed that M_{max} varies as $(\alpha H)^3$ and we make the approximation

$$M_{max} = (\alpha H/0.2)^3 (M_{max})_{\alpha H=0.2} \tag{39}$$

the results are a close approximation to the exact values, as shown in Fig. 7.

Finally, the results for the finite-length coating will be considered. It was shown in Section 3 that the instability in this case is divergence. The length of the coatings on Kramer's models was about 990 mm. From Eqn. (38) it is found that $n_m \simeq 170$ corresponding to a wave-length, $\lambda_m = 2\ell/n_m \simeq 11.7$ mm, which is very slightly longer than for the travelling-wave flutter instability predicted by the infinite-length theory. Eqn. (37) gives a critical velocity of 16.6 m/s which is very slightly less than for travelling-wave flutter. For practical purposes the critical velocities for the two types of instability are identical. The growth rates for divergence (i.e. D_I) are plotted in Fig. 8 for various values of M. It turns out that $D_I \times 10^{-3}$ corresponds closely to $\bar{c}_I$ so the growth rates in Fig. 8 can be compared directly to those in Figs. 4 and 5. The most striking difference is that a rise in viscosity always reduces the growth rate for divergence.

A comparison of the two types of instability may provide an explanation for Kramer's observation that a 200 cSt substrate fluid gave the best drag reduction. In the past this empirical result has caused a certain amount of puzzlement since both Brooke Benjamin (Ref. 5) and Landahl (Ref. 6) have shown that an increase in substrate damping destabilises Tollmein-Schlichting waves and it was also known, e.g. see Kornecki (Ref. 7), that damping destabilises travelling-wave flutter. Consequently, it appeared that a low substrate viscosity would give the best results. For a 200 cSt ($M = 0.0065$) substrate fluid, however, $\bar{c}_I \simeq 0.05$ and $D_I \times 10^{-3} \simeq 0.10$ at $V = 1.1$. So for Kramer's optimum viscosity value both types of instability have roughly similar growth rates. At lower viscosities the growth rate for divergence would be larger and at higher viscosities the growth rate for travelling-wave flutter and Tollmein-Schlichting waves would be greater.

5. CONCLUSIONS

Flow-induced instabilities of Kramer-type compliant coatings have been studied theoretically. The main practical conclusions are as follows:
  (i)   There are two possible types of instability namely, travelling-wave flutter and divergence.
  (ii) The critical velocity and wavelength are virtually the same for both types of instability. The predicted wavelength agrees reasonably well with experimental observation.
  (iii) The growth rate for travelling-wave flutter is greatest for a fluid substrate viscosity of about 400 cSt for a Kramer coating.
  (iv) This worst value of viscosity varies as the cube of substrate depth.
  (v)   The growth rate for divergence falls monotonically with increasing substrate viscosity.
  (vi) At Kramer's empirically observed optimum value of 200 cSt for substrate viscosity the growth rates for travelling-wave flutter and divergence are roughly similar in magnitude.

REFERENCES

1. Kramer, M.O.: "Boundary layer stabilization by distributed damping". Journal of the American Society of Naval Engineers, 72, 1960, pp. 25-33.

2. Puryear, F.W.: "Boundary layer control drag reduction by compliant surfaces". U.S. Department of Navy, David Taylor Model Basin, 1962, Report 1668.
3. Ritter, H. and Messum, L.T.: "Water tunnel measurements of turbulent skin friction on six different compliant surfaces of one foot length". Admiralty Research Laboratory, 1964, Report ARL/G/N9.
4. Ritter, H. and Porteous, J.S.: "Water tunnel measurements of skin friction on a compliant coating". Admiralty Research Laboratory, 1965, Report ARL/N3/G/HY/9/7.
5. Brooke Benjamin, T.: "Effects of a flexible boundary on hydrodynamic stability". Journal of Fluid Mechanics, 9, 1960, pp. 513-532.
6. Landahl, M.T.: "On the stability of a laminar incompressible boundary layer over a flexible surface". Journal of Fluid Mechanics, 13, 1962, pp. 609-632.
7. Kornecki, A.: "Aeroelastic and hydroelastic instabilities of infinitely long plates. I". Solid Mechanics Archives, 3, 1978, pp. 381-440.
8. Garrad, A.D. and Carpenter, P.W.: "A theoretical investigation of flow-induced instabilities in compliant coatings". Journal of Sound and Vibration, 1982, (to be published).
9. Garrad, A.D. and Carpenter, P.W.: "On the aerodynamic forces involved in aeroelastic instability of two-dimensional panels in uniform incompressible flow". Journal of Sound and Vibration, 80, 1982, pp. 437-439.

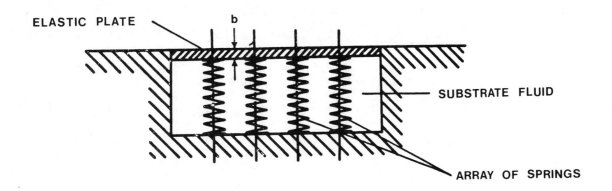

Figure 1.   Compliant coating.

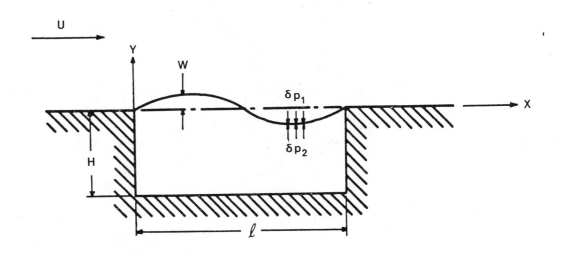

Figure 2.   Notation for coating.

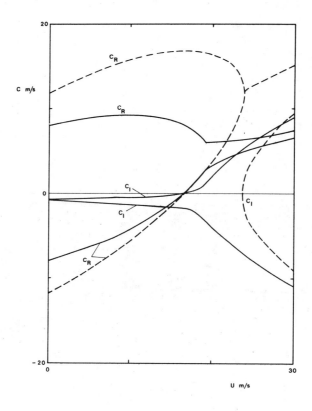

Figure 3.   Complex wave speed versus main stream velocity.
_____ 200 cSt fluid substrate, _ _ _ no substrate.

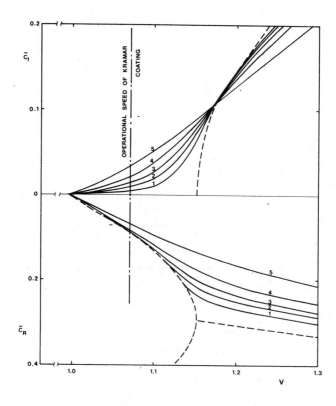

Figure 4.   Effect of low substrate viscosity on complex wave speed.
___ inviscid fluid substrate, 1 ~ M = 0.001, 2 ~ M = 0.002,
3 ~ M = 0.003, 4 ~ M = 0.005, 5 ~ M = 0.01.

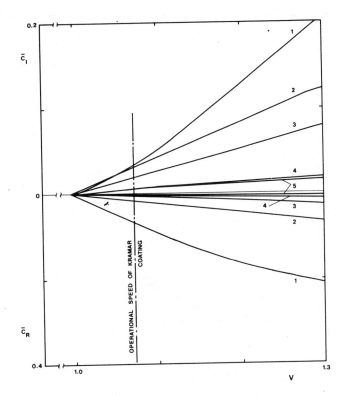

Figure 5.   Effect of high substrate viscosity on complex wave speed.
1 ~ M = 0.01, 2 ~ M = 0.05, 3 ~ M = 0.1, 4 ~ M = 0.5,
5 ~ M = 1.0.

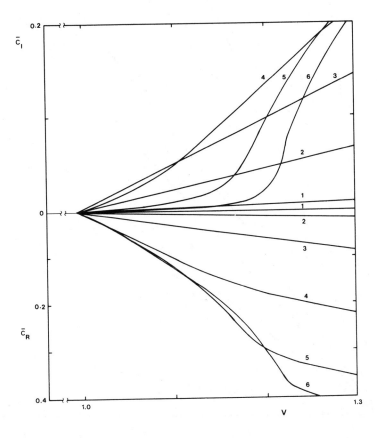

Figure 6.   Effect of substrate depth on complex wave speed.
M = 0.0065 (ν_S = 200 cSt).   1 ~ α H = 0.1, 2 ~ α H = 0.2,
3 ~ α H = 0.3, 4 ~ α H = 0.5, 5 ~ α H = 1.0, 6 ~ α H = ∞.

381

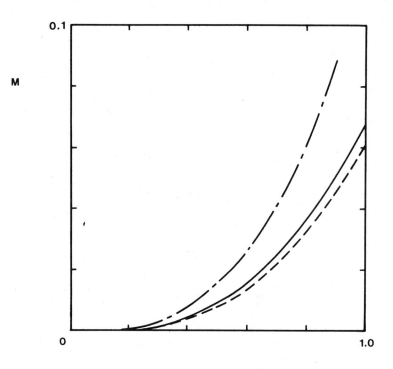

Figure 7.   Effect of substrate depth on M_{max}.

_____ exact theory, ___ _ ___ small depth approximate theory, ___ Eqn. (39).

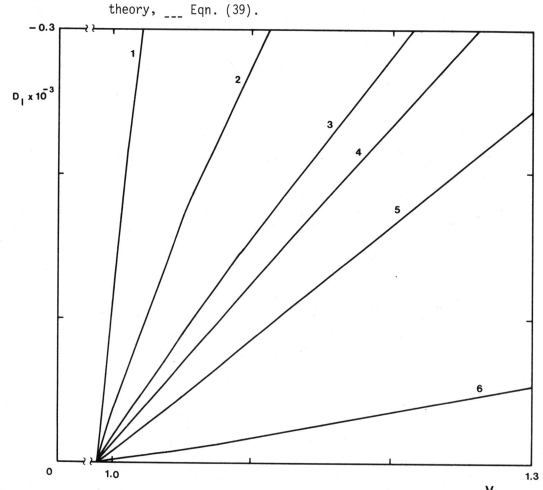

Figure 8.   Effect of substrate viscosity on growth rate for divergence.
1 ~ M = 0.0, 2 ~ M = 0.002, 3 ~ M = 0.005, 4 ~ M = 0.0065, 5 ~ M = 0.01, 6 ~ M = 0.05.